The Eastern San Juan Mountains

The Eastern

San Juan

Mountains

THEIR GEOLOGY, ECOLOGY,
AND HUMAN HISTORY

edited by

Rob Blair *and* **George Bracksieck**

UNIVERSITY PRESS OF COLORADO
MOUNTAIN STUDIES INSTITUTE / SAN JUAN COLLABORATORY

© 2011 by the University Press of Colorado

Published by the University Press of Colorado
5589 Arapahoe Avenue, Suite 206C
Boulder, Colorado 80303

 The University Press of Colorado is a proud member of
the Association of American University Presses.

The University Press of Colorado is a cooperative publishing enterprise supported, in part, by Adams State College, Colorado State University, Fort Lewis College, Metropolitan State College of Denver, Regis University, University of Colorado, University of Northern Colorado, and Western State College of Colorado.

∞ The paper used in this publication meets the minimum requirements of the American National Standard for Information Sciences—Permanence of Paper for Printed Library Materials. ANSI Z39.48-1992

Library of Congress Cataloging-in-Publication Data

The eastern San Juan Mountains : their geology, ecology, and human history / Rob Blair and George Bracksieck, editors.
    p. cm.
  Includes bibliographical references and index.
  ISBN 978-1-60732-084-5 (pbk. : alk. paper) — ISBN 978-1-60732-085-2 (ebook) 1. San Juan Mountains (Colo. and N.M.)—Description and travel. 2. Geology—San Juan Mountains (Colo. and N.M.) 3. Ecology—San Juan Mountains (Colo. and N.M.) 4. San Juan Mountains (Colo. and N.M.)—Environmental conditions. 5. San Juan Mountains (Colo. and N.M.)—History. 6. Natural history—San Juan Mountains (Colo. and N.M.) I. Blair, Rob, 1943– II. Bracksieck, George.
  F782.S18.E37 2011
  978.8'3—dc23
                                                                                            2011017144

Design by Daniel Pratt

20 19 18 17 16 15 14 13 12 11   10 9 8 7 6 5 4 3 2 1

Contents

Foreword by Governor Bill Ritter   *vii*

Preface   *ix*

Acknowledgments   *xi*

PART 1: PHYSICAL ENVIRONMENT OF THE SAN JUAN MOUNTAINS

CHAPTER 1   A Legacy of Mountains Past and Present in the San Juan Region   3
David A. Gonzales and Karl E. Karlstrom

CHAPTER 2   Tertiary Volcanism in the Eastern San Juan Mountains   17
Peter W. Lipman and William C. McIntosh

CHAPTER 3   Mineralization in the Eastern San Juan Mountains   39
Philip M. Bethke

CHAPTER 4   Geomorphic History of the San Juan Mountains   61
Rob Blair and Mary Gillam

CHAPTER 5   The Hydrogeology of the San Juan Mountains   79
Jonathan Saul Caine and Anna B. Wilson

CHAPTER 6   Long-Term Temperature Trends in the San Juan Mountains   99
Imtiaz Rangwala and James R. Miller

PART 2: BIOLOGICAL COMMUNITIES OF THE SAN JUAN MOUNTAINS

CHAPTER 7 Mountain Lakes and Reservoirs   113
Koren Nydick

CHAPTER 8 Fens of the San Juan Mountains   129
Rodney A. Chimner and David Cooper

CHAPTER 9 Fungi and Lichens of the San Juan Mountains   137
J. Page Lindsey

CHAPTER 10 Fire, Climate, and Forest Health   151
Julie E. Korb and Rosalind Y. Wu

CHAPTER 11 Insects of the San Juans and Effects of Fire on Insect Ecology   173
Deborah Kendall

CHAPTER 12 Wildlife of the San Juans: A Story of Abundance and Exploitation   185
Scott Wait and Mike Japhet

PART 3: HUMAN HISTORY OF THE SAN JUAN MOUNTAINS

CHAPTER 13 A Brief Human History of the Eastern San Juan Mountains   203
Andrew Gulliford

CHAPTER 14 Disaster in La Garita Mountains   213
Patricia Joy Richmond

CHAPTER 15 San Juan Railroading   231
Duane Smith

PART 4: POINTS OF INTEREST IN THE EASTERN SAN JUAN MOUNTAINS

CHAPTER 16 Eastern San Juan Mountains Points of Interest Guide   243
Rob Blair, Hobie Dixon, Kimberlee Miskell-Gerhardt, Mary Gillam, and Scott White

Glossary   299

Contributors   311

Index   313

Foreword

As a COLORADO NATIVE AND LIFELONG OUTDOORSMAN, it doesn't take much to get me to boast about my home state. The duties of governor have taken me to the four corners of Colorado and around the country, so I can say with authority that the San Juan Mountains epitomize what makes "Colorful Colorado" one of the best places in the world in which to live and work.

The San Juans have a tumultuous history. Huge volcanic eruptions and the relentless grinding of glacial ice have created a rugged landscape known as the "Switzerland of America." With thirteen peaks reaching 14,000 feet into the sky and more than a few streams where an angler can wet a fly and marvel at Creation, the San Juans are quite simply a paradise for outdoor lovers. In the space of a few dozen miles, a traveler can range from the austere San Luis Valley in the east to some of the highest peaks in the Rockies, traversing a diverse array of life zones from grassy steppe to alpine tundra amid the most spectacular scenery anywhere.

Volcanoes—one of which was the largest ever documented on the planet—created the rich lodes of gold and silver that convinced miners to endure privation and hardship seeking their fortune. They dug in deposits dumped by the glaciers below steep and forbidding slopes that still echo with the sounds of avalanches and rockfall. The San Juans are geology in action.

The Eastern San Juan Mountains explores the important geologic, biologic, and historic events that have made this geologic province one of the most storied places in Colorado. Published by the University Press of Colorado and the Mountain Studies Institute, it stands as a worthy companion to *The Western San Juan Mountains* on the bookshelf of every naturalist, researcher, resident, student, and tourist who wants to probe more deeply into this marvelous place and its history.

I would suggest that to get even more from this book, toss it into the car as you travel along the Silver Thread Scenic Byway between South Fork and Lake City, which is detailed in a road guide featured in this book.

The San Juans are Colorado at its best. The geology, biology, and history of the San Juans are the intellectual threads we can follow to better understand this amazing place. As governor of Colorado, I want to personally thank the two dozen authors—all knowledgeable, experienced researchers in their fields—who have so artfully brought the San Juans that much closer to all of us.

GOVERNOR BILL RITTER JR.
COLORADO'S 41ST GOVERNOR

Preface

THIS VOLUME IS THE COMPANION TO *The Western San Juan Mountains: Their Geology, Ecology, and Human History*, which was published in 1996. The earlier volume includes chapters that focus on all of the San Juan Mountains, and the same is true for this book. Although both volumes address the same general geographic area, and there is some overlap, we made a concerted effort not to duplicate content. One of the primary distinctions between the eastern and western volumes lies in their respective points of interest guides: highways through and around the eastern two-thirds of the range are covered in this book, while the San Juan Skyway and Alpine Loop were documented in the western volume.

No satisfactory boundary line divides the eastern and western San Juan Mountains, only lines of convenience. Although the Continental Divide is a natural boundary, in the San Juans it wanders as much from east to west as from north to south. A geographically appropriate, easy-to-visualize division between eastern and western aspects is a north-south line that runs approximately through Lake City, south to the crossing of the Piedra River by US Highway 160.

This book, as was its companion volume, is divided into four parts: (1) physical environment (geology, hydrology, and climate), (2) ecology (animals, fungi, insects, forests, fens, and lakes), (3) human histories, and (4) a highway points of interest guide. Some of the chapters (especially those that address geology and biology) are not an easy read and may require background study. Because of the use of technical language, a general glossary is included at the end of the book.

The purpose of these two volumes is to inform residents, researchers, educators, and visitors about the workings of a complex mountain system. Thus geology, ecology, climate, hydrology, and human activity are addressed. Neither the western volume nor this text is considered an "end-all" reference to this magnificent mountain range. Indeed, it is our objective to challenge the reader toward a greater understanding of complex mountain systems and to stimulate further research. As a consequence of the success of the western volume, the nonprofit Mountain Studies Institute (MSI) was established in 2002, headquartered in Silverton, Colorado. In 2008 the MSI, along with Fort Lewis College and the University of Colorado, joined together to form the San Juan Collaboratory, dedicated to the study of the San Juan Mountains and other mountain systems.

The editors and authors of this book volunteered their time and their own expense to assemble this text. As with the western volume, the royalties generated will be donated to the San Juan Mountains Undergraduate Research Fund, which provides economic assistance to students conducting research associated with this mountain range.

Acknowledgments

IN ADDITION TO THE TWENTY-EIGHT CONTRIBUTORS TO THIS TEXT, numerous other people behind the scenes have contributed to this publication through reviews, edits, time provided, and donated images. If we miss anyone, please accept our apologies. We especially thank these reviewers: Cathy Ager, Olivier Bachmann, Kip Bossong, Peter Brown, Jeff Coe, Cathy Cripps, Skip Cunningham, Emmett Evanoff, Allen Farnsworth, Tony Gurzick, Shaul Hurwitz, David Jamieson, Donald E. Kendall, Bob Kirkham, Joe Lewandowski, Vince Mathews, Robert Mathiasen, John M. Smethurst III, Donna M. Steele, and Jim White. We are particularly indebted to the anonymous reviewers provided by the University Press of Colorado who gave us insight and allowed us to improve the final product. Mark Williams, in association with the San Juan Collaboratory, was instrumental in providing funds for color photo reproductions. The editors are keenly aware that errors might have occurred, and we accept ultimate responsibility for such oversights. But like other publications of a scientific nature, we view this text as a stepping stone to better understanding and treat it as a current "progress" report.

The Eastern San Juan Mountains

Part 1
Physical Environment of the San Juan Mountains

CHAPTER ONE

A Legacy of Mountains Past and Present in the San Juan Region

···

David A. Gonzales and Karl E. Karlstrom

THROUGHOUT TIME, PEOPLE HAVE BEEN DRAWN TO MOUNTAINS for inspi-ration, recreation, and scientific exploration. Mountains are also vast ware-houses of natural resources and libraries of geologic history.

Mountains form in response to the dynamic forces of our planet. The life spans of mountain belts, from initial uplifts to erosion to base levels, run from tens of mil-lions to hundreds of millions of years. Ancient and active mountain belts are part of the fabric of continental crust and provide clues to events that build and reshape continents. The concept of plate tectonics (Condie 1989) provides a framework within which to investigate and explain mountain building. (For a summary of plate tectonics, refer to *The Western San Juan Mountains*, chapter 2.)

The San Juan Mountains are part of the extensive Southern Rocky Mountains (figure 1.1) and are dominated by some of the highest and most jagged summits in the continental United States. The San Juans reveal a fascinating geologic story of the creation and demise of many mountain ranges in this region during the past 1.8 billion years, including probable current uplift from active mountain-building processes.

The history of any mountain belt is deciphered from its modern landscape and the remaining rock record. An understanding of the geologic evolution of the San Juan Mountains comes from many studies done over the past 125 years. Field

·······

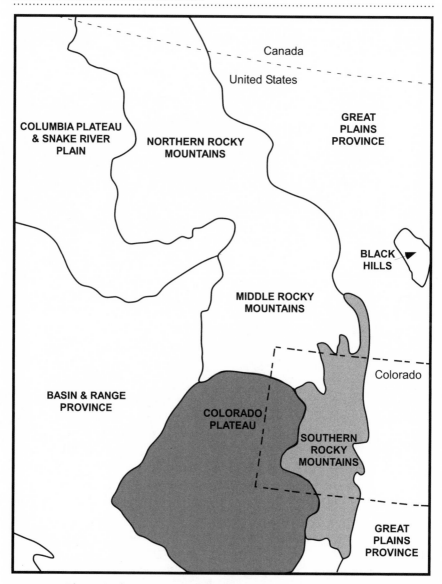

FIGURE 1.1 Physiographic provinces of the Four Corners region showing the extent of the Colorado Plateau and the Southern Rocky Mountains. From http://www2. nature.nps.gov/GEOLOGY/usgsnps/province/INDEXrockyMtnSUBS.gif.

studies of rocks exposed in the region are supported by satellite images, geophysical probes of the deep earth, measurements of rock ages, and analyses of rock composi- tions. These data give geologists the ability to reconstruct the geologic history in the San Juan Mountains over a span of nearly two billion years, almost half the age of the Earth.

The modern San Juan Mountains preserve a history of many geologic events. In this chapter, we provide a basic overview of our understanding of these events within the context of models geologists have proposed.

ASSEMBLY OF A CONTINENT

Evidence of the oldest mountain-building events in the San Juan Mountains is preserved in ancient rocks that are exposed mostly in the Needle Mountains, along the southwestern edge of the San Juan Volcanic Field (plate 1), and on some of the area's isolated peaks and in some of its deeply eroded canyons. The Precambrian rocks in the San Juan Mountains were formed during ancient mountain-building events, between 1.8 and 1.0 billion years ago (Ga). Long ago, these rocks were eroded and beveled to sea level, leaving only their contorted mid-crustal roots to be studied. In the discussion that follows, we use the present location of North America as a reference point, although the continents have shifted positions over time and North America's old rocks did not form at the latitudes and longitudes at which they are now exposed.

The continental crust that forms the foundation of the San Juan Mountains was assembled in a series of continent-scale interactions along the previous edge of North America, during the early to middle Proterozoic (plate 2). Formation of these oldest mountains involved landmass collisions accompanied by magmatism and widespread contortion of rocks. This hypothesized scenario is similar to the tectonic processes active in modern Indonesia. In Indonesia, the Australian plate, which is analogous to the continental nucleus of proto–North America, is colliding with numerous volcanic-arc terranes (e.g., New Guinea and Sumatra), all of which are being sutured together to form a larger continental mass.

It is hypothesized that prior to 1.8 Ga, an ancient landmass known as Laurentia had been assembled by the accretion of rocks older than 2.5 Ga in a series of regional tectonic events (Van Schmus et al. 1993) (plate 2). The margin of this landmass was located along the present-day Colorado-Wyoming border. A volume of evidence indicates that between 1.8 and 1.7 Ga, large tracts of new continental crust were generated in subduction zones on or near the edge of Laurentia. This new crust collided with and was added to the margin of Laurentia in one or more regional tectonic events, forming the crustal foundation of Colorado and the Four Corners region (plate 2) (Gonzales and Van Schmus 2007; Karlstrom et al. 2004; Whitmeyer and Karlstrom 2007).

An alternative hypothesis to explain the creation of ancient crust in Colorado has been proposed in the past ten years (Bickford and Hill 2006; Hill and Bickford 2001). In this model, the edge of Laurentia, consisting mostly of rocks that are now ~1.85 Ga, extended into what is now central Colorado. This margin of Laurentia then underwent extensive rifting and magmatism between 1.8 and 1.7 Ga, producing large volumes of mafic-to-felsic volcanic and plutonic rocks. The key point in this model is that the edge of Laurentia was recycled and modified by later tectonic and magmatic events rather than composed of new additions of Proterozoic magmatic-arc crust.

In the Needle Mountains of southwestern Colorado (plate 2), the Irving Formation and Twilight Gneiss are interpreted as the remnants of ancient amalgamated oceanic volcanic-arc mountains that formed between 1.8 and 1.75 Ga (Barker 1969; Gonzales 1997; Gonzales and Van Schmus 2007). This block of crust was added to Laurentia between 1.73 and 1.7 Ga in a regional tectonic event known as the Yavapai orogeny (Karlstrom et al. 2004) (plate 2).

Following erosion and beveling of ancient volcanic-arc mountains in what is now southwestern Colorado, the region was covered by marine and river deposits sometime after 1.69 Ga. Deposited in tectonic basins that developed during continued regional convergence and compression (Karlstrom et al. 2004), these rocks were later buried during thrusting events and metamorphosed.

The period between 1.4 and 1.0 Ga was marked by episodes of magma generation over the present-day region of the San Juan Mountains. These events involved the emplacement of large masses of mostly granitic magma beneath the margin of Laurentia and were possibly related to renewed convergent tectonic events (subduction or continent collision). The granitic rocks formed during these events have been eroded into some of the majestic and rugged peaks in the Needle Mountains, such as the 14,000-foot Eolus, Sunlight, and Windom Peaks.

Regionally, tectonic events between 1.8 and 1.0 Ga in the San Juan Mountain region (plate 2) generated and modified crustal provinces over an extended period (Gonzales 1997; Gonzales and Van Schmus 2007). Mountain building during this period was a product of the assembly of crustal blocks along the entire edge of Laurentia. These events are collectively responsible for the formation of the crust beneath Colorado and are evident only in the vestiges of the now-eroded Precambrian mountain belts. The crustal-scale fabrics (i.e., faults, folds, and weak zones) created in these ancient mountain-building events, however, appear to have had important controls on the location and formation of many younger events, such as magmatism and mineralization along the Colorado Mineral Belt (Karlstrom et al. 2005; McCoy et al. 2005a).

THE ANCESTRAL ROCKY MOUNTAINS

Geologists have established many lines of evidence to indicate that, starting about 300 Ma, all of the world's major landmasses were assembled into a supercontinent called Pangaea (plate 3). Formation of this supercontinent is hypothesized as a series of major plate collisions that occurred between two large continental masses—the Northern Hemisphere continent, Laurasia (Laurussia in plate 3), and the Southern Hemisphere continent, Gondwana. Laurasia includes modern-day Europe, Asia exclusive of the Indian subcontinent, Greenland, and North America. Gondwana includes modern-day South America, India, Australia, Africa-Arabia, and Antarctica.

The collision of Laurasia and Gondwana progressed from what is now north to south, along what is now the East Coast of North America, causing the uplift and formation of North America's Appalachians and Europe's Caledonian Mountains during the early phases of collision. Stresses from the collisions were transferred throughout the North American Plate. This led to uplift of blocks of crust in a currently northwest-southeast pattern, throughout the region that makes up present-day central and southwestern Colorado and southeastern Utah (plates 3 and 4). This created an ancient mountain range known as the Ancestral Rocky Mountains. These mountains were formed largely by uplift of large blocks of mountains, relative to other large blocks (deep sedimentary basins), along deep, near-vertical faults and, in places, thrust faults.

Remnants and evidence of the Ancestral Rocky Mountain uplifts are found in southwestern Colorado, as well as in Colorado's Front, Elk, and Sangre de Cristo Ranges. Numerous fractures and faults throughout the San Juan Mountains trend roughly parallel to the structural fabrics created during the uplift and development of the Ancestral Rocky Mountains. For example, the Snowdon Mountain block, in the Grenadier Range, was persistently a high area because of the resistant metamorphosed quartz-rich rocks exposed in this fault block.

The main evidence for the Ancestral Rocky Mountains, however, comes not from the mountains, because they were rather quickly eroded and covered by deposits of Permian–Jurassic age. Nor does it come from the faults, which tended to be reactivated and obscured by later mountain-building events. Instead, the best evidence comes from the very thick sedimentary deposits that filled basins adjacent to the uplifts (plate 4). As the fault-bound blocks of the Ancestral Rocky Mountains rose, erosion incised and wore away these highlands. This generated a vast accumulation of river and lake deposits in basins that flanked the highlands, such as the Paradox Basin. These deposits are preserved in rocks that are exposed in the region, such as the maroon-colored rocks of the Cutler Formation. The basins accumulated sediments near sea level, so these mountains might have been like modern-day mountainous areas of Greece, Italy, and the Aegean Sea—rugged uplifts cored by basement rocks but not as high above sea level as the present Rockies.

THE LARAMIDE ROCKY MOUNTAINS

During the Cretaceous, southwestern Colorado was covered by a vast inland seaway (plate 5). This seaway began a retreat as the Four Corners region once again underwent uplift, starting about 80 Ma and extending to perhaps 36 Ma (Cather 2004). During this pulse of mountain building, the Laramide orogeny, the western United

States was compressed and uplifted. In the Southwest, regional deformation of the crust caused broad uplift of the Colorado Plateau and an extensive north-south belt of ranges that form the Laramide Rocky Mountains. The Rocky Mountain belt attained some of its present elevation during the Laramide orogeny, although these mountains were extensively eroded before about 35 Ma and have experienced later periods of uplift.

In the Southern Rocky Mountains (figure 1.1), the Laramide orogeny created a series of mountain ranges and related structural sedimentary basins. In many areas, the mountain units are bound by upturned sedimentary rocks and bordered by steep reverse and thrust faults. The San Juan Mountains rest on part of the fault-truncated Sangre de Cristo Range to the east, the Sawatch Range to the north, and the San Juan Uplift (Needle Mountains Uplift) to the south (plate 1).

Along the southern edge of the San Juan Mountains, the Needle Mountains block was uplifted during the Laramide, as a result of compression and shortening, to form a domal structure that is elongated to the northeast. Cather (2004) discusses current ideas regarding the structural evolution of the San Juan Uplift and San Juan Basin. He summarizes recent evidence for dextral-transpressive deformation along the eastern margin of the Colorado Plateau in the development of Laramide uplifts and basins in the Southern Rocky Mountains, although the tectonics models for the Laramide orogeny in this region remain controversial (Seager 2004).

Paleozoic to Cenozoic sedimentary rocks are bent and tilted on the flanks of San Juan Uplift and are relatively flat-lying in the central part of the uplifted block. A deep bend on the southern edge of the Laramide uplift in southwestern Colorado created an asymmetrical syncline that defines the San Juan Basin (plate 1). This basin presents about 3,000 feet of structural relief and ~18,000 feet of structural relief between the Needle Mountains Uplift and the deepest part of the San Juan Basin (Kelley 1955). Cather (2004) reported that the evolution of the San Juan Basin involved diachronous subsidence and deposition, with at least three distinct periods of basin development.

The San Juan Basin is bound by abrupt and pronounced structural margins to the northwest (Hogback Monocline), northeast (southwest limb of the Archuleta Anticlinorium), and east (Nacimiento Uplift). In southwestern Colorado, tilted sedimentary strata of the Hogback define a flexure, partly fault-controlled, on the southern and eastern edges of the San Juan Uplift. As uplift of the Needle Mountains block continued into the Paleocene and Eocene, it was dissected by erosion that shed vast amounts of detritus as stream and fan deposits (Ojo Alamo Sandstone, Animas Formation, Nacimiento Formation, and San Jose Formation) into the San Juan Basin. Rock fragments in these deposits show a progressive uncovering of the uplifted block, down to its Proterozoic core. The Animas Formation also contains abundant fragments of volcanic rock, indicating the existence of volcanic centers to the north of the San Juan Basin prior to ~55 Ma.

At about 70 Ma, intermediate to felsic magmas rose into the crust, forming numerous mushroom-shaped bodies called laccoliths along the western edge of Colorado. Eroded remnants of Laramide laccolithic mountains in this part of the world include the La Plata, Abajo, and La Sal Mountains and smaller masses exposed on the fringes of the eastern San Juan Volcanic Field. These Laramide-intrusive complexes form the older component of a belt of Tertiary magmatism that extends from north-central into southwestern Colorado and that provided the loci of many mineralized deposits during the generation and development of the modern Rocky Mountains. Convincing geologic evidence (McCoy et al. 2005a; Tweto and Sims 1963) suggests that this belt, referred to as the Colorado Mineral Belt, has lineage in fractures and cracks that developed during the formation of Proterozoic mountain belts on the North American craton.

The widely accepted model (Dickinson 1981; Dickinson and Snyder 1978; Lipman, Prostka, and Christiansen 1971) for the Laramide orogenic event involves subduction that extended far inland, beneath North America's West Coast. In this "flat-slab," or shallow-subduction, model, the rate of subduction increased over time. This is thought to have caused the subducted slab to become more buoyant and rise to a lower angle beneath the North American Plate. This process caused "scraping" of the subducted plate 1,000 km inboard of the western edge of North America, leading to crustal compression and uplift (Dickinson 1981; Dickinson and Snyder 1978). Gutscher and colleagues (2000) offer convincing evidence for the flat-slab subduction in segments of the active subduction system of western South America. They argue that this process is caused by the subduction of buoyant oceanic plateaus, which allows the subducted slab to be driven up to 800 km from the active trench. Further deformation, up to 200 km inboard, is documented as block-type uplifts, or transpressional faulting for oblique convergence.

Other explanations for the formation of the Laramide Rocky Mountains have been proposed, but they have not provided substantial evidence to gain the support of most of the geosciences community. For example, Gilluly (1971, 1973) proposed that the Laramide orogeny was related to heterogeneities in the crust-mantle structure. Some recent studies provide evidence that forces in the earth at that time might have caused reactivation and movement along old fractures, which facilitated uplift and formed pathways for melted rock (Karlstrom and Humphreys 1998; Karlstrom et al. 2005; McCoy et al. 2005a, 2005b; Mutschler, Larsen, and Bruce 1987). In this model, the need for subduction as the catalyst for uplift and magmatism is less critical because magma production is dependent upon the release of pressure on the mantle to initiate melting rather than on the melting of the subducted oceanic slab. This is further supported by geophysical and geochemical studies that have concluded that the Laramide event did not have a substantial impact on the thermal state of the lithosphere; such an impact would be expected as a result of magmatism related to the proposed low-angle subduction that accompanied mountain building (Karlstrom et al. 2005; Livaccari and Perry 1993; Riter and Smith 1996).

McCandless and Johnson (2000) and McMillan (2004) note that although most of the geochemical and isotopic signatures of igneous rocks generated during this period are consistent with volcanic-arc systems, the temporal patterns of magmatism do not fully support this model. Furthermore, Gutscher and colleagues (2000) used geophysical data to argue that low-angle subduction creates a "cool thermal structure" in the lithosphere, which reduces heat flow and inhibits melting and arc magmatism. The more realistic model for Laramide magmatism may therefore involve some mechanism other than arc subduction.

THE SAN JUAN VOLCANIC FIELD

According to the flat-slab model, the rate of subduction involved in the formation of the modern Rocky Mountains slowed down at ~35 Ma. Foundering and sinking of the subducted plate to the west from its previous contact with the North American plate induced inflow of asthenospheric mantle, leading to widespread melting of the lithosphere and magmatism (Lipman, Prostka, and Christiansen 1971). This pulse of magmatism generated some of the largest volcanic eruptions in the Earth's geologic record (plate 6; see also Lipman and McIntosh, chapter 2, this volume).

Oligocene magmatic events in the Four Corners region were linked to mantle-magma sources (Carlson and Nowell 2001; Farmer, Bailey, and Elkins-Tanton 2007; Johnson 1991; Lipman 2007; Lipman et al. 1978; Riciputi et al. 1995; Roden 1981). Evidence supports a model in which upwelling mantle magmas invaded the lithosphere and caused partial melting of the crust and extensive production of intermediate to felsic magmas. These magmas were the catalyst for the production of eruptive phases in the San Juan Volcanic Field (Farmer, Bailey, and Elkins-Tanton 2008; Johnson 1991; Lipman 2007). This regional magmatic-volcanic event is commonly called the "ignimbrite flare-up," which was initiated farther west and north of the San Juan Mountains, although the San Juan Volcanic Field is a spectacular easterly expression of this process.

The ignimbrite flare-up resulted in the eruptions of large volumes of viscous, gas-rich magmas from about twenty caldera complexes preserved in the San Juan Volcanic Field (e.g., the Silverton–Lake City Calderas). Each caldera represents the collapse of land surface as a result of the massive expulsion of lava and ash. The calderas are inferred, on the basis of geophysical data (McCoy et al. 2005b; Plouff and Pakiser 1972), to be underlain by large volumes of granite that were emplaced at this time.

While mountain building by erupting magmas and accumulating volcanic deposits was somewhat localized, the ignimbrite flare-up also caused regional uplift of the Southern Rocky Mountains. Evidence for this is provided by analysis of the cooling histories of rocks in the Great Plains and in the Rockies, indicating that the

Southern Rocky Mountains were tilted up in a broad dome that caused additional erosion and uplift of the Rockies (Roy et al. 2004a, 2004b). This Oligocene contribution to the uplift of the Rockies was driven by deep-seated magmatism, and it resulted in broad doming of the region.

Current studies along the southern and western edges of the San Juan Mountains lend evidence for further jostling and uplift of blocks of the earth's crust at the onset of extensive volcanic eruptions of the San Juan Volcanic Field. During the early pulses of volcanic activity, as convincingly evident from field studies (Gonzales, Kray, and Gianniny 2005; Harraden and Gonzales 2007), the western San Juan Mountains, at least, were undergoing a renewed period of uplift. This uplift may reflect continued southwest-to-northeast compression (Cather 2004), but the mechanism of uplift is not well understood. Uplift was manifested in the reactivation of large blocks of crust that were eroded, leaving river gravels. These river gravels, called the Telluride Conglomerate, contain fragments of intermediate to felsic volcanic rocks similar to those found in the San Juan Volcanic Field, suggesting a contemporaneous series of events that must somehow be linked. The mountains were being uplifted and eroded as volcanic activity waned and rifting began in the region. Uplift between 40 and 30 Ma might have contributed to further elevation of the Needle Mountains and related subsidence in the San Juan Basin.

About 5 million years after this Early-Oligocene uplift and volcanism, crustal stretching and thinning began near the western and eastern boundaries of the San Juan Mountains. Incipient rifting was manifested in a southwest-to-northeast belt of potassium-rich mafic dikes and explosive diatreme volcanoes that formed as early as 28 Ma (Laughlin et al. 1986; Nowell 1993; Roden and Smith 1979). The eroded remnants ("necks") of these diatremes are exemplified by towers of dark-colored rock that dot the Navajo Volcanic Field. Ship Rock is the best-known of these iconic features. East of the San Juan Mountains, basin development along the Rio Grande rift indicates the onset of rifting between 35 and 30 Ma (Smith 2004) (plate 6). A series of generally north-south–trending mafic dikes located between Durango and Pagosa Springs (Gonzales 2009; Gonzales et al. 2006; Gonzales et al. 2010) may also provide further evidence for regional extension just after the onset of volcanism in the San Juan Volcanic Field, although the ages of these dikes are not constrained.

MID- TO LATE-TERTIARY ROCKY MOUNTAINS: LOCAL COLLAPSE OF HIGH TOPOGRAPHY

The Southern Rocky Mountains of New Mexico continued to undergo extension in an east-west direction between 25 and 10 Ma. This continental rifting overlaps in timing with the formation of extensional structures of the Basin and Range province, although the location of the north-south rift—parallel to the north-south Rocky Mountains uplift—and the fact that the rifting decreases in magnitude northward suggest that it was driven in part by gravitational collapse of the high mountains that had formed earlier. The Rio Grande Rift is not well-developed northward in Colorado, although the active extension is still ongoing, as indicated by the hot springs, high heat flow, and young faulting along the Arkansas Valley and in other areas east of the San Juan Mountains.

Is uplift of the San Juan Mountains ongoing? This region is underlain by some of the lowest-velocity crust found anywhere—the Aspen anomaly (Karlstrom et al. 2005). The Aspen anomaly is characterized by a mantle zone similar in shape and character to each of those found in the Yellowstone and Jemez (Valles) hotspot anomalies. These other areas are associated with recent (< 1 Ma) caldera eruptions, whereas the Aspen anomaly is not—yet. Mantle gases detected in hot springs throughout and adjacent to the northern San Juan Mountains (Newell et al. 2005) suggest young and active mantle upwelling beneath the region. Reiter (2008) provides evidence from thermal- and gravity-anomaly sources for a contribution to current uplift in the San Juan Mountains, from upper-mantle buoyancy. One view is that the very rugged nature of the San Juan Mountains reflects erosion responding to this ongoing uplift; another model is that relatively wet climates during the last few million years have caused increased erosion of a previously elevated region.

THE FUTURE OF THE SAN JUAN MOUNTAINS

In the past several million years, the San Juan Mountains have been worn and sculpted by the active erosive forces of glaciers and rivers. Removal of crustal mass may be facilitating minor uplift. This erosion has also exposed windows into the crust, allowing geologists to study older rocks and gain insight into the mountain-building processes responsible for (1) formation of the continent, as recorded by the contorted deep-crustal roots of a series of planed Precambrian mountain belts; (2) formation of the Ancestral Rocky Mountains by Paleozoic plate collisions; (3) formation of the modern Rocky Mountains by Laramide compression and uplift; (4) the San Juan Volcanic Field and concurrent uplift; and (5) ongoing uplift related to continued mantle buoyancy, the Rio Grande Rift, and mantle degassing associated with the Aspen anomaly.

The southwestern United States is undergoing widespread stretching and extension as the Pacific and North American plates interact along the San Andreas Fault in California. In tens of millions of years, the San Juan Mountains may also be dismembered by this process, forming a landscape similar to that of the Basin and Range country in Nevada and in western and southern Arizona. The Rio Grande rift may be unzipping northward. Alternatively, the partially melted mantle zone known as the Aspen anomaly may herald a future caldera eruption similar to Yellowstone, the Jemez Caldera, and the San Juan calderas before them. If so, uplift and building of mountains may continue. The question of whether the San Juan Mountains are presently going up, as a result of mantle uplift, or down, as a result of erosion, is unresolved. Eventually, the San Juan Mountains will be eroded to base level, as were the Precambrian arc mountains and the Ancestral Rockies.

Whatever the future holds for the San Juan Mountains, cycles of mountain building and beveling will continue as plate tectonics and erosion transform layers of crust into the yet-unwritten pages of Earth's history.

REFERENCES

Barker, F. 1969. *Precambrian Geology of the Needle Mountains*. US Geological Survey Professional Paper 644-A, 35.

Bickford, M. E., and B. M. Hill. 2006. Is the Arc-Accretion Model for the 1780–1650 Ma Crustal Evolution of Southern Laurentia Correct? Let's Look at the Rocks. *GSA Abstracts with Programs* 38(6):12.

Carlson, R. W., and G. M. Nowell. 2001. Olivine-Poor Sources for Mantle-Derived Magmas: Os and Hf Isotopic Evidence from Potassic Magmas of the Colorado Plateau. Paper no. 2000GC000128. *Geochemistry, Geophysics, Geosystems* 2(6):17 pp.

Cather, S. M. 2004. Laramide Orogeny in Central and Northern New Mexico and Southern Colorado. In G. H. Mack and K. A. Giles, eds., *The Geology of New Mexico: A Geologic History*. Special Publication 11. Printed in Canada: New Mexico Geological Society, 203–248.

Condie, K. C. 1989. *Plate Tectonics and Crustal Evolution*, 3rd ed. Oxford: Pergamon.

Dickinson, W. R. 1981. Plate Tectonic Evolution of the Southern Cordillera. In W. R. Dickinson and W. D. Payne, eds., *Relations of Tectonics to Ore Deposits in the Southern Cordillera*. Tucson: Arizona Geological Society Digest 14:113–135.

Dickinson, W. R., and W. S. Snyder. 1978. *Plate Tectonics of the Laramide Orogeny*. Memoir 151. Boulder: Geological Society of America, 355–366.

Farmer, G. L., T. Bailey, and L. T. Elkins-Tanton. 2008. Mantle Source Volumes and the Origin of the Mid-Tertiary Ignimbrite Flare-up in the Southern Rocky Mountains, Western U.S. *Lithos* 102:279–294.

Gilluly, J. 1971. Plate Tectonics and Magmatic Evolution. *Geological Society of America Bulletin* 82(9):2382–2396.

———. 1973. Steady Plate Motion and Episodic Orogeny and Magmatism. *Geological Society of America Bulletin* 84(2):499–514.

Gonzales, D. A. 1997. Crustal Evolution of the Needle Mountains Proterozoic Complex, Southwestern Colorado. PhD dissertation, University of Kansas, Lawrence.

———. 2009. Insight into Middle Tertiary Crustal Extension and Mantle Magmatism along the Northern Margin of the San Juan Basin at the Onset of Volcanism in the San Juan Mountains. Paper no. 10-4. *Geological Society of America Abstracts with Programs* 41(6):18. RMS 61st Annual Meeting, May 11–13.

Gonzales, D. A., R. T. Burgess, M. R. Critchley, and B. E. Turner. 2006. New Perspectives on the Emplacement Mechanisms Involved in Diatreme Formation in the Northeastern Navajo Volcanic Field, Southwestern Colorado. Abstract 202. *Geological Society of America Abstracts with Programs* 38(6):4. RMS 58th Annual Meeting, May 17–19.

Gonzales, D. A., B. Kray, and G. Gianniny. 2005. Insight into the Timing of Cenozoic Uplift in Southwestern Colorado. *Geological Society of America Abstracts with Programs* 37(6):16.

Gonzales, D. A., B. E. Turner, R. T. Burgess, C. C. Holnback, and M. R. Critchley. 2010. New Insight into the Timing and History of Diatreme-Dike Complexes of the Northeastern Navajo Volcanic Field, Southwestern Colorado. In J. E. Fassett, K. Zeigler, and V. W. Lueth, eds., *Geology of the Four Corners Country*. New Mexico Geological Society 60th Field Conference Guidebook, 163–172, Socorro, New Mexico.

Gonzales, D. A., and W. R. Van Schmus. 2007. Proterozoic History and Crustal Evolution in Southwestern Colorado: Insight from U/Pb and Sm/Nd Data. *Precambrian Research* 154(1–2):31–70.

Gutscher, M., W. Spakman, H. Bijwaard, and R. R. Engdahl. 2000. Geodynamics of Flat Subduction: Seismicity and Tomographic Constraints from the Andean Margin. *Tectonic* 19(5):814–833.

Harraden, C. L., and D. A. Gonzeles. 2007. An Assessment of the Depositional History of the Telluride Conglomerate and Implications for Mid-Tertiary Tectonic-Volcanic Events in the Western San Juan Mountains, Colorado. *Geological Society of America Abstracts with Programs* 39(5):10.

Hill, B. M., and M. E. Bickford. 2001. Paleoproterozoic Rocks of Central Colorado: Accreted Arcs or Extended Crust. *Geology* 29:1015–1018.

Johnson, C. M. 1991. Large-Scale Crust Formation and Lithosphere Modification beneath Middle- to Late-Cenozoic Calderas and Volcanic Fields, Western North America. *Journal of Geophysical Research* 96(B8):13,485–507.

Karlstrom, K. E., J. M. Amato, M. L. Williams, M. Heizler, C. A. Shaw, A. S. Read, and P. Bauer. 2004. Proterozoic Tectonic Evolution of the New Mexico Region: A Synthesis. In G. H. Mack and K. A. Giles, eds., *The Geology of New Mexico: A Geologic History*. Special Publication 11. Printed in Canada: New Mexico Geological Society, 1–34.

Karlstrom, K. E., and E. D. Humphreys. 1998. Persistent Influence of Proterozoic Accretionary Boundaries in the Tectonic Evolution of Southwestern North America: Interaction of Cratonic Grain and Mantle Modification Events. *Rocky Mountain Geology* 33(2):161–179.

Karlstrom, K. E., S. J. Whitmeyer, K. Dueker, M. L. Williams, S. A. Bowring, A. Levander, E. D. Humphreys, G. R. Keller, and the CD-ROM Working Group. 2005. Synthesis of Results from the CD-ROM Experiment: 4-D Image of the Lithosphere beneath the Rocky Mountains and Implications for Understanding the Evolution of Continental Lithosphere. In K. E. Karlstrom and G. R. Keller, eds., *The Rocky Mountain Region—An Evolving Lithosphere: Tectonics, Geochemistry, and Geophysics*. Geophysical Monograph Series 154. Washington, DC: American Geophysical Union, 421–441.

Kelley, V. C. 1955. *Regional Tectonics of the Colorado Plateau and Relationships to the Origin and Distribution of Uranium*. Publications in Geology 5. Albuquerque: University of New Mexico.

Laughlin, A. W., M. J. Aldrich, M. Shafiqullah, and J. Husler. 1986. Tectonic Implications of the Age, Composition, and Orientation of the Lamprophyre Dikes, Navajo Volcanic Field, Arizona. *Earth and Planetary Science Letters* 76:361–374.

Lipman, P. W. 2006. Geologic Map of the Central San Juan Caldera Complex, Southwestern Colorado. US Geologic Investigation Series I–2799.

———. 2007. Incremental Assembly and Prolonged Consolidation of Cordilleran Magma Chambers: Evidence from the Southern Rocky Mountain Volcanic Field. *Geosphere* 3(1):42–70.

Lipman, P. W., B. R. Doe, C. E. Hedge, and T. A. Steven. 1978. Petrologic Evolution of the San Juan Volcanic Field, Southwestern Colorado: Pb and Sr Isotope Evidence. *Geological Society of America Bulletin* 89:59–82.

Lipman, P. W., H. J. Prostka, and R. L. Christiansen. 1971. Evolving Subduction Zones in the Western United States, as Interpreted from Igneous Rocks. *Science* 174:821–825.

Livaccari, R. F., and F. V. Perry. 1993. Isotopic Evidence for Preservation of the Cordilleran Lithospheric Mantle during the Sevier-Laramide Orogeny, Western United States. *Geology* 21:719–722.

McCandless, T. E., and D. A. Johnson. 2000. Trends and Non-Trends in Laramide Magmatism. *Geological Society of America Abstract with Programs* 32(7):107.

McCoy, A., K. E. Karlstrom, M. L. Williams, and C. A. Shaw. 2005a. Proterozoic Ancestry of the Colorado Mineral Belt: 1.4 Ga Shear Zone System in Central Colorado. In K. E. Karlstrom and G. R. Keller, eds., *The Rocky Mountain Region—An Evolving Lithosphere: Tectonics, Geochemistry, and Geophysics.* Geophysical Monograph 154. Washington, DC: American Geophysical Union, 71–90.

McCoy, A., R. Mousoumi, L. Trevino, and R. G. Keller. 2005b. Gravity Modeling of the Colorado Mineral Belt. In K. E. Karlstrom and G. R. Keller, eds., *The Rocky Mountain Region—An Evolving Lithosphere: Tectonics, Geochemistry, and Geophysics.* Geophysical Monograph 154. Washington, DC: American Geophysical Union, 99–106.

McMillan, N. J. 2004. Magmatic Record of Laramide Subduction and the Transition to Tertiary Extension: Upper Cretaceous through Eocene Igneous Rocks of New Mexico. In G. H. Mack and K. A. Giles, eds., *The Geology of New Mexico: A Geologic History.* Special Publication 11. Printed in Canada: New Mexico Geological Society, 249–267.

Mutschler, F. E., E. E. Larsen, and R. M. Bruce. 1987. Laramide and Younger Magmatism in Colorado. *Colorado School of Mines Quarterly* 82:1–45.

Newell, D. L., L. J. Crossey, K. E. Karlstrom, and T. B. Fischer. 2005. Continental-Scale Links between the Mantle and Groundwater Systems of the Western United States: Evidence from Travertine Springs and Regional He Isotope Data. *Geological Society of America Today* 15(12):1–10.

Nowell, G. M. 1993. Cenozoic Potassic Magmatism and Uplift of the Western United States. PhD dissertation, Open University, Milton Keynes, UK.

Plouff, D., and L. C. Pakiser. 1972. *Gravity Study of the San Juan Mountains, Colorado.* United States Geological Survey Professional Survey 800-B:183–190.

Reiter, M. 2008. Geothermal Anomalies in the Crust and Upper Mantle along Southern Rocky Mountain Transitions. *Geological Society of America Bulletin* 120(3–4):431–441.

Riciputi, L. R., C. M. Johnson, D. A. Sawyer, and P. W. Lipman. 1995. Crustal and Magmatic Evolution in a Large Multicyclic Caldera Complex: Isotopic Evidence from the Central San Juan Volcanic Field. *Journal Volcanology and Geothermal Research* 67:1–28.

Riter, J.C.A., and D. Smith. 1996. Xenolith Constraints on the Thermal History of the Mantle below the Colorado Plateau. *Geology* 24:267–270.

Roden, M. F. 1981. Origin of Coexisting Minette and Ultramafic Breccia, Navajo Volcanic Field. *Contributions to Mineralogy and Petrology* 77:195–206.

Roden, M. F., and D. Smith. 1979. Field Geology, Chemistry, and Petrology of Buell Park Minette Diatreme, Apache County, Arizona. In F. R. Boyd and H.O.A. Meyer, eds., *Kimberlites, Diatremes, and Diamonds: Their Geology, Petrology, and Geochemistry.* Washington, DC: American Geophysical Union, 364–381.

Roy, M., S. Kelley, F. Pazzaglia, S. Cather, and M. House. 2004a. Middle Tertiary Buoyancy Modification and Its Relationship to Rock Exhumation, Cooling, and Subsequent Extension at the Eastern Margin of the Colorado Plateau. *Geology* 32:925–928.

———. 2004b. Middle Tertiary Buoyancy Modification and Its Relationship to Rock Exhumation, Cooling, and Subsequent Extension at the Eastern Margin of the Colorado Plateau. *Geological Society of America Abstracts with Programs* 36(5):119.

Seager, W. 2004. Laramide (Late Cretaceous–Eocene) Tectonics of Southwestern New Mexico. In G. H. Mack and K. A. Giles, eds., *The Geology of New Mexico: A Geologic History*. Special Publication 11. Printed in Canada: New Mexico Geological Society, 183–202.

Smith, G. 2004. Middle to Late Cenozoic Development of the Rio Grande Rift and Adjacent Regions in Northern New Mexico. In G. H. Mack and K. A. Giles, eds., *The Geology of New Mexico: A Geologic History*. Special Publication 11. Printed in Canada: New Mexico Geological Society, 1–34.

Tweto, O., and P. K. Sims. 1963. Precambrian Ancestry of the Colorado Mineral Belt. *Geological Society of America Bulletin* 74:991–1014.

Van Schmus, W. R., and 24 others. 1993. Transcontinental Proterozoic Provinces. In J. C. Reed Jr., M. E. Bickford, R. S. Houston, P. K. Link, D. W. Rankin, P. K. Sims, and W. R. Van Schums, eds., *Precambrian: Conterminous U.S.* The Geology of North America C-2. Geological Society of America, 171–334.

Whitmeyer, S., and K. E. Karlstrom. 2007. Tectonic Model for the Proterozoic Growth of North America. *Geosphere* 3(2):220–259; doi 10.1130/GES00055.1.

Tertiary Volcanism in the Eastern San Juan Mountains

Peter W. Lipman and William C. McIntosh

ANDESITIC TO RHYOLITIC VOLCANIC ROCKS of the San Juan Mountains, along with associated epithermal ores, were studied intensively during the twentieth century (e.g., Larsen and Cross 1956), culminating with detailed study of the Creede Mining District by Steven and Ratté (1965) as well as regional field and volcanological studies (Lipman, Steven, and Mehnert 1970; Steven and Lipman 1976; Steven et al. 1974). While much had previously been learned about the evolution of several complex caldera clusters from which at least twenty-two major ignimbrite (ash-flow) sheets (each 150–5,000 km³) were erupted at 30–26 Ma, research in support of the Creede Scientific Drilling Project (Bethke and Hay 2000; Lipman 2000, 2006) and more recent mapping, petrologic study, and high-resolution geochronology have provided new insights regarding the stratigraphy, duration of volcanism, eruptive processes, magmatic evolution, and regional structure in eastern parts of this widely known volcanic region.

REGIONAL SETTING

The San Juan Mountains are the largest erosional remnants of a composite volcanic field (figure 2.1) that covered much of the Southern Rocky Mountains in mid-Tertiary time (Lipman 2007; Steven 1975). As explosive volcanism migrated

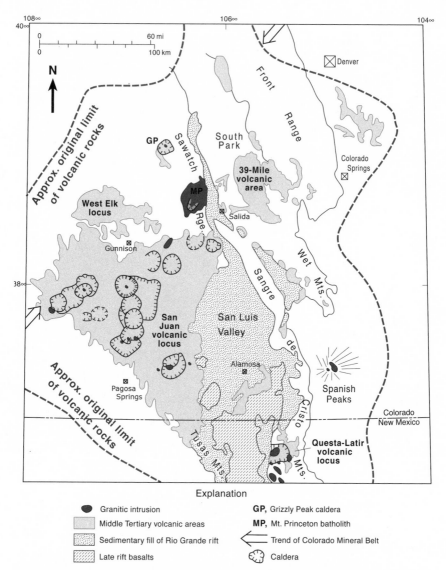

FIGURE 2.1 Map of the Southern Rocky Mountain Volcanic Field, showing ignimbrite calderas, major erosional remnants, and inferred original extent of once nearly continuous mid-Tertiary volcanic cover, caldera-related granitic intrusions, and later Tertiary sedimentary fill of the Rio Grande Rift Zone. Arrows indicate trend of Late-Cretaceous–Early-Tertiary (Laramide) intrusions of the Colorado Mineral Belt. Modified from McIntosh and Chapin (2004).

southward along the Sawatch Range Trend toward the San Juan region between 37 and 33 Ma (McIntosh and Chapin 2004), widely scattered centers of intermediate composition (dominantly andesite, with lesser volumes of dacite and minor volumes of rhyolite) erupted lava flows and flanking volcaniclastic breccias. Volcanic rocks in the San Juan region consist mainly of intermediate-composition lavas and breccias erupted at about 35–30 Ma from scattered central volcanoes (Conejos Formation) and overlying voluminous ignimbrites erupted from caldera sources. At about 26 Ma, volcanism shifted to a bimodal assemblage dominated by trachybasalt and silicic rhyolite, concurrent with the inception of regional extension during establishment of the Rio Grande Rift (Lipman, Steven, and Mehnert 1970).

Tertiary volcanic rocks preserved in the San Juan region occupy an area of more than 25,000 km^2 and have a volume of about 40,000 km^3. They cover a varied basement of Precambrian to lower-Tertiary rocks along the uplifted and eroded western margin of the Late-Cretaceous–Early-Tertiary (Laramide) uplifts of the Southern Rocky Mountains and adjoining portions of the San Juan Basin, on the eastern Colorado Plateau (figure 2.1). The San Juan region is one of many loci of Tertiary volcanic activity that developed along the eastern Cordilleran margin of the North American plate, probably in a complex response to changing subduction geometry along its western margin. Other such loci include the Sierra Madre Occidental, Trans-Pecos, Mogollon-Datil, Absaroka, Challis, and Lowland Creek fields.

In the eastern and central San Juan Mountains (figure 2.1), eruptions of at least 10,000–12,000 km^3 of dacitic-rhyolitic magma as fifteen major ignimbrite sheets (individually 150–5,000 km^3) were accompanied by recurrent caldera subsidences between 29.5 and 26.9 Ma. Voluminous andesitic-dacitic lavas and breccias erupted from central volcanoes prior to the ignimbrite eruptions, and similar lava eruptions continued within and adjacent to the calderas during silicic explosive volcanism (Lipman, Doe, and Hedge 1978). Exposed calderas vary from 10 to 75 km in maximum dimension, with the largest calderas associated with the most voluminous eruptions (table 2.1).

IGNIMBRITE AND CALDERA PROCESSES

The ignimbrite sheets and calderas of the San Juan region (table 2.1) have been widely recognized as exceptional sites for the study of explosive volcanic processes (Lipman 2000; Steven and Lipman 1976; Steven and Ratté 1965). Lavas, volcaniclastic sedimentary rocks, and intrusions emplaced concurrently with ignimbrites in the San Juan Mountains define multiple caldera cycles, based on affinities of geographic distribution and stratigraphic sequence, isotopic age, paleomagnetic pole directions, and petrologic character (table 2.1).

Briefly, many large ignimbrite calderas, such as those in the San Juan region, form at sites of previous volcanism that conditions the Earth's crust for shallow

Table 2.1. Ignimbrites and calderas of the San Juan region

	Ignimbrite					Caldera
Site	Name	% SiO^2	Rock, phenocrysts	Vol, km3	Age, Ma	Name
Western San Juan	Sunshine Peak	Zoned, 76–68	Qtz, sodic san	200–500	23.0	Lake City
Central San Juan	Snowshoe Mountain	62–66	Xl-rich dacite	> 500	26.9	Creede
San Luis complex	Nelson Mountain	Zoned, 74–63	Xp rhy; xl dacite	> 500	26.9	San Luis–Cochetopa
	Cebolla Creek	61–64	Xl dacite, hbl, no san	250	26.9	San Luis complex
	Rat Creek	Zoned, 74–65	Xp rhy; xl dacite	150	26.9	San Luis complex
Central San Juan cluster	Wason Park	Zoned, 72–63	Xl rhyolite; xl dacite	>500	27.4	South River
	Blue Creek	64–68	Xl dacite, no san	250	27.45	[concealed]
	Carpenter Ridge	Zoned, 74–66	Xp rhy; xl dacite	> 1,000	27.5	Bachelor
Western San Juan	Crystal Lake	72–74	Xp rhyolite	50–100	27.6	Silverton
Central San Juan	Fish Canyon	66–68	Xl dacite, san, hbl, qtz	5,000	28.0	La Garita
Southeast San Juan	Chiquito Peak	64–67	Xl dacite, sanidine	500–1,000	28.35	Platoro
	Sapinero Mesa	72–75	Xp rhyolite	> 1,000	28.35	Uncompahgre–San Juan
	Dillon Mesa	72–75	Xp rhyolite	50–100	28.4	Uncompahgre?
Western San Juan	Blue Mesa	72–74	Xp rhyolite	200–500	28.5	Lost Lakes (buried)
	Ute Ridge	66–68	Xl dacite, sanidine	250–500	28.6	Ute Creek
Central San Juan	Masonic Park	62–66	Xl dacite, no san	500	28.7	[concealed]
	South Park	68–70	Xl dacite, sanidine	50–100	28.8	Platoro/Summitville?
	Ra Jadero	64–66	Xl dacite, sanidine	150	28.8	Summitville?
Southeastern San Juan	Ojito Creek	67–70	Xl dacite, no san	100	29.1	Summitville?
	La Jara Canyon	66–68	Xl dacite, no san	500–1,000	29.3	Platoro
	Black Mountain	67–69	Xl dacite, no san	200–500	29.4	Platoro
	(Barret Creek)	65–73	(Xl dacite-rhy lavas)	50?	29.8	[failed?]
Northeastern San Juan	Saguache Creek	73–75	Alkali rhyolite, no bio	250–500	32.2	North Pass
	Bonanza	74–63	Zoned: dac-rhy-dac	250?	33.0	Bonanza-Gribbles

Abbreviations: rhy, rhyolite; dac, dacite; san, sanidine; qtz, quartz; hbl, hornblende; xl, crystal; xp, crystal-poor
Note: Ages are best estimates, most with an uncertainty of < 0.1 million years.
Source: Modified from Lipman 2007.

accumulations of silicic, gas-rich magma. Large eruptions (> 50–100 km^3 of ignimbrite magma) cause caldera collapse concurrently with venting, as indicated by thick intra-caldera ignimbrite fill and interleaved collapse-slide breccias from oversteepened crater walls. Volumes of intra-caldera and outflow tuffs tend to be subequal; correlation between them is commonly complicated by contrasts in the abundance and size of phenocrysts and lithic fragments, degree of welding, devitrification, alteration, and chemical composition. Structural boundaries of calderas are commonly single ring faults or composite ring-fault zones that dip steeply. Scalloped topographic walls beyond the structural boundaries of most calderas are a result of secondary gravitational slumping during subsidence. The area and volume of caldera collapse are roughly proportional to the amount of erupted material. Postcollapse volcanism may occur from varied vent geometries within ignimbrite calderas; ring-vent eruptions are most common in resurgent calderas and reflect renewed magmatic pressure. Resurgence within a caldera may result in a symmetrically uplifted dome or more geometrically complex forms. In addition to resurgence within individual calderas, broader magmatic uplift occurs within some silicic volcanic fields, reflecting isostatic adjustment to emplacement of associated subvolcanic batholiths. Large intrusions related to resurgence are exposed centrally or along the margins of some deeply eroded calderas. Hydrothermal activity and mineralization accompany all stages of ignimbrite magmatism, becoming dominant late in caldera evolution. Some mineralization occurs millions of years after caldera collapse, where the caldera served primarily as a structural control for late intrusions and associated hydrothermal systems.

EARLY CENTRAL VOLCANOES

Construction of large stratocones, dominantly composed of intermediate-composition lavas and breccias, marked the inception of volcanism widely in the San Juan region, generally commencing several million years prior to initial ignimbrite eruptions in the area. These lavas and breccias have long been included in the Conejos Formation (Larsen and Cross 1956; Steven and Lipman 1976). Composite volumes of the early volcanoes are large; stratigraphic sequences are commonly more than a kilometer thick, and total volume, estimated at 25,000 km^3 (Lipman, Steven, and Mehnert 1970), exceeds that of the later-erupted ignimbrites by at least 50 percent. High-K calc-alkaline andesite and dacite are dominant compositions; true basalt is almost absent. Conspicuous eroded remnants of such stratocones are clustered in the areas that subsequently became sites of ignimbrite eruption and caldera collapse; this is especially clear in the Bonanza and Platoro centers in the eastern San Juan region. Between these clusters is a north-south alignment of compositionally diverse volcanoes (the Tracy Creek, Beidell Creek, and Summer Coon centers) along the eastern margin of the mountains.

Small hypabyssal stocks, commonly exposed in cores of these volcanoes and loci for outward-radiating dikes, are similar in composition to the dominant eruptive products; large silicic plutons are absent. Along with flanking lavas, these widely scattered high-level intrusive centers have been inferred to record initial phases of assembly of upper-crustal magma chambers that subsequently coalesced and enlarged to become sites for ignimbrite eruptions (Lipman, Doe, and Hedge 1978). Phenocryst-poor rhyolitic lavas, although regionally rare, occur locally as late eruptive products from some central volcanoes, especially those clustered close to sites of subsequent ignimbrite eruptions. These may mark increasing efficiency of fractionation processes as magmatic power enlarged the size and energetics of the initial subvolcanic magma chambers.

NORTHEAST CALDERAS

The northeastern San Juans, the least-studied portion of this well-known volcanic region, are the site of several recently identified and reinterpreted ignimbrite-caldera systems (plate 7). The long-recognized Bonanza Caldera (McIntosh and Chapin 2004; Steven and Lipman 1976; Varga and Smith 1984), source of the compositionally zoned Bonanza Tuff (dacite-rhyolite-dacite) at 33.0 Ma, structurally lies within the San Juan erosional remnant of mid-Tertiary volcanic deposits yet is probably more appropriately interpreted as the youngest and southernmost system along the Sawatch trend (McIntosh and Chapin 2004). The Bonanza Caldera subsided within the topographical high of several clustered andesitic stratocones, from which preserved flank lavas are known as the Rawley Andesite (equivalent to older parts of the Conejos Formation to the south).

The previously unrecognized North Pass Caldera (Lipman and McIntosh 2008), along the Continental Divide in the Cochetopa Hills, was the source of the 32.25-Ma Saguache Creek Tuff (~250 km^3). This regionally distinctive crystal-poor alkalic rhyolite helps fill an apparent gap in the southwestward migration from older explosive activity, from calderas along the north-south Sawatch locus in central Colorado (youngest: Bonanza Tuff at 33 Ma), to the culmination of Tertiary volcanism in the San Juan region, where large-volume ignimbrite eruptions started at ~29 Ma and peaked with the enormous Fish Canyon Tuff (5,000 km^3), at 28.0 Ma. The entire North Pass cycle, including small precursor tuffs, caldera-sourced Saguache Creek Tuff, thick caldera-filling lavas, and a late small-volume tuff sheet, is tightly bracketed at 32.5–32.2 Ma.

No large tuff sheets were erupted in the San Juan region during 32–29 Ma. However, a previously unmapped cluster of dacite to rhyolite lava flows and small tuffs associated with a newly recognized intermediate-composition intrusion, 5 × 10 km across (the largest subvolcanic intrusion in the San Juan region) and centered 15 km north of the North Pass Caldera, marks a near-caldera-size silicic system active at 29.8 Ma.

SOUTHEAST CALDERAS (PLATORO COMPLEX)

The Platoro Caldera complex has been interpreted as recording the initial rise of inter-mediate-composition magma bodies to shallow crustal levels preparatory to ignim-brite volcanism (Lipman, Doe, and Hedge 1978). Conejos lavas in the vicinity of the Platoro Caldera complex have been subdivided into the Horseshoe Mountain, Rock Creek, and Willow Mountain members, based on differences in composition and phenocryst mineralogy (Colucci et al. 1991). Potassium-argon and $^{40}Ar/^{39}Ar$ ages indicate activity from before 33 Ma to about 29.5 Ma.

Ignimbrite sheets erupted from the Platoro Caldera complex between 29.4 and 28.35 Ma were deposited widely across the southeastern San Juan region; those have been described as formations within the Treasure Mountain Group (Lipman et al. 1996). Although all six regional units of the Treasure Mountain Group (Chiquito Peak, South Fork, Ra Jadero, Middle, La Jara Canyon, and Black Mountain Tuffs; table 2.1) were unequivocally erupted from within the Platoro Caldera complex, intra-caldera facies of only the two largest eruptions, La Jara Canyon and Chiquito Peak, are exposed (figure 2.2). Thick sequences of locally derived andesitic lavas recurrently filled the Platoro complex to overflowing—and also intertongue with many of the ignimbrite sheets.

The Platoro complex is an extreme example, for the San Juan region and else-where, in terms of successive voluminous ash flows that alternated with more quies-cent lavas from an areally restricted source. Geologically recent multicyclic caldera complexes are relatively few, and none record as many successive eruptions from a single source. While the evidence is clear in support of major subsidence accompa-nying the Chiquito Peak eruption, as indicated by the nearly 1 km of exposed thick-ness of intra-caldera tuff on the uplifted intra-caldera block on Cornwall Mountain, the subsidence history during earlier eruptions is obscure. Based on their outflow volumes (at least 2,600 km³; table 2.1), eruptions of the Black Mountain, Ojito Creek, Ra Jadero, and South Fork Tuffs were also likely accompanied by substan-tial caldera collapses. However, the volumes of tuff ponded within their associated calderas can be inferred only by analogy with better-exposed systems. For each large caldera, about half of the total eruptive volume is, on average, deposited as the outflow tuff sheet, and about half ponds within the concurrently subsiding caldera (Lipman 1984).

Postcaldera lava domes of dacite and rhyolite, along with several large gran-itoid intrusions (Alamosa River, Cat Creek, and Crater Creek stocks; figure 2.2), occupy an east-west zone across the northern caldera area and are associated with base- and precious-metal mineralization at Summitville. Broadly similar mineral-ization near the town of Platoro is structurally controlled by faults of the trapdoor resurgent uplift within the caldera and probably by a largely concealed underlying intrusion.

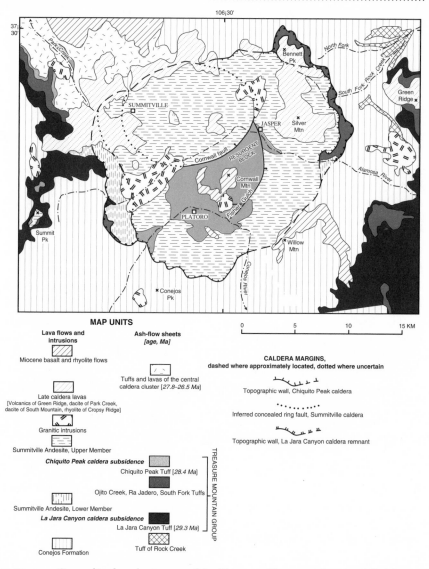

FIGURE 2.2 Generalized geologic map of the Platoro caldera complex. Modified from Lipman et al. 1996, figure 2.

CENTRAL CALDERA CLUSTER

Ignimbrite eruptions became progressively focused in the central San Juan region between 28.8 and 26.9 Ma (figure 2.3), leading to the discharge of at least 8,800 km³ of dacitic-rhyolitic magma and emplacement of nine major ignimbrite sheets

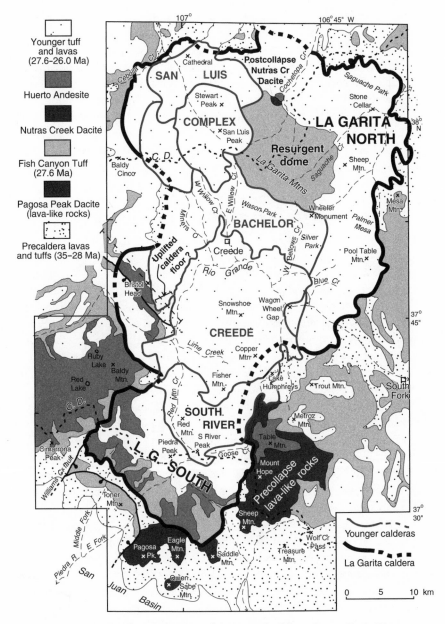

Legend:

Younger tuff and lavas (27.6–26.0 Ma)

Huerto Andesite

Nutras Creek Dacite

Fish Canyon Tuff (27.6 Ma)

Pagosa Peak Dacite (lava-like rocks)

Precaldera lavas and tuffs (35–28 Ma)

107° 106° 45' W

Cebolla Cr

Cathedral

SAN   LUIS

Postcollapse
Nutras Cr
Dacite

Cochetopa Cr

Saguache Park

Stewart
Peak

Stone
Cellar

COMPLEX

San Luis
Peak

LA GARITA
NORTH

38°
N

Resurgent
dome

Saguache Cr

Sheep
Mtn.

C. D.

Baldy
Cinco

La Garita Mtns

Mesa
Mtn.

W. Willow Cr

E. Willow Cr

Wason Park

Wheeler
Monument

Palmer
Mesa

Miners Cr

BACHELOR

Silver
Park

Pool Table
Mtn.

Creede

W. Bellows Cr

Uplifted
caldera
floor ?

Rio   Grande

Blue Cr

Bristol
Head

Snowshoe
Mtn.

Wagon
Wheel
Gap

37°
45'

CREEDE

Lime Creek

Copper
Mtn.

Ruby
Lake

Baldy
Mtn.

Fisher
Mtn.

Lake
Humphreys

Trout Mtn.

South
Fork

Red
Lake

Red Mtn. Cr

Metroz
Mtn.

C. D.

SOUTH.
RIVER

Red
Mtn.

S River
Peak

Table
Mtn.

Cimarrona
Peak

L. G. SOUTH

Piedra
Peak

Goose Cr

Mount
Hope

Precollapse
lava-like rocks

Williams Cr. fault

Toner
Mtn.

37°
30'

Piedra R. Middle Fork

E Fork

Pagosa
Pk.

Eagle
Mtn.

Sheep
Mtn.

Wolf Cr
Pass

Saddle
Mtn.

Treasure
Mtn.

Younger calderas

San

Juan   Basin

Quien
Sabe
Mtn.

La Garita caldera

0   5   10 km

FIGURE 2.3 Generalized geologic map of the central caldera cluster. From Lipman 2000, figure 3.

(individually 150–5,000 km³) accompanied by recurrent caldera subsidence. In the central cluster, erosional dissection to depths of as much as 2.5 km has exposed diverse features of intra-caldera ignimbrite and interleaved caldera-collapse land-slide deposits that accumulated to multi-kilometer thicknesses within concurrently subsiding caldera structures. Exposed calderas vary in size from 10 to 75 km in maximum diameter, with the largest calderas associated with the most voluminous eruptions. Here also, voluminous andesitic-dacitic lavas and breccias were erupted from central volcanoes prior to the ignimbrite eruptions. Similar lava eruptions continued within and adjacent to the calderas during the period of explosive volca-nism, making the central San Juan caldera cluster an exceptional site for the study of caldera-related volcanic processes.

MASONIC PARK TUFF AND ITS SOURCE

The 28.8-Ma Masonic Park Tuff—the earliest ignimbrite erupted from a central San Juan source—is transitional in age, petrology, and caldera location between tuffs of the Treasure Mountain Group, erupted from the Platoro Caldera complex ~29.5 to 28.4 Ma, and younger tuffs from the central cluster. The Masonic Park Tuff is a crystal-rich mafic dacite (62–65% SiO_2; 30–40% plagioclase, biotite, and clinopyroxene), characterized by exceptionally developed, compound cooling zona-tions and numerous flow-unit partings. The Masonic Park is a typical "monoto-nous intermediate tuff," as are other phenocryst-rich dacite tuffs that lack obvious magmatic compositional variation (Ute Ridge, Chiquito Peak, Fish Canyon, Blue Creek, Cebolla Creek, and Snowshoe Mountain Tuffs; table 2.1).

Until recently, a widespread ignimbrite sheet in the southeastern San Juans was mapped as part of the Masonic Park Tuff (Steven and Lipman 1976). This deposit is now recognized as a separate unit, the younger Chiquito Peak Tuff, product of the youngest large eruption from the Platoro complex at 28.4 Ma (Lipman et al. 1996).

The Mount Hope area was formerly interpreted as a caldera source for the Masonic Park Tuff by Steven and Lipman (1976), based on multiple criteria that remain partly valid as indicators of eruptive proximity, but the existence of any Mount Hope Caldera now appears unlikely (Lipman, Dungan, and Bachman 1997). The thick "Fish Canyon Tuff" in the Mount Hope area, previously interpreted as a mas-sive tuff that ponded within a preexisting Mount Hope caldera (Steven and Lipman 1976), is now recognized as a large lava-like mass of Pagosa Peak Dacite, a precursor to eruption of the overlying Fish Canyon Tuff (see discussion in the next section). Alternatively, a Masonic Park caldera probably lies to the northwest of Mount Hope and is completely buried within the southern sector of La Garita Caldera.

FISH CANYON TUFF (28.02 MA) AND LA GARITA CALDERA

The Fish Canyon Tuff, long recognized as among the world's largest ignimbrite sheets, consists of uniform phenocryst-rich dacite (Bachmann, Dungan, and Lipman

2002; Lipman, Steven, and Mehnert 1970) that spread widely beyond its source caldera and ponded within it. This tuff is an exceptional example of a large "monotonous intermediate" eruption from a chamber that lacked compositional gradients. Among the twenty-two large (100–>1,000 km³) Tertiary ignimbrite sheets in the San Juan region, the Fish Canyon is unique in its enormous volume (greater than 5,000 km³), compositional uniformity (67.5%–68.5% SiO_2), high phenocryst content (35%–50%), and near-solidus phenocryst assemblage (plagioclase, sanidine, quartz, biotite, hornblende, sphene, apatite, zircon, and Fe-Ti oxides). It is the only San Juan ignimbrite that contains phenocrystic hornblende without augite. Despite its large volume, the entire Fish Canyon Tuff is a single ignimbrite sheet, generally having a simple cooling zonation. Its bulk composition is dacite because of high phenocryst content, even though the matrix is silicic rhyolite (75%–76% SiO_2). A large precaldera, lava-like body (Pagosa Peak Dacite: 30 km across, locally more than 800 m thick, and 200–300 km³ in volume) along the recently identified south margin of La Garita Caldera and a small postcaldera lava flow in the northern moat (figure 2.2) are compositionally indistinguishable from the main Fish Canyon Tuff, documenting variable eruptive processes from an enormous homogeneous magma body (Bachmann, Dungan, and Lipman 2002).

The overall geometry of La Garita Caldera, which subsided in response to eruption of the main Fish Canyon Tuff, has long been challenging to decipher because of its great size (35 × 75 km), concealment by later eruptive deposits and caldera subsidences, physical difficulties of access, rugged terrain, and vegetative cover. The elongate La Garita Caldera collapsed in three successive northward-migrating segments during eruption of the enormous Fish Canyon Tuff at 28.02 Ma (Lipman 2000). The northern segment is inferred to have been the last to subside or to have completed its subsidence, based on confinement of distinctive late-erupted granophyric fragments of the Fish Canyon magma type to this sector. The southern sector is thought to have been the first segment to subside, based on proximity to the accumulations of early-erupted Pagosa Peak Dacite. Even though the segmented caldera shape implies a multistage origin for La Garita Caldera, the outflow Fish Canyon Tuff is a single ignimbrite sheet characterized by a simple to weakly compound welding zonation, indicating deposition from a single sustained eruption. In each sector, outflow tuff is truncated along the caldera walls, against which the intra-caldera tuff wedges out depositionally.

In the northern caldera sector, La Garita Mountains (figure 2.2) are a resurgently uplifted block of densely welded intra-caldera Fish Canyon Tuff > 1,200 m thick, without exposed basal contacts (Steven and Lipman 1976). The intra-caldera tuff is strongly indurated and oxidized red-brown, in comparison to the light-gray outflow, and it contains larger and more coarsely porphyritic pumice lenses (10–20 cm). Erosional levels in the southern sector, as in the northern, expose thick intra-caldera tuff (> 800 m) without reaching the caldera floor. No marked resurgence is evident in the southern sector; instead, linear northwest-trending faults recurrently disrupted the caldera fill, which includes andesitic lavas (Huerto Andesite) that flooded the southern sector after collapse (figure 2.3). The thick Fish Canyon Tuff within La Garita Caldera and similarly thick tuff fills within other San Juan calderas provide clear evidence that these calderas collapsed concurrently with eruption of the tuffs, a relationship first documented in the central San Juan Mountains (Steven and Ratté 1965).

After the collapse of La Garita Caldera, seven additional large ignimbrites erupted from inside La Garita depression, within about one million years (table 2.1).

CARPENTER RIDGE TUFF (27.55 MA) AND BACHELOR CALDERA

The outflow Fish Canyon Tuff and Huerto Andesite are widely overlain by the 27.55-Ma Carpenter Ridge Tuff, the second-largest ignimbrite sheet (1,000 km^3) erupted from the central caldera cluster. The Carpenter Ridge Tuff is a laterally and vertically complex ignimbrite characterized by magmatic compositional gradients, intricate welding and crystallization zones, variably intense potassic alteration, and interdigitated wedges of lithologically diverse landslide breccia deposits.

The elliptical Bachelor Caldera is located centrally within La Garita Caldera, while its southern and northern margins have been concealed beneath younger subsidence structures. The core of Bachelor Caldera was broadly uplifted to form a resurgent dome, with its crest eccentrically north of the center of subsidence, although detailed geometric interpretation is uncertain because of truncation by the Creede and San Luis Calderas and widespread cover by younger units. Faults along the crest of the Bachelor Dome define a keystone graben that has had a complex history of later recurrent movement and provided the dominant structural control for mineralization in the Creede Mining District (Steven and Lipman 1976).

BLUE CREEK TUFF (27.45 MA) AND ITS SOURCE

Distribution of the Blue Creek Tuff, a uniformly phenocryst-rich dacite (65%–68% SiO$_2$; 30%–40% phenocrysts of plagioclase, biotite, and augite), is largely controlled by ponding within the moat of Bachelor Caldera and the central segment of La Garita Caldera. No remnants of a caldera source for the Blue Creek Tuff are exposed at present erosional levels, but the widespread distribution (at least 1,000 km^2) and sizable volume of outflow tuff (> 150 km^3) strongly suggest that caldera subsidence accompanied its eruption. A concealed caldera source is inferred to lie beneath southern parts of Creede Caldera, based on distribution of the tuff. Andesitic and dacitic lavas that overlie the Blue Creek Tuff near Wagon Wheel Gap (volcanics of McClelland Mountain) also provide some suggestion that the source of this ignimbrite is nearby, to the west, within the younger Creede Caldera.

WASON PARK TUFF (27.40 MA) AND SOUTH RIVER CALDERA

Typical Wason Park Tuff is a distinctive phenocryst-rich rhyolite and silicic dacite (68%–72% SiO_2), characterized where densely welded by large light-gray, vapor-phase–crystallized pumice and brick-red/brown groundmass (Steven and Ratté 1965). Bulk samples contain from 15 to 25 percent phenocrysts of plagioclase, sanidine, and biotite. The outflow tuff sheet accumulated largely within La Garita and Bachelor Calderas, where it ponded as much as 300 m thick, forms rugged cliffs, and constitutes a distinctive marker along the walls of the younger Creede Caldera.

The Wason Park Tuff, long thought to have come from a source concealed beneath Creede Caldera (Steven and Lipman 1976), was erupted from the recently recognized South River Caldera, which was filled to overflowing by younger andesitic and dacitic lavas (Lipman 2000, 2006). At exposed levels, exhumed margins of the South River Caldera are strikingly defined by unconformities along two arcuate drainages that expose parts of the original caldera walls: Red Mountain Creek to the west, and Goose Creek to the southeast (figure 2.2).

SAN LUIS CALDERA COMPLEX AND ITS RELATION TO COCHETOPA PARK CALDERA (ALL 26.9 MA)

The San Luis complex (plate 8), a cluster of three overlapping calderas at the northern end of the central caldera cluster, was the source (in eruptive sequence) of the Rat Creek, Cebolla Creek, and Nelson Mountain Tuffs (figure 2.4). All three tuff sheets were erupted in rapid succession at 26.9 Ma (Lipman and McIntosh 2008). Outflow sheets of the Rat Creek and Nelson Mountain Tuffs are strikingly compositionally zoned, from non- to weakly welded crystal-poor rhyolite upward into densely welded crystal-rich dacite. The main phenocrysts are plagioclase, sanidine, biotite, and augite. In contrast, the intervening Cebolla Creek Tuff is a monotonously uniform, crystal-rich dacite that contains abundant hornblende and lacks sanidine. Lying stratigraphically between these tuffs, especially within and near their source calderas, are thick sequences of compositionally diverse lava. The Nelson Mountain Tuff, the youngest of the three San Luis units, is the most voluminous (> 500 km³) but was erupted from a relatively small ("underfit") caldera of only 8 × 10 km (plate 8), seemingly concurrently with larger-scale caldera subsidence at Cochetopa Park 30 km to the northeast.

Cochetopa Park is morphologically the most striking caldera in the San Juan Volcanic Field, yet its origin and associated erupted deposits have remained confusing and obscure. The original topographic rim of the caldera is defined by old volcanic rocks along the Continental Divide, to the south and east, and by the topographic highs of Razor Creek Dome and Sawtooth Mountain, to the north. The interior of the caldera is marked by a constructional pile of rhyolitic lavas that form geographic Cochetopa Dome, which is surrounded by the low basin of Cochetopa

Nelson Mtn. Tuff
(dw caprock)

Nelson Mtn. Tuff (nw base)

Cebolla Creek Tuff

Rat Creek Tuff
(upper part)

FIGURE 2.4 View of three tuff sheets erupted from the San Luis Caldera complex at Wheeler Geologic Monument. Lowest sheet is Rat Creek Tuff, which is non-welded throughout and grades upward from light-tan rhyolite (~74% SiO_2) into pale-brown dacite (~66% SiO_2) that contains sparse dark-brown andesitic scoria. Distinctive hornblende-rich Cebolla Creek Tuff contains basal surge beds overlain by vitrophyre of uniform mafic dacite that becomes less welded upward. Uppermost Nelson Mountain Tuff consists of non-welded crystal-poor rhyolite, which grades upward to a densely welded caprock of crystal-rich dacite (~68% SiO_2). White arrows show contacts between units. Photo by P. W. Lipman.

Park underlain by readily eroded, weakly indurated bedded tuffs. Previous interpretations (based on recon mapping by Steven in the early 1970s) that the Cochetopa Caldera formed in response to eruption of the "tuff of Cochetopa Creek" (Steven and Lipman 1976) have become untenable; this tuff is not a separate emplacement unit but rather the northeastern outflow of the 500-km³ Nelson Mountain Tuff, erupted from the San Luis Caldera complex (Lipman 2000). Recent field and lab studies have led to the somewhat novel interpretation that no tuff erupted directly from this large obvious caldera structure. Rather, magma (1) drained from beneath Cochetopa Park as it subsided passively, (2) flowed laterally to the southwest along graben-related dikes, and (3) erupted as Nelson Mountain Tuff from within the San Luis complex (Lipman and McIntosh 2008).

SNOWSHOE MOUNTAIN TUFF (26.9 MA) AND CREEDE CALDERA

Most of the central San Juan calderas have been deeply eroded, and their identification is dependent upon detailed geologic mapping. In contrast, the primary volcanic morphology of the symmetrically resurgent Creede Caldera, source of the Snowshoe Mountain Tuff, has been exceptionally preserved because of rapid infilling by moat sediments of the Creede Formation, which were preferentially eroded during the past few million years (figure 2.5). The Snowshoe Mountain Tuff, another crystal-rich dacite, erupted at 26.9 Ma and is more than 1.4 km thick (with no base exposed) on the steep-sided resurgent dome within Creede Caldera, but outflow tuff is preserved only locally. Local eruption of thick flows of Fisher Dacite accompanied resurgence of Creede Caldera, especially along its south side. No direct stratigraphic relations are preserved between tuffs erupted from Creede Caldera and from the San Luis complex; all yield high-resolution ages that are analytically indistinguishable at 26.9 Ma, although geologic relations indicate that Creede Caldera is the youngest of the central cluster (Lipman 2000).

LATE INTRUSIONS AND MINERALIZATION

Central San Juan calderas display a variety of postcollapse resurgent uplift structures, and caldera-forming events produced complex fault geometries that localized

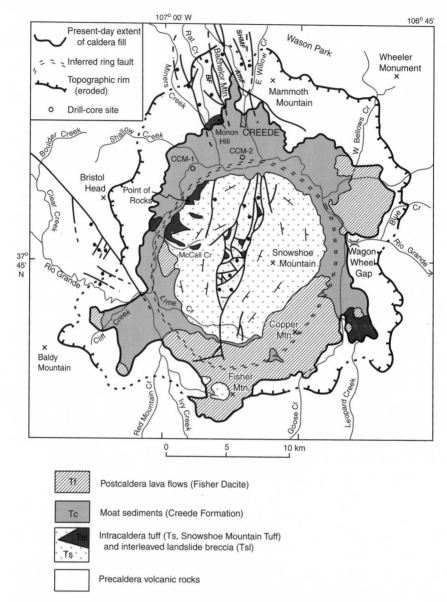

FIGURE 2.5 Generalized geologic map of Creede Caldera, showing approximate location of eroded topographic caldera rim, present-day extent of caldera-fill deposits, inferred buried ring fault, and late normal faults during resurgent doming (Deep Creek Graben) and mineralization (Creede Graben). Faults of Creede Graben: BF = Bulldog Mountain Fault; AmF = Amethyst Fault; SHMF = Solomon-Holy Moses Faults. From Lipman 2000, figure 12.

late mineralization, including the epithermal base- and precious-metal veins of the well-known Creede Mining District. Subvolcanic intrusions are present within several of the calderas and typically consist of relatively fine-grained granodiorite.

While surface exposures provide only limited perspectives regarding the scale of the subsurface magmatic system active during San Juan volcanism, regional Bouguer gravity data define a steep-sided, flat-floored negative anomaly extending over an elliptical area of about 75 × 100 km and having an amplitude of about −50 mGal. The San Juan anomaly has long been modeled as recording the presence of an upper-crustal granitic batholith that has much of its upper contact within a few kilometers of the present surface (Plouff and Pakiser 1972). The negative anomaly encloses most of the calderas in the San Juan region, although individual calderas have little or no gravity expression and sizable central portions of the deepest gravity low lie outside any caldera, indicating that caldera fills contribute little to the anomaly. Because cumulative subsidence at multiple nested calderas, especially in the central San Juan cluster, is interpreted to exceed 10–15 km in some places (Lipman 2000, figure 16), growth of the subvolcanic batholith to its present shallow crustal level must have continued after caldera subsidence. The San Juan anomaly and a similar gravity low of comparable areal extent, linking the Mount Princeton Batholith and Grizzly Peak Caldera areas in the Sawatch trend, are the most negative gravity lows (> −355 mGal) anywhere in the United States.

Consistent with this timing of sustained batholith assembly and consolidation are the resurgent uplift at many of the San Juan calderas and sustained intervals of postcaldera lava eruptions. Thus the general pattern of mid-Tertiary ignimbrite volcanism in the San Juan region, and elsewhere in the Southern Rocky Mountains, is a several-million-year period of multistage, open-system magmatism. It began with premonitory eruptions of predominantly intermediate-composition lavas from central volcanoes, led to ignimbrite eruptions at times of peak magma input, and was followed by waning surface magmatism as subvolcanic plutons of a composite batholith consolidated beneath the volcanic loci.

STRUCTURES

Structural features of the central San Juan Mountains (figure 2.6) involve complex interactions between diverse localized faulting associated with volcanism, especially at the large calderas, and effects of west-southwest–directed regional extension associated with inception of the Rio Grande Rift Zone. Although many faults are exposed within the region, erosion levels are insufficient to expose any ring faults directly related to caldera collapse, such as those well exposed within the Lake City and Silverton calderas in the western San Juan Mountains.

A distinctive area of rectilinear faulting within and adjacent to the southern segment of the enormous La Garita Caldera appears to have accommodated early piecemeal-style collapse, probably initiated during precursor eruption of the

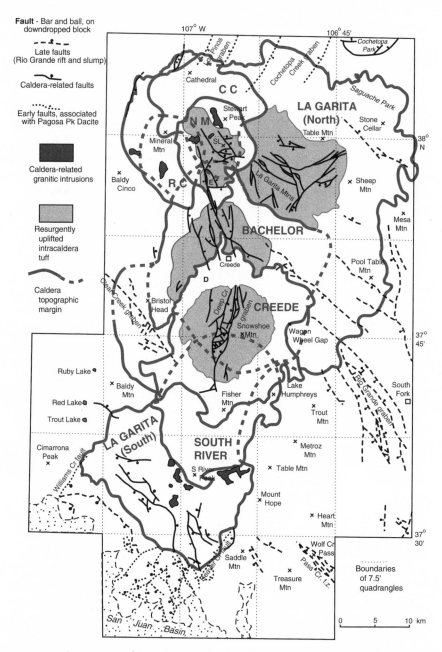

FIGURE 2.6 Structures and intrusions in vicinity of the central caldera complex. Abbreviated names: SL = San Luis Peak. Calderas of the San Luis complex: RC = Rat Creek; CC = Cebolla Creek; NM = Nelson Mountain. Modified from Lipman 2000, figure 14.

Pagosa Peak Dacite (Lipman 2000); some of these faults indicate localized and continued subsidence in the southern segment during subsequent eruptions of Fish Canyon Tuff. Several fault clusters are related to resurgent uplift of caldera floors. Especially conspicuous is the Deep Creek Graben along the keystone crest of the steep-sided Snowshoe Mountain Dome, within Creede Caldera (Steven and Ratté 1965). Other resurgent structures include the graben faults of the Creede Mining District that were also initiated as keystone faults on the elliptical resurgent uplift within the Bachelor Caldera (Steven and Lipman 1976), faults bounding trapdoor-style uplift of the San Luis Peak block within the caldera associated with eruption of the Nelson Mountain Tuff (plate 8), and probably also the faults that cut the intra-caldera Fish Canyon Tuff in the uplifted block of La Garita Mountains.

Another group of structures includes linear grabens and other faults (figure 2.6) adjacent to calderas that appear largely to have been initially established during segmented subsidence of La Garita Caldera, then passively buried by younger tuff sheets and lava flows. These include the Los Pinos and Cochetopa Grabens that connect the northern segment of La Garita Caldera to the Cochetopa Caldera; some faults of the Clear Creek Graben, to the west of the central segment; and perhaps initial faulting along the Rio Grande Graben, to the southeast of the central segment. The southwestern bounding faults of the Clear Creek Graben appear to have controlled a subparallel segment of La Garita Caldera–wall unconformity, along which the Carpenter Ridge Tuff and younger units are depositionally banked against steep slopes without major later faulting. In contrast, northeastern bounding faults from Bristol Head to Spring Creek Pass had continued movement after eruption of the Nelson Mountain Tuff. To the southeast, the Pass Creek Fault Zone runs between the central caldera cluster and the Platoro Caldera complex, displacing younger than the Fish Canyon Tuff and associated with modest hydrothermal alteration.

RIFT-RELATED BIMODAL VOLCANISM

During late stages of San Juan volcanism, the Rio Grande Rift Zone became active within the present-day San Luis Valley, to the east of the San Juan Mountains (figure 2.1). The style of volcanism changed to a bimodal assemblage of widely distributed flows of weakly alkalic basalt and basaltic andesite, accompanied by small lava flows of silicic rhyolite, especially in the vicinity of the older Platoro Caldera complex. Only a few north- and northwest-trending faults within the eastern San Juans have appropriate geometry and timing to reflect such regional tectonism. Northwest-trending faults of the Rio Grande Graben cut basaltic lava flows of the Hinsdale Formation dated at about 24 Ma, south of South Fork. Late movement along the Bristol Head master fault of the Clear Creek Graben suggests that faults localized by, and initially active during formation of, La Garita Caldera were also influenced by regional stresses associated with initial southwestward-directed extension along the Rio Grande Rift Zone. North-trending normal faults along the east margin of the North Pass Caldera also displace lavas of the Huerto Andesite, thus documenting rift-related extension much more recent than the caldera-related structures.

REFERENCES

Bachmann, O., M. A. Dungan, and P. W. Lipman. 2002. The Fish Canyon Magma Body, San Juan Volcanic Field, Colorado: Rejuvenation and Eruption of an Upper Crustal Batholithic Magma Chamber. *Journal of Petrology* 43:1469–1503.

Bethke, P. M., and R. L. Hay, eds. 2000. *Ancient Lake Creede: Its Volcano-Tectonic Setting, History of Sedimentation, and Relation to Mineralization in the Creede Mining District.* Special Paper 346. Washington, DC: Geological Society of America.

Colucci, M. T., M. A. Dungan, K. M. Ferguson, P. W. Lipman, and S. Moorbath. 1991. Precaldera Lavas of the Southeast San Juan Volcanic Field: Parent Magmas and Crustal Interactions. *Journal of Geophysical Research* 96:13, 412–434.

Larsen, E. S., and W. Cross. 1956. *Geology and Petrology of the San Juan Region, Southwestern Colorado.* Professional Paper 258. Boulder: US Geological Survey.

Lipman, P. W. 1984. The Roots of Ash-Flow Calderas in Western North America: Windows into the Tops of Granitic Batholiths. *Journal of Geophysical Research* 89:8801–8841.

———. 2000. The Central San Juan Caldera Cluster: Regional Geologic Framework. In P. M. Bethke and R. L. Hay, eds., *Ancient Lake Creede: Its Volcano-Tectonic Setting, History of Sedimentation, and Relation to Mineralization in the Creede Mining District.* Special Paper 346. Boulder: Geological Society of America, 9–70.

———. 2006. Geologic Map of the Central San Juan Caldera Complex, Southwestern Colorado. Map I–2799, scale 1:50,000, 4 sheets. Reston, VA: US Geological Survey.

———. 2007. Incremental Assembly and Prolonged Consolidation of Cordilleran Magma Chambers: Evidence from the Southern Rocky Mountain Volcanic Field. *Geosphere* 3:42–70.

Lipman, P. W., B. Doe, and C. Hedge. 1978. Petrologic Evolution of the San Juan Volcanic Field, Southwestern Colorado: Pb and Sr Isotope Evidence. *Geological Society of America Bulletin* 89:59–82.

Lipman, P. W., M. A. Dungan, and O. Bachman. 1997. Eruption of Granophyric Granite from a Large Ash-Flow Magma Chamber: Implications for Emplacement of the Fish Canyon Tuff and Collapse of the La Garita Caldera, San Juan Mountains, Colorado. *Geology* 25:915–918.

Lipman, P. W., M. A. Dungan, L. D. Brown, and A. Deino. 1996. Recurrent Eruption and Subsidence at the Platoro Caldera Complex, Southeastern San Juan Volcanic Field, Colorado: New Tales from Old Tuffs. *Geological Society of America Bulletin* 108: 1039–1055.

Lipman, P. W., and W. C. McIntosh. 2008. Eruptive and Non-Eruptive Calderas, Northeastern San Juan Mountains, Colorado: Where Did the Ignimbrites Come From? *Geological Society of America Bulletin* 120:771–795; doi: 10.1130/B26330.1.

Lipman, P. W., T. A. Steven, and H. H. Mehnert. 1970. Volcanic History of the San Juan Mountains, Colorado, as Indicated by Potassium-Argon Dating. *Geological Society of America Bulletin* 81:2327–2352.

McIntosh, W. C., and C. E. Chapin. 2004. Geochronology of the Central Colorado Volcanic Field. *New Mexico Bureau of Mines and Mineral Resources Bulletin* 160:205–237.

Plouff, D., and L. C. Pakiser. 1972. *Gravity Study in the San Juan Mountains, Colorado*. Professional Paper 800. Washington, DC: US Geological Survey, B183–190.

Steven, T. A. 1975. Middle Tertiary Volcanic Field in the Southern Rocky Mountains. In B. F. Curtis, ed., *Cenozoic History of the Southern Rocky Mountains*. Memoir 144. Boulder: Geological Society of America, 75–94.

Steven, T. A., and P. W. Lipman. 1976. *Calderas of the San Juan Volcanic Field, Southwestern Colorado*. Professional Paper 958. Washington, DC: US Geological Survey.

Steven, T. A., P. W. Lipman, W. J. Hail Jr., F. Barker, and R. G. Luedke. 1974. Geologic Map of the Durango Quadrangle, Southwestern Colorado. Miscellaneous Geologic Investigations, Map I–764, scale 1:250,000. Washington, DC: US Geological Survey.

Steven, T. A., and J. C. Ratté. 1965. *Geology and Structural Control of Ore Deposition in the Creede District, San Juan Mountains, Colorado*. Professional Paper 487. Washington, DC: US Geological Survey.

Varga, R. J., and B. M. Smith. 1984. Evolution of the Early Oligocene Bonanza Caldera, Northeast San Juan Volcanic Field, Colorado. *Journal of Geophysical Research* 89: 8679–8694.

CHAPTER THREE

Mineralization in the Eastern San Juan Mountains

Philip M. Bethke

THE EASTERN SAN JUAN MOUNTAINS were less intensely mineralized than the western San Juans; in addition, the styles of mineralization were less diverse and the time interval of mineralization was shorter. Nevertheless, mines in the eastern San Juans produced more than $375 million of mineral wealth (more than a billion dollars at 2007 metal prices). The two most important mining districts, Creede and Summitville, have been the subjects of extensive scientific study and stand as classic examples, or archetypes, of their styles of mineralization.

All of the important mineralization in the eastern San Juans occurred during the caldera cycle of volcanism (see Lipman and McIntosh, chapter 2, this volume). A few minor deposits were formed on the eastern flank of the mountains during the earlier period of intermediate-composition volcanic eruptions, but they have received little study and have produced only minor amounts of metals. Those are not discussed herein. Of the fourteen eastern San Juan calderas identified by Lipman and McIntosh, only the Creede, Platoro-Summitville, and Bonanza Calderas are known to be significantly mineralized, although some mineralization may be related to the San Luis Caldera complex.

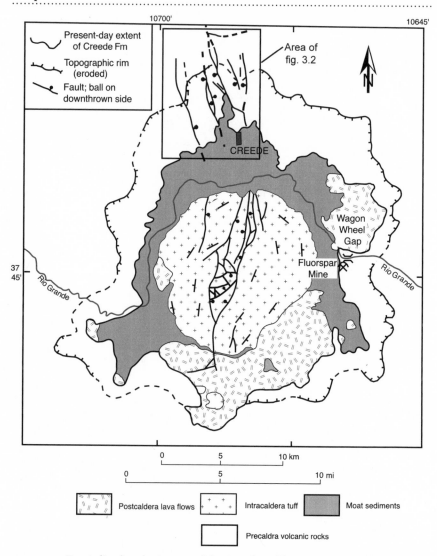

FIGURE 3.1 Generalized geologic map of the Creede Caldera, showing the locations of the Creede Mining District and the Wagon Wheel Gap fluorspar deposit. Modified from Bethke and Hay 2000.

MINERALIZATION ASSOCIATED WITH THE CREEDE CALDERA

The Creede Caldera (figure 3.1) is the youngest in the central San Juan Caldera complex (see Lipman and McIntosh, chapter 2, this volume). The caldera was strongly resurgent, and the moat between its topographic wall and resurgent dome contains volcanic-ash-rich sediments deposited in a saline lake. The nature of the lake and its relation to ore formation in the Creede Mining District are described in a Special Paper of the Geological Society of America (Bethke and Hay 2000). The most important ore deposits related to the Creede Caldera were deposited as open-space vein fillings in fissures extending northward from its northern margin. These deposits were exploited in the Creede Mining District. Much less important is the fluorspar deposit at Wagon Wheel Gap, along the eastern margin of the Creede Caldera.

Creede Mining District

The Creede Mining District is an archetypical example of a style of epithermal (near-surface hydrothermal) mineralization produced by near-neutral ore fluids and characterized by sericite ± adularia wall-rock alteration and the presence of adularia in the vein filling. This type of mineralization was termed *adularia-sericite* by Heald-Wetlaufer and others (1987) and is often also referred to as *low-sulfidation* (Hedenquist 1987). The Creede district and its geologic setting have long been the subject of scientific studies, beginning with those of Emmons and Larsen (1913a, 1923) and Larsen (1930). A more recent district study by the US Geological Survey (USGS) (Steven and Ratté 1965) led to extensive mineralogical, geochemical, and volcanological studies by other USGS researchers and by mining company and university geoscientists, recently summarized by Barton and others (2000).

In the 1870s, the supply route to the mines in the western San Juans ran from Del Norte up the Rio Grande Valley, over aptly named Stony Pass, and into Silverton. The mountains surrounding the valley were tempting prospecting grounds, and in the early 1870s claims were staked at Sunnyside, near the junction of Miners Creek and Rat Creek (figure 3.2), approximately 2 miles (3.2 km) west of the present-day town of Creede (Emmons and Larsen 1923). Although some rich ore was found, the venture was not profitable. Some low-grade ore was taken from the Bachelor claim at the south end of the Amethyst Vein in 1884, but it was not until Nicholas C. Creede (after whom the town of Creede is named) discovered rich ore and staked the Holy Moses claim above East Willow Creek in 1889 that the real prospecting rush began. In 1891, several claims, including the Amethyst, Last Chance, and Commodore, were staked on the Amethyst Vein along the western slope of West Willow Creek. By December of that year, a rail line was completed to Creede, and ore was being shipped in large quantities. The Creede district was in continuous production until 1985, except for the Depression years of 1931 through 1933—a remarkable run of continuous mining activity.

As the last of the great Colorado bonanza mining districts, Creede attracted its share of nefarious characters who frequented the mining camps of the western United States. They included Bat Masterson, Soapy Smith, Poker Alice Tubbs, and Bob Ford, the killer of Jesse James. Creede's colorful early history has been described by a number of authors, including Wolle (1949), Feitz (1969, 1973), and Schroder

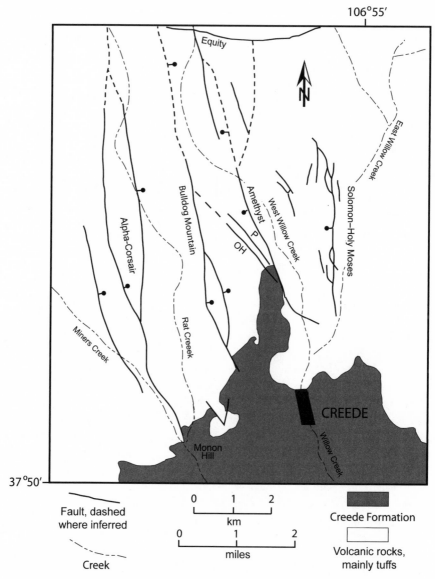

FIGURE 3.2 Generalized geologic map of the Creede Mining District, showing the locations of mineralized structures and the Monon Hill area. Modified from Steven and Eaton 1975.

(2004). Richard Huston's extensively researched and richly detailed *A Silver Camp Called Creede* (2005) is an outstanding record of Creede's mining days. It has served as the basis for much of the mining history briefly summarized in this section.

By far the greatest production from the Creede district was from the Amethyst and related OH and P Veins and the Bulldog Mountain Vein System (figure 3.2). Although the early discoveries along the Alpha-Corsair and Solomon–Holy Moses Fault Systems led to the intense prospecting that resulted in the discovery of mineralization along the Amethyst Vein, mines along those vein systems have contributed only a small percentage of the district's total production. The Alpha-Corsair and Solomon–Holy Moses Fault Systems are the westernmost and easternmost boundaries, respectively, of the Creede Graben, a fault-bounded depression that trends north-northwest from the northern margin of the Creede Caldera. The Creede Graben formed because of reactivation of the keystone graben of the earlier, moderately resurgent Bachelor Caldera. The Amethyst and Bulldog Mountain Faults form the central core of the graben. The wall rocks enclosing the veins of the Creede district are the intra-caldera ash-flow tuffs that erupted from the Bachelor Caldera and accumulated in its subsided basin. The Creede mineralization has been dated at 25.1 Ma (Bethke et al. 1976, as modified by Hon and Mehnert 1983), a million years later than the last dated volcanism in the central San Juan Caldera complex (Lanphere 2000).

Amethyst Vein System. The Amethyst Vein was mined continuously for a horizontal distance of over 2 miles (3.2 km) and appears to have been mineralized for another 3 miles (4.9 km), to the Equity Mine, although the continuity of mineralization has not been fully tested. The Bulldog Mountain Vein system has been exploited for a lateral distance of 1.5 miles (2.4 km). In contrast, the maximum vertical dimension of the ore zone along each of the vein systems was less than 1,000 feet (305 m).

Exploitation of the ores along the Amethyst Vein in the Bachelor, Commodore, Last Chance–New York–Del Monte, Amethyst, Happy Thought, White Star, and Park Regent Mines (noted in south-to-north order) was responsible for the early bonanza days of the Creede district. The boom began with the first shipments of ore from the Amethyst and Last Chance Mines in the fall of 1891. In 1893, the mines produced the largest number of ounces of silver ever extracted in a single year from the Creede district (nearly 5 million ounces). The repeal of the 1890 Sherman Silver Purchase Act in the fall of 1893 led to a precipitous drop in silver prices and a corresponding drop in production from the Creede district. The great bonanzas of the first two years of production were never seen again. By 1920, the Amethyst Vein had been essentially mined out, and further production from the mines along its strike came from veins in the hanging wall, as discussed in the next section.

Except for the Bachelor and Commodore Claims, mines along the Amethyst Vein were initially developed through shafts sunk from the outcrop area high on the slopes of West Willow Creek. The miners lived just over the crest of the slope in the town of Bachelor, a short walk to the mine workings. Abandoned in the early 1900s, Bachelor is said to have had a population of over 2,000 residents in its heyday. The Commodore and Bachelor Mines were accessed by horizontal tunnels driven into the southern end of the vein system. The view of the service buildings of the Commodore 3, 4, and 5 levels (from top to bottom) from a parking area just north of the junction of East and West Willow Creeks provides a popular photo opportunity.

Initially, ore was transported to the various mills by pack animals, over narrow trails. However, roads were soon constructed, and the pack animals were replaced by horse-drawn wagons. The lower portion of the descent from the upper workings down West Willow Creek, over what became known as the Black Pitch, was particularly perilous, and runaways—with the losses of loads and deaths of horses—were common. Soon, aerial tramways (seven in all) were constructed along East and West Willow Creeks to transport the ore more safely and efficiently. The driving of the Nelson-Wooster-Humphreys Tunnel beneath the ore bodies along the Amethyst Vein drained the workings and allowed ore from most of the workings to be dropped down raises for transport in ore cars to the Humphreys Mill. The concrete foundations of the mill can be seen at the junction of East and West Willow Creeks. Farther north, the remains of the rail tramway from the tunnel to the mill can be seen clinging to the steep slopes on the west side of West Willow Creek.

Early production from the Amethyst Vein was from a shallow zone containing ores enriched by downward-percolating, oxidizing groundwaters that leached metals from the overlying vein materials and deposited them deeper in the vein system when the waters' oxidizing capacity declined. The minerals of the enriched zone included native silver (much of it as wires and leaves), acanthite, cerargyrite, pyrolusite, anglesite, pyromorphite, smithsonite, malachite, and barite, along with various iron oxides and clay minerals.

Beneath this enriched zone, the vein material consisted mainly of sulfide ores in a gangue of white and amethystine quartz, iron-rich chlorite, barite, and minor rhodochrosite and adularia. The main sulfide minerals included sphalerite, galena, chalcopyrite, and pyrite. Gold was always a very minor part of the value of the Amethyst Vein and the district as a whole, although it was found in unusually high amounts in the sulfide ores of the Happy Thought Mine. Barite was most abundant in the upper levels and south end of the vein, mainly occurring with manganese-oxide minerals that were probably produced by the oxidation of rhodochrosite. Large vein fillings that consist of intricately banded amethyst and white quartz are characteristic of the upper parts of the Amethyst Vein, particularly in the region of the Last Chance and Amethyst Mines. This so-called Creede sowbelly is highly prized by rock hounds.

The vein material in the Amethyst Vein system was brecciated and crushed by repeated movement along the fault, during and following mineralization. This crushing undoubtedly contributed to enrichment, although it made the mineralogy and mineral relations of the ores more difficult to study. The crushed nature of the vein filling and the fact that the amethyst was essentially mined out by 1920, making most of the workings inaccessible, precluded modern study of the ores of the Amethyst Vein.

Modern concepts of the nature of the Creede ore-forming system are based on detailed studies of the OH, P, Bulldog Mountain, and Equity Vein Systems cited by Barton and others (2000). These studies are supplemented by those of the southern end of the Amethyst Vein (Hemingway 1986; Robinson and Norman 1984) and of the area of a plexus of hanging-wall veins at the projected intersection of the OH Vein with the Amethyst (Giudice 1980). However, observations along those vein systems are entirely consistent with those by early workers on the main part of the Amethyst Vein (Emmons and Larsen 1913a, 1923; Larsen 1930).

Hanging-Wall Veins. As noted earlier, when the main ores on the Amethyst Vein were mined out, attention was directed to a plexus of minor fractures in the hanging wall. Although small and, in cases, discontinuous, many of these fractures contained high silver values, and their exploitation carried mining in the Creede district through the 1920s until 1930, when the mines were shut down because of low metal prices during the Great Depression. The near-doubling of the price of silver by presidential proclamation, followed by the Silver Purchase Act of 1934 and consequent reopening of the mines, encouraged increased exploration in the hanging wall of the Amethyst Vein, which led to the discovery of the OH Vein in 1938 or 1939.

The OH Vein occupies a pronounced, northwest-trending, near-vertical struc-ture antithetic to the Amethyst Vein. It does not truly intersect the Amethyst Vein where seen in the workings; instead, it dies out in the plexus of minor hanging-wall veins. From shortly after its discovery, the OH Vein was the principal supplier of ores along the Amethyst Vein system until discovery of the P Vein in 1961. The P Vein is a shorter vertical structure, subparallel to the OH and lying between it and the Amethyst Vein. It gradually became the main producer as the ores of the OH Vein were mined out. Both the OH and P Vein structures were open fractures, and the ores were deposited on their walls. The OH Vein was mined for a distance of 1.5 miles (2.4 km) over a vertical interval of 700 feet (213 m); the P Vein was mined for 0.7 miles (1.1 km) over the same vertical interval. Mineralization on both veins terminates in a capping of argillic wall-rock alteration, at elevations ranging from 9,900 feet (3,018 m) in the north to 10,000 feet (3,048 m) in the south. Neither the OH nor the P Vein shows evidence of significant post-mineralization move-ment. Consequently, the ores deposited in the open spaces between their walls were well preserved, and many beautiful mineral specimens with large crystals have been collected from their workings. Such specimens provided the basis for the detailed mineralogical and geochemical studies that, along with similar studies on the Bulldog Mountain Vein System and the Equity area, have made the Creede district an archetypical example of the adularia-sericite, or low-sulfidation, type of base- and precious-metal epithermal (near-surface hydrothermal) ore deposits.

Sulfide ores dominated by sphalerite and galena, with lesser amounts of chalco-pyrite and pyrite, were mined from both the OH and P Veins. Fine-grained milky quartz with intergrown adularia was the earliest material deposited in the veins. This was followed by relatively fine-grained sulfides that contained most of the sil-ver values as tetrahedrite-tennantite intergrown with sphalerite. Later sulfide min-eralization produced large crystals up to 6 inches (15 cm) in diameter. Fluorite was an important component of the mineral assemblage, but later hydrothermal fluids leached much of it away, leaving only remnants. Major gangue minerals consisted of quartz (much of it amethystine), hematite, and iron-rich chlorite. Chlorite became increasingly abundant as mining progressed northward. The last stage of mineral deposition consisted of fibrous, botryoidal pyrite, with some marcasite, primarily in the upper parts of the vein system. The closing of zinc smelters in the western United States in the early 1970s and the weakening of the northern extension of the OH and P Veins shut down mining operations in 1976.

Bulldog Mountain Vein System. The Homestake Mining Company's Bulldog Mountain Mine, which exploited the ores along the Bulldog Mountain Fault System, was a late addition to the Creede district's mining history. In the vicinity of the Bulldog Mountain Mine, the north-northwest-trending fault system consists of two main east-dipping zones, the East and the West strands. Several claims were filed along the Bulldog Mountain Fault in the early 1900s, and the Nickel Plate exploration tunnel was driven at an elevation of 9,940 feet (3,030 m), sometime prior to 1920. However, no economic mineralization was found, and the fault was virtually ignored until a US Geological Survey report (Steven and Ratté 1960a) suggested that, along with several other areas, the Bulldog Mountain Fault was a favorable area for exploration. Very quickly, many claims along the fault were filed, and several exploration holes were drilled. In 1963 Homestake acquired the mining rights and began intensive exploration, culminating in the driving of a horizontal tunnel at an elevation of 9,700 feet (2,957 m) in 1964. The tunnel encountered the East strand of the Bulldog Mountain Fault System at 800 feet (244 m) and followed it northward for a similar distance until encountering a rich, high-grade pocket of leaf silver and pyrargyrite. In 1966, encouraged by the results of explora-tion on the 9,700-foot level, Homestake drove another tunnel at an elevation of 9,360 feet (2,853 m) to test the vertical continuity of the ore. With that continu-ity confirmed, Homestake constructed a mill, completed in 1969, and the Bulldog Mountain Mine was commissioned. The Bulldog Mountain Mine produced over 25 million ounces of silver and 48 million pounds of lead during its lifetime. From 1971 to 1984, it was Colorado's major silver producer.

The East strand of the Bulldog Mountain Fault is a zone of anastomosing veins called the Puzzle System. Most of the mine production came from this system. The

West strand is less anastomosing and less well-mineralized. Veins along the Puzzle System vary considerably in width and may be as wide as 9 feet (3 m). The principal vein minerals are barite, rhodochrosite, galena, sphalerite, and quartz, some of which is amethystine. Silver occurs as wire and leaf silver and as microscopic tetrahedrite-tennantite blebs in sphalerite. Less abundant minerals include chalcopyrite, fluorite (much of it removed by hydrothermal leaching), silver-bearing minerals such as acanthite, pyrargyrite-proustite, and polybasite-pearceite, and a variety of silver-copper minerals. As in the OH Vein, a late generation of fibrous botryoidal pyrite, up to 10 inches thick, coats earlier material, principally in the upper parts of the ore body (Plumlee and Whitehouse-Veaux 1994).

The mineral assemblages vary in a complex manner, both vertically and horizontally, along the Bulldog Mountain Vein System, but the overall pattern is similar to that of the Amethyst Vein System, where the sulfide-rich assemblage in a quartz-chlorite gangue occurs more northerly and deeper in the system and the southern and upper parts of the vein system are characterized by a rhodochrosite-barite-rich assemblage. The rhodochrosite-barite-rich assemblage is much better developed along the Bulldog Mountain Vein System and occupies a larger proportion of the vein filling. Sulfide assemblages with chlorite as a gangue mineral appear only in the most northerly and deeper parts of the workings. As in the Amethyst system, silver values along the Bulldog Mountain Vein System are higher in the shallower parts of the system.

The wide barite-filled veins presented mining difficulties. The heavy and weak barite created dangerous mining conditions. Because of this, Homestake went to a mining system whereby the tailings from the mill were pumped back into the open workings to stabilize the stopes. This method of mining obviates the need to place waste material in dumps on the surface, although it is very expensive. In January 1985, the low prices for silver and lead, combined with the high costs of mining, forced the secession of operations at the Bulldog Mountain Mine. The mine was decommissioned and the site reclaimed in 1994, closing the final chapter in the history of mining in the Creede district—at least to date.

Equity–North Amethyst Vein System. The Equity Mine was developed through an adit at an elevation of 11,200 feet (3,353 m) on the eastern slope of West Willow Creek, about 6 miles (9.7 km) north of the town of Creede. The mine was developed at the intersection of the west-dipping Amethyst Fault with the steeply north-dipping Equity Fault. The portal was sealed for safety reasons following cessation of mining activity. Development was mainly along the east-trending Equity Fault, a major structure representing the southern structural margin of the San Luis Caldera complex (Lipman and McIntosh, chapter 2, this volume). The Equity Fault is marked by a series of rusty outcrops running up the steep east slope of West Willow Creek. The original discovery was along these outcrops, about 500 feet (152 m) above the Equity portal. The first shipment of ore was accomplished in 1912, and the mine was operated intermittently until 1967. Its total production was just over 5,000 tons of ore, containing 643 ounces of gold and 153,971 ounces of silver. The ratio of gold to silver was more than twice that of the district as a whole.

In 1971 the Homestake Mining Company initiated an exploration project to evaluate the mineral potential of the Equity and Amethyst Veins at depth. Diamond drilling from the surface encountered some rich intercepts carrying high gold values. An inclined tunnel was driven from the portal level to an elevation of ~10,000 feet (~3,048 m), with exploration drifts driven at intermediate levels. Rumors of high values created great excitement in the minerals industry and the press, but lack of continuity and a continuing decline in silver prices discouraged Homestake from bringing the property into production. The exploration program was discontinued in 1988. Ore mined during underground exploration was stockpiled at the Bulldog Mountain Mine and milled in 1988, when the Homestake Mill was reopened for a brief period. Approximately $500,000 in metal values was recovered. The Equity property was reclaimed in 1993.

Detailed studies of the ores by Foley (1990), Foley and others (1993), and Foley and Ayuso (1994) showed that there were two periods of mineralization, closely related in time, in the Equity–North Amethyst area. The earlier manganese-gold assemblage contained native gold-silver alloys, gold-silver sulfides, silver sulfosalts, and base-metal sulfides in a quartz, manganese-silicate, and carbonate gangue. The later assemblage contained base-metal sulfides and silver sulfosalts in a quartz-chlorite-hematite gangue, similar to that of the deeper and northern parts of the OH and P Veins. Associated argillic wall-rock alteration is also similar to that in the southern part of the district. Adularia from the later assemblage has been dated at 25.1 Ma, the same age as that from the OH Vein (Foley 1990). The earlier assemblage has not been dated; because it occurs along the Equity Fault, it must be no older than 26.9 Ma, the age of the San Luis Caldera complex.

Solomon–Holy Moses Vein System. The Solomon–Holy Moses vein system, along East Willow Creek, was the site of Nicholas Creede's original discovery (the Holy Moses) in 1889, which led to the development of the Creede camp. Mines along the vein system operated episodically from 1891 to 1971, yet have contributed only a small percentage of the district's total production. The Ridge, Ethel, Solomon, and Mexico ore bodies, at the southern end of the Solomon–Holy Moses Vein System, were mined as one. Dump materials and old portals representing those ore bodies can be seen along the west side of East Willow Creek. The workings were developed along the lower part of the vein system and produced galena and sphalerite ores in a green chlorite matrix. Silver content was relatively low. The upper portions of the vein system were developed through the Holy Moses and Phoenix Mines, upslope and north of the lower mines. The material in the Holy Moses and Phoenix ore bodies was oxidized and contained some high silver values. The ore bodies, however, proved to be small. The lower, base-metal-rich ore bodies were somewhat larger, yet still very small relative to those along the Amethyst and Bulldog Mountain Vein Systems.

Alpha-Corsair Vein System. The Alpha-Corsair Fault System, located in the Sunnyside area near the junction of Miners and Rat Creeks, forms the western margin of the Creede graben. Production from mines along the Alpha-Corsair Fault System contributed little to the total district production, even less than that from mines along the Solomon–Holy Moses Vein System. Dump material from the Alpha and Corsair Mines can be seen just north of the junction of the Miners Creek and Rat Creek roads. The vertical extent of workings on the Alpha-Corsair Vein was only a few hundred feet, and only a few miles of tunnel were driven. The ore extracted from these workings was oxidized similar to that from the upper levels of the Amethyst Vein and contained low values of lead and zinc, although some pockets contained high silver values. Mines along the Alpha-Corsair Fault were worked only sporadically, and the workings have mainly been inaccessible for recent geologic study.

Replacement Ores in the Moat Sediments

As initially pointed out by Steven and Ratté (1960a), all veins in the Creede district were mineralized subsequent to deposition of the lacustrine and clastic rocks in the moat of the Creede Caldera. Ore fluids exited the southern end of the veins and permeated the sedimentary rocks in two places. Ore fluids exited from the Amethyst Vein System and entered tuffaceous sands and conglomerates deposited in an ancient river channel cut into the northern wall of the Creede Caldera, just west of the present town of Creede. Silver minerals and barite were deposited in these beds, along with manganese-oxide minerals. One such bed of tuffaceous sandstone contained abundant carbonized wood fragments that are especially silver-rich. The ancient channel was prospected in the 1970s and 1980s by surface drilling and by a 680-foot-long (207 m) exploration tunnel. Although a significant tonnage of ore-grade silver mineralization was found, the high manganese-oxide content presented metallurgical problems, and tests indicated that silver recovery would be low. Therefore, the project was terminated.

The situation was more favorable in the Sunnyside area. There, ore fluids exiting the Alpha-Corsair Fault Zone deposited silver ores along the contact between the lake sediments and a buried hill of underlying volcanic rocks at Monon Hill and centered about a half-mile east-southeast of the dumps of the Alpha-Corsair. The earliest recorded mining activity here was in 1894, but significant production did not begin until after the discovery of a large ore body in 1915. The main productive period was between 1918 and 1922, yet some ore was produced in 1934 and 1941. Total dollar value of minerals from the Monon Hill ore bodies was nearly $1 million, mostly in silver and with some contribution from lead. The Monon Hill ore bodies were mined through six tunnels over a vertical range of about 400 feet (133 m), the lowest of which (the Silver Horde) can be seen just east of and slightly above the Miners Creek road.

WAGON WHEEL GAP FLUORSPAR MINE

The Wagon Wheel Gap fluorspar deposit is not part of the Creede Mining District per se. It is located along Goose Creek on the eastern margin of the Creede Caldera, approximately 8 miles (13 km) southwest of Creede and directly across from the 4UR Guest Ranch. The deposit was first described by Emmons and Larsen (1913b) and more recently by Korzeb (1993). It was discovered around 1891 by prospectors who mistook the purple fluorite for amethyst and thought the deposit might be an extension of the highly productive Amethyst Vein in the Creede district. Actually, the mineralization at Wagon Wheel Gap is not related to that in the Creede district and appears to have occurred several million years later. In 1911, S. B. Collins, a well-known mining engineer, recognized the fluorite and formed the American Fluorspar Company to exploit the deposit. Production began in 1914 and continued intermittently until 1925, when the property was purchased by the Colorado Fuel and Iron Corporation (CF&I), which operated the Wagon Wheel Gap Fluorspar Mine until 1950, except for a few years during the depth of the Great Depression. Steven (1968) estimated that 120,000 tons of fluorite concentrate was produced from the mine.

The mine was originally developed through two main tunnels, several shafts, and open cuts located up the eastern slope of Goose Creek. The ore was transported to a mill just above the creek by an aerial tramway. CF&I abandoned the old workings and exploited the ore through a new tunnel driven at the level of the mill that was connected to the upper working by shafts and ore passes, thereby eliminating the need for the aerial tramway. The mill buildings remain standing and can be seen from the road to the 4UR ranch. The ore consisted mainly of fluorite and barite. A fluorite concentrate was made by gravity separation. The concentrate was shipped to the CF&I steel mills in Pueblo, for use as a flux, and to St. Louis, Missouri, for the manufacture of hydrofluoric acid. The rare mineral Creedite was first described by Larsen and Wells (1916) from the Wagon Wheel Gap deposit and was named after the Creede Quadrangle, in which the mine is located.

MINERALIZATION ASSOCIATED WITH THE PLATORO-SUMMITVILLE CALDERA COMPLEX

The Platoro and Summitville Calderas were the sites of six ash-flow eruptions between 29.4 and 28.35 Ma (Lipman and McIntosh, chapter 2, this volume). Flooding of each of the calderas by later andesite flows has obscured most of the intra-caldera equivalents of the outflow facies, making reconstruction of the erup-

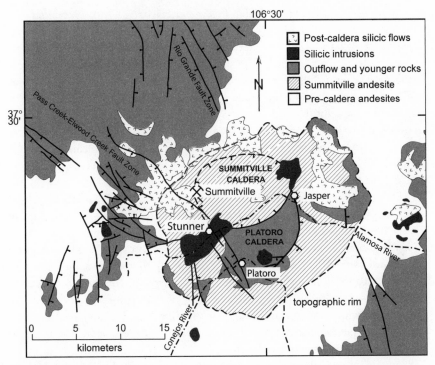

FIGURE 3.3 Generalized geologic map of the Platoro-Summitville Caldera complex, showing the locations of the Summitville, Platoro, Stunner, and Jasper districts and the Pass Creek–Elwood Creek and Rio Grande Fault Zones. Modified from Bethke et al. 2005.

tive history difficult. The Summitville Caldera is nested within the northern part of the larger Platoro Caldera. The caldera complex and its adjacent areas were intruded by granitic stocks of slightly younger age. Voluminous silicic lavas were erupted and several volcanic domes emplaced along the northern border of the complex between 29 and 20 Ma. The caldera complex lies along the trends of two major regional northwest-trending graben systems: the Rio Grande Graben to the north and the Pass Creek–Elwood Creek Fault Zone, which transects the western margin of the complex (figure 3.3). These regional graben systems reflect crustal doming above a large batholith that underlies the San Juan Volcanic Field (Lipman 1975).

SUMMITVILLE DISTRICT

The gold-silver-copper deposit at Summitville is located approximately 26 miles (42 km) southwest of the town Del Norte, between 11,300 and 12,300 feet (3,444 and 3,749 m) elevation. It is an archetypical example of a type of mineralization variously termed *acid-sulfate* (Heald-Wetlaufer, Foley, and Hayba 1987) or *high-sulfidation* (Hedenquist 1987). Hills (1885) and Patton (1917) first described the Summitville deposit, although the classic description is that of Steven and Ratté (1960b). More recent and detailed studies by Stoffregen (1987), Gray and Coolbaugh (1994), and Bethke and others (2005)—based on observations of diamond-drill core samples, bench mapping, and sampling made possible by recent open-pit mining—have added to the descriptions and addressed the origin of the alteration and mineralization. Contamination of the Wightman Fork of the Alamosa River, as a result of mining activity, has focused seminal studies of environmental problems engendered by the development of such acid-sulfate deposits (see articles in Posey, Pendleton, and Van Zyl 1995).

Placer gold was discovered along Wightman Fork in 1870, attracting hundreds of prospectors to the area. By 1874, lode deposits on the northern slope of South Mountain were being developed. The small town of Summitville grew up along Wightman Fork, and the remains of some of the original buildings can still be seen just west of the mine entrance. Placer mining continued sporadically until the late 1880s and contributed only slightly to production. The main mining activity was focused on the rich, oxidized, near-surface lode deposits. South Mountain was studded with numerous mines and eleven mills. Initially, those deposits were exploited by open cuts; underground workings were soon developed, becoming the principal means of extracting ore until 1985, when open-pit operations began. By 1888, the oxidized surficial ores were depleted, and later production exploited the underlying lower-grade sulfide ores that were more difficult to beneficiate. Most operators closed their mines, and production declined precipitously. Between 1888 and 1984, production waxed and waned, with peak periods being 1926–1928 and 1934–1947; during some years, there was no recorded production.

Several mining companies conducted exploration programs at Summitville between 1948 and 1983 and outlined a large volume of low-grade ore. In 1984, the Summitville Consolidated Mining Company, a subsidiary of Galactic Resources, acquired the property. In 1985, it began a large-tonnage, low-grade, open-pit mining operation, using a cyanide-heap-leach recovery system. In December 1992, the company abandoned the property and declared bankruptcy. The total metal value of Summitville ore production is estimated to exceed $125 million, based on metal prices at the time of production. More than 99 percent of metal values came from gold (Gray and Coolbaugh 1994; Koschman and Bergendahl 1968; Steven and Ratté 1960b).

The Summitville deposit occurs in a quartz-latite volcanic dome emplaced along the coincident margins of the Platoro and Summitville Calderas, near their intersections with the Pass Creek–Elwood Creek Fault Zone (figure 3.3). Emplacement of the dome and mineralization were essentially contemporaneous at 22.5 Ma. Intense acid leaching along fractures in the quartz latite produced irregular pods and mainly northwest-trending, sub-vertical, highly silicified lenses containing a small percent of fine-grained pyrite. These lenses are termed *vuggy silica* because of the prominent voids left by the total dissolution of large potassium-feldspar phenocrysts in the host quartz latite. Alteration zones of quartz-alunite, quartz-kaolinite, and clay enclose the vuggy silica, sequentially. The intense leaching was produced by sulfuric acid formed by the dissolution of SO_2-rich magmatic vapors into the overlying groundwater. The succession of surrounding alteration zones represents increasing neutralization of the acid fluids. Gold values are highest in the vuggy silica zones and decrease outward. Main-stage mineralization consists of pyrite, enargite, luzonite, covellite, and native gold, or electrum, accompanied in places by native sulfur. Spectacular specimens of large covellite blades, an inch (2.54 cm) or more in diameter, were recovered by underground mining in the early 1970s. Narrow barite-rich veins containing pyrite, base-metal sulfides, and visible gold cut earlier alteration and mineralization. At higher levels, this vein material has been oxidized to barite, goethite, jarosite, and native gold by descending surface waters that might have been heated during dissolution of the magmatic vapors. These barite veins were the source of the famous Summitville Boulder, a 114-pound (51.7 kg) barite boulder that contains abundant visible gold and is on display at the Denver Museum of Nature and Science.

The mining at Summitville, along with natural acidic drainage, caused serious environmental damage downstream. Acidic runoff from waste dumps, leakage from the cyanide heap-leach pads, and discharge from mine tunnels were the causative processes (Gray et al. 1994; papers in Posey, Pendelton, and Van Zyl 1995). Leakage from the leach pads did not present a serious long-term problem because the cyanide entering the drainage was quickly oxidized, and the amount of cyanide in the leach pads was limited. The acid mine drainage from waste dumps and tunnels was more serious and of longer term. These problems were caused by the exposure of sulfide minerals, particularly fine-grained pyrite, to atmospheric oxygen by mining operations. The oxidation of the sulfides produced sulfuric acid, which increased the dissolution of metal-bearing minerals. The sulfuric acid could not be neutralized by the host rocks, as happens in many mining districts, because the intense pre-mineralization alteration—which is characteristic of this type of deposit—destroyed the host-rock neutralizing capacity. This process had been going on for millions of years prior to mining as erosion exposed the altered and mineralized rocks, although it was greatly exacerbated by mining, particularly open-pit operations.

The acid mine waters and runoff from the Summitville mine and dumps carried high amounts of metals, such as copper, aluminum, iron, zinc, manganese, and heavy metals, which were discharged into the Wightman Fork and transported downstream to the Alamosa River. The Alamosa River is a major source of irrigation water for the agricultural fields of the Lower San Luis Valley, raising concerns about possible accumulation of metals in crops. (This problem was later shown to be insignificant.) Metals such as zinc are toxic to fish, leading to concerns about effects of metal-contaminated water on animal species. Further, with increasing acidity in the river, metal irrigation structures tended to corrode more rapidly. Immediately following abandonment of the property, the state of Colorado asked for emergency assistance from the US Environmental Protection Agency (EPA). In May 1994, Summitville was declared a Superfund site, and remediation was undertaken cooperatively by the state and the EPA. Remedial actions included detoxifying and re-vegetating the heap-leach pad, removing material from the Cropsy and North waste dumps, backfilling the North and South open pits, and enlarging the water-runoff ponds. The water-treatment plant used during mining operations continues to treat the acid mine waters. In January 2005, the EPA turned over management of site operations to the state of Colorado. As a result of these actions, the metal concentrations in Wightman Fork and the Alamosa River have been greatly reduced. Remediation at Summitville is expected to be ongoing for the foreseeable future, and the water-treatment plant and holding ponds will need to be enlarged to guard against pond overflow in years of high precipitation.

PLATORO, STUNNER, AND JASPER DISTRICTS

The discovery of gold at Summitville spurred prospecting in the surrounding region, resulting in the location of a number of claims along the Alamosa and Conejos Rivers, south of Summitville (Patton 1917). The most important of these discoveries was the Mammoth-Revenue Mine just west of the town of Platoro, along the Conejos River. As described by Bird (1972), the Mammoth-Revenue Vein lies in the South Mountain–Platoro Fault Zone, at the southern end of the Pass Creek–Elwood Creek Fault Zone (Lipman 1975). The vein consists of massive quartz with arsenopyrite and marcasite, along with silver sulfide and sulfosalt minerals. The mine has been worked sporadically since the 1880s and has produced less than $200,000 in mining revenue. Mining activity ceased in the mid-1970s, yet the town of Platoro remains a popular, rustic summer vacation locality.

A number of prospects and small mines were developed around the old mining camps of Stunner and Jasper, along the Alamosa River, in the mid-1870s and early 1880s. Neither district has been the subject of modern scientific studies. Much of what is known is based on the work of Patton (1917). Although some very rich ore was reportedly found, little production was reported from these camps. Stunner lies in the South Mountain–Platoro Fault Zone, approximately 3 miles (5 km) north of Platoro. Jasper, located about 13 miles (21 km) east of Summitville, does not appear related to any regional faulting but instead to structures related to the southeast rim of the Summitville Caldera. Both Stunner and Jasper are located in areas of intense hydrothermal wall-rock alteration of granitic stocks (Lipman 1975). The ores at both camps consisted of quartz-pyrite veins, with copper and silver sulfides and sulfosalts and silver-gold telluride minerals. The town site of Stunner is now the location of a US Forest Service campground, and only a few of the original log houses at Jasper survive as summer homes.

MINERALIZATION ASSOCIATED WITH THE BONANZA CALDERA

The ore deposits of the Bonanza district lie mainly within the subsided block of the 33-Ma Bonanza Caldera, one of several early calderas in the northeastern San Juan Mountains (Lipman and McIntosh, chapter 2, this volume). Although the caldera has been known for some time, it has not been studied in the same detail as other calderas in the San Juan Volcanic Field. Early studies by Patton (1915) and Burbank (1932) described the geology and ore deposits of the district but did not recognize the caldera as such, although Burbank recognized a collapse feature centered near the mining district. Gravity studies by Karig (1965) provided evidence for a caldera, and later authors accepted its probable existence (Steven and Lipman 1976). Supported by geochemical, isotopic, and geochronological studies, more recent study by Varga and Smith (1984) confirmed its existence and provided a modern interpretation of the geology. Using high-precision radiometric dating of the ash-flow tuff erupted from the caldera, McIntosh and Chapin (2004) established the age of caldera formation at 33 Ma.

The town of Bonanza is situated 15 miles (24 km) northwest of Villa Grove, on Saguache County Road LL56. Prospectors, including Nicolas Creede, established it as Bonanza City in 1881. It was a thriving boomtown in the 1880s, and in 1880–1881, 40,000 people, including Ulysses S. Grant, passed through the town. An estimated 1,500 prospect holes and mines were opened by 1900. By that time, most of the rich and easily mined ore had been exploited, and the less-rich, deeper ores were less profitable, if at all. The boom had ended, but some mining persisted. The town was nearly destroyed by fire in 1927, although several of the original buildings remain. Today, Bonanza has the distinction of being Colorado's least-populated incorporated town, with a permanent population of 14, according to the 2000 Census.

The early investigations by Patton (1915) and Burbank (1932) are the only comprehensive studies of the Bonanza district's ore deposits. Few mines were accessible at the time of Burbank's investigation, and none have been so in recent years. Burbank recognized two areas with different mineralogy in the Bonanza district. North of the town of Bonanza, the veins are relatively high in sulfide minerals, and quartz is the principal gangue mineral. South of town, the veins are low in sulfides, and quartz is joined by rhodochrosite and fluorite as the principal gangue minerals. Mines in the southern portion were generally unproductive; only the Eagle Mine shipped a significant amount of ore. The Bonanza district is characterized by numerous small mines, each of which generally produced fewer than 1,000 tons of ore. Patton (1915) lists forty-seven mines that produced ore between 1881 and

1900, and Burbank (1932) adds twenty-one producers through 1930, the end of significant mining in the district. Through 1970, the Bonanza district produced just over $11.5 million in gold, silver, copper, lead, and zinc (Marsh and Queen 1974).

Mineralization in the Bonanza district was of the adularia-sericite, or low-sulfidation, type, as at Creede. However, there are significant differences between the districts. The ores at Bonanza contained gold and silver telluride minerals, unknown at Creede, and the Bonanza Veins filled relatively short structures in a network of variously oriented faults in the subsided block of the caldera, as opposed to the horizontally continuous fault systems of the Creede Mining District. The mines at Bonanza exploited quartz-pyrite veins that contained sphalerite, galena, chalcopyrite, bornite, tennantite, and gold and silver telluride minerals as the principal ore components. Wall-rock alteration associated with mineralization consisted of seritization (a general term that encompasses the formation of fine-grained muscovite, or sericite, and clay minerals such as illite). An earlier period of acid-sulfate alteration, similar to that at Summitville, affected the district's volcanic rocks—particularly a volcanic dome at Porphyry Peak—but apparently had no related mineralization.

Of the many mines in the Bonanza district, the Rawley Mine was by far the most important. Ore was originally discovered in 1880 on the north slope of Rawley Gulch, approximately 2 miles (3.2 km) northeast of the town of Bonanza, where the mine exploited a north-south-trending vein. A small mill was constructed in 1902, and a small amount of concentrate was shipped in that year and again in 1905. Until 1910, most work was devoted to development and the locations of ore bodies. In 1911–1912, a 6,000-foot-long (1.8 km) tunnel was driven perpendicular to the vein, approximately 1,200 feet (366 m) below the uppermost workings, to more economically drain the workings and transport the broken ore to the mill in Squirrel Gulch. A new mill was completed in Squirrel Gulch in 1923, and a 7.5-mile-long (12 km) aerial tramway was constructed from the mill to the town of Shirley on the Denver and Rio Grande Railway, directly north of the mill. The years 1923 through 1930, when the mine was shut down, were the mine's most productive. In total, the Rawley Mine produced 480,644 tons of ore between 1902 and 1930. The next-most-productive mines in the district were the Cocomongo and Bonanza Mines (operated as one), located on Kerber Creek about 1.25 miles (2 km) north of town. Together, these mines produced about 160,000 tons of ore between 1902 and 1927, when operations ceased. The Empress Josephine Mine, located in 1881 in Copper Gulch, three-quarters of a mile (1.2 km) northeast of Bonanza, produced some high-grade ores that contained a variety of telluride minerals including the gold telluride mineral Empressite, first described from that mine and named after it. The Empress Josephine was the largest producer in the district in the early years, but the ore body turned out to be small, and total production was apparently limited to 5,000 tons of ore. Some mining activity continued at Bonanza until 1970, but it is not clear that there was any production from the district after 1930.

Acid mine drainage from waste piles and workings adversely affected Kerber Creek and some of its tributaries, particularly Squirrel Creek. The greatest damage resulted from waters draining waste at the Rawley tunnel and mill site. In 1991, ASARCO, the main recent operator in the district, and other private parties historically involved in mining operations in the district formed the Bonanza Group to undertake a voluntary cleanup of lands contaminated by their operations, in cooperation with local governments, the Colorado Department of Public Health and Environment (CDPHE), the US Forest Service (USFS), and the EPA. The USFS was responsible for managing activities on public lands and the CDPHE for work on private lands, as well as oversight for construction on both public and private lands. Remediation efforts included moving the waste pile from the Rawley mill site to a more secure on-site repository, plugging the Rawley tunnel, and reclaiming riverbanks in the Kerber Creek Watershed. As a result of these efforts, the metal content and acidity of the waters of Kerber Creek were significantly lowered, and its waters now support aquatic life. The Bonanza reclamation efforts stand as exemplary of a voluntary and cooperative industry-government response to environmental damage caused by mining activities.

REFERENCES

Barton, P. B., Jr., R. O. Rye, and P. M. Bethke. 2000. Evolution of the Creede Caldera and Its Relation to Mineralization in the Creede Mining District. In P. M. Bethke and R. L. Hay, eds., *Ancient Lake Creede, Its Volcano-Tectonic Setting: History of Sedimentation and Relation to Mineralization in the Creede Mining District*. Special Paper 346. Boulder: Geological Society of America, 301–326.

Bethke, P. M., P. B. Barton Jr., M. A. Lanphere, and T. A. Steven. 1976. Environment of Ore Deposition in the Creede Mining District, San Juan Mountains, Colorado, Part 2: Age of Mineralization. *Economic Geology* 71:1006–1011.

Bethke, P. M., and R. L. Hay. 2000. *Ancient Lake Creede: Its Volcano-Tectonic Setting, History of Sedimentation and Relation to Mineralization in the Creede Mining District*. Special Paper 346. Boulder: Geological Society of America.

Bethke, P. M., R. O. Rye, R. E. Stoffregen, and P. G. Vikre. 2005. Evolution of the Magmatic-Hydrothermal Acid-Sulfate System at Summitville, Colorado: Integration of Geological, Stable Isotope, and Fluid Inclusion Evidence. *Chemical Geology* 215:281–315.

Bird, W. H. 1972. Mineral Deposits of the Southern Portion of the Platoro Caldera Complex, Southeast San Juan Mountains, Colorado. *Mountain Geologist* 9(4):379–387.

Burbank, W. S. 1932. *Geology and Ore Deposits of the Bonanza Mining District, Colorado*. Professional Paper 169. Washington, DC: US Geological Survey.

Emmons, W. H., and E. S. Larsen. 1913a. A Preliminary Report on the Geology and Ore Deposits of Creede, Colorado. In *Contributions to Economic Geology (Short Papers and Preliminary Reports, 1911. US Geological Survey Bulletin* 530:42–65.

———. 1913b. The Hot Springs and the Mineral Deposits of Wagon Wheel Gap, Colorado. *Economic Geology* 8:235–246.

————. 1923. Geology and Ore Deposits of the Creede District, Colorado. *US Geological Survey Bulletin* 718:198 pp.

Feitz, L. 1969. *A Quick History of Creede.* Denver: Golden Bell.

————. 1973. *Soapy Smith's Creede.* Colorado Springs: Little London.

Foley, N. K. 1990. Petrology and Geochemistry of Precious- and Base-Metal Mineralization, North Amethyst Vein System, Mineral County, Colorado. Unpublished PhD dissertation, Virginia Polytechnic Institute and State University, Blacksburg.

Foley, N. K., and R. Ayuso. 1994. Lead-Isotope Compositions as Guides to Early Gold Mineralization; The North Amethyst Vein System, Creede District, Colorado. *Economic Geology* 89:1842–1859.

Foley, N. K., S. W. Caddey, C. B. Byington, and D. M. Vardiman. 1993. *Mineralogy, Mineral Chemistry, and Paragenesis of Gold, Silver, and Base-Metal Ores of the North Amethyst Vein System, San Juan Mountains, Mineral County, Colorado.* Professional Paper 1537. Washington, DC: US Geological Survey.

Giudice, P. M. 1980. Mineralization at the Convergence of the Amethyst and OH Fault Systems, Creede District, Mineral County, Colorado. Unpublished MSc thesis, University of Arizona, Tucson.

Gray, J. E., and M. F. Coolbaugh. 1994. Geology and Geochemistry of Summitville, Colorado: An Epithermal Acid-Sulfate Deposit in a Volcanic Dome. *Economic Geology* 89: 1906–1923.

Gray, J. E., M. F. Coolbaugh, G. S. Plumlee, and W. W. Atkinson. 1994. Environmental Geology of the Summitville Mine, Colorado. *Economic Geology* 89:2006–2014.

Heald-Wetlaufer, P. H., N. K. Foley, and D. O. Hayba. 1987. Comparative Anatomy of Volcanic-Hosted Epithermal Deposits: Acid-Sulfate and Adularia-Sericite Types. *Economic Geology* 82:1–26.

Hedenquist, J. W. 1987. Mineralization Associated with Volcanic-Related Hydrothermal Systems in the Circum-Pacific Basin. In M. K. Horn, ed., *Transactions of the Fourth Circum-Pacific Energy and Mineral Resources Conference*, August 1986, Singapore. *American Association of Petroleum Geologists*:513–524.

Hemingway, M. P. 1986. Mineralogy and Geochemistry of the Southern Amethyst Vein System, Creede Mining District, Colorado. Unpublished MSc thesis, New Mexico Institute of Mining and Technology, Socorro.

Hills, R. C. 1885. Ore Deposits of Summit District, Rio Grande County, Colorado. *Colorado Scientific Society Proceedings* 1:20–37.

Hon, K., and H. H. Mehnert. 1983. Compilation of Revised Ages of Volcanic Units in the San Juan Mountains, Colorado. *US Geological Survey Open-File Report* 83:668.

Huston, R. C. 2005. *A Silver Camp Called Creede: A Century of Mining.* Montrose, CO: Western Reflections.

Karig, D. E. 1965. Geophysical Evidence of a Caldera at Bonanza, Colorado. US Geological Survey Professional Paper 525B. In *Geological Survey Research 1965*, 9–12.

Korzeb, S. L. 1993. The Wagon Wheel Gap Fluorspar Mines, Mineral County, Colorado. *Mineralogical Record* 24:23–24.

Koschman, A. H., and M. H. Bergendahl. 1968. *Principal Gold-Producing Districts of the United States.* Professional Paper 610. Washington, DC: US Geological Survey.

Lanphere, M. A. 2000. Duration of Sedimentation of the Creede Formation, from ^{39}Ar/^{40}Ar Ages. In P. M. Bethke and R. L. Hay, eds., *Ancient Lake Creede: Its Volcano-Tectonic*

Setting, History of Sedimentation and Relation to Mineralization in the Creede Mining District. Special Paper 346. Boulder: Geological Society of America, 301–326.

Larsen. E. S. 1930. Recent Mining Developments in the Creede Mining District, Colorado. *US Geological Survey Bulletin* 811B:89–112.

Larsen, E. S., and R. C. Wells. 1916. Some Minerals from the Fluorite-Barite Vein near Wagon Wheel Gap, Colorado. In *Proceedings of the National Academy of Sciences,* vol. 2. Washington, DC: US Government Printing Office.

Lipman, P. W. 1975. *Evolution of the Platoro Caldera Complex and Related Volcanic Rocks, Southeastern San Juan Mountains, Colorado.* Professional Paper 852. Washington, DC: US Geological Survey.

Marsh, W. R., and R. W. Queen. 1974. Map Showing Localities and Amounts of Metallic Mineral Production in Colorado. Map MR-58 and accompanying text and tables. Washington, DC: US Geological Survey.

McIntosh, W. C., and C. E. Chapin. 2004. Geochronology of the Central Colorado Volcanic Field. *New Mexico Bureau of Mines and Mineral Resources Bulletin* 160:205–237.

Patton, H. B. 1915. Geology and Ore Deposits of the Bonanza District, Saguache County, Colorado. *Colorado Geological Survey Bulletin* 1(9):122 pp.

———. 1917. Geology and Ore Deposits of the Platoro-Summitville Mining District, Colorado. *Colorado Geological Survey Bulletin* 13:122 pp.

Plumlee, G. S., and P. H. Whitehouse-Veaux. 1994. Mineralogy, Paragenesis, and Mineral Zoning of the Bulldog Mountain Vein System, Creede District, Colorado. *Economic Geology* 89:1883–1905.

Posey, H. H., J. A Pendleton, and D.J.A Van Zyl, eds. 1995. *Proceedings: Summitville Forum '95.* Special Publication—Colorado Geological Survey Report 38. Denver: Colorado Department of Natural Resources.

Robinson, R. W., and D. I. Norman. 1984. Mineralogy and Fluid Inclusion Study of the Southern Amethyst Vein System, Creede Mining District, Colorado. *Economic Geology* 79:439–447.

Schroder, M. F. 2004. *Nicholas Creede and the Amethyst Vein.* Creede, CO: Bonanza Press.

Steven, T. A. 1968. Ore Deposits in the Central San Juan Mountains, Colorado. In J. D. Ridge, ed., *Ore Deposits of the United States 1933–1967, Graton-Sales Volume.* New York: American Institute of Mining and Metallurgical Engineers, 706–713.

Steven, T. A., and G. P. Eaton. 1975. Environment of Ore Deposition in the Creede Mining District, San Juan Mountains, Colorado: Geologic, Hydrologic and Geophysical Setting. *Economic Geology* 70:1023–1037.

Steven, T. A., and P. W. Lipman. 1976. *Calderas of the San Juan Volcanic Field, Southwestern Colorado.* Professional Paper 958. Washington, DC: US Geological Survey.

Steven, T. A., and J. C. Ratté. 1960a. Relation of Mineralization to Caldera Subsidence in the Creede District, San Juan Mountains. Colorado. In *Short Papers in the Geological Sciences.* Professional Paper 400B. Washington, DC: US Geological Survey, 14–17.

———. 1960b. *Geology and Ore Deposits of the Summitville District, San Juan Mountains, Colorado.* Professional Paper 343. Washington, DC: US Geological Survey.

———. 1965. *Geology and Structural Control of Ore Deposition in the Creede District, San Juan Mountains, Colorado.* Professional Paper 487. Washington, DC: US Geological Survey.

Stoffregen, R. E. 1987. Genesis of Acid-Sulfate Alteration and Cu-Au-Ag Mineralization at Summitville, Colorado. *Economic Geology* 82:1575–1591.

Varga, R. J., and B. M. Smith. 1984. Evolution of the Early Oligocene Bonanza Caldera, Northeast San Juan Volcanic Field, Colorado. *Journal of Geophysical Research* 89: 8679–8694.

Wolle, M. S. 1949. *Stampede to Timberline: The Ghost Towns and Mining Camps of Colorado.* Chicago: Sage Books/Swallow.

....

Geomorphic History of the San Juan Mountains

....

Rob Blair and Mary Gillam

THE SAN JUAN MOUNTAINS encompass 6,000–7,000 mi² (15,000–18,000 km²) and thirteen peaks that rise above 14,000 feet (4,267 m). Some of these "fourteeners" are flanked by valley floors as low as 7,000 feet (2,133 m). The surface of the eastern San Juans is dominated by Tertiary volcaniclastic breccias and ignimbrites (ash-flow tuffs) derived from more than two dozen volcanic centers. The volcanic pile is hundreds to several thousand meters thick. At the surface, the western San Juans are composed mostly of resistant Precambrian metamorphic and plutonic rocks, Paleozoic through Cenozoic sedimentary layers, and smaller areas of Tertiary volcanic rocks (see geologic map, plate 37) This chapter briefly outlines the dominant geomorphic processes (such as uplift, volcanism, and erosion) that have shaped (and continue to shape) the rugged San Juans during the last 70 million years.

EARLY WORK

The earliest publications that addressed the San Juan Mountains landscape came from the work of the *Geological Survey of the Territories* by F. V. Hayden (in 1869–1876), followed by the work of Whitman Cross (in 1895–1909) and others (Larsen and Cross 1956). Cross recognized the uniqueness of the San Juan landscape and in 1909 arranged for Wallace Atwood to lead an investigation into the mountains'

....

physiography and Quaternary geology (Atwood and Mather 1932). This later work emphasized the importance of erosional surfaces and the influence of glaciation. Glen Scott, another US Geological Survey geologist, later identified six stages of geomorphic development of the Southern Rocky Mountain region (Scott 1975): (1) Laramide uplift, (2) beveling of an erosional surface during the Eocene, (3) Oligocene deposition, (4) uplift during the Miocene through the Pliocene, (5) Pliocene incision, and (6) Quaternary glaciation. Although the San Juan Mountains record evidence of all these stages, volcanism dominated the area between 35 and 15 Ma (Oligocene through Miocene), and uplift is now thought to have occurred mostly between 5 and 2.5 Ma (Late Miocene–Pliocene). The volcanic overprint makes these mountains unique. Steven, Hon, and Lanphere (1995) documented the influence of volcanism, tectonism, and erosion in the San Juans and refined the key stages on which this chapter is based.

LARAMIDE OROGENY

The Laramide orogeny (70–50 Ma, Late Cretaceous through Early Eocene) was a compressional tectonic event that resulted in thrust faulting, crustal thickening, and mountain uplift across portions of western North America, forcing the eastward retreat of the Cretaceous mid-continent inland sea. In the Rocky Mountain West, this orogeny formed intra-foreland basins and basement-block uplifts (Cather 2004; Suppe 1985). In southwest Colorado, the preceding broad coastal plain gave way to the San Juan Uplift (a broad dome centered in today's Needle Mountains, southeast of Silverton) and adjacent troughs to the south (the San Juan Basin, which lies mostly in northwest New Mexico) and to the east (the San Juan Sag, between Pagosa Springs and Saguache). The conglomerates found in the Animas Formation (near Durango and from a deep well drilled near South Fork) and the Ridgway Conglomerate (west of Ridgway) attest to the uplift and vigorous erosion that occurred during the Paleocene and Early Eocene as a result of the Laramide orogenic event (Baars 1992; Cather 2004; Van Houten 1957).

LATE-EOCENE EROSIONAL SURFACE

The clearest evidence of pre-volcanic topography is seen where unconformities are exposed, such as at the erosional contacts between the Telluride Conglomerate (Late Eocene) and the Cutler Formation (Permian), west of Molas Pass, and between the Blanco Basin (Eocene?) and Morrison formations (Jurassic), north of Chama, New Mexico. These unconformities are indicative of erosion that accompanied the San Juan Uplift during the Paleocene and Early Eocene. This erosional cycle removed much of the Phanerozoic sedimentary cover around the perimeter of the San Juans and completely removed the sedimentary cover from the newly exposed Precambrian formations (see geologic map, plate 37; Gonzales and Karlstrom, chapter 1 this volume). Later portions of the erosion surface were buried beneath conglomerates shed from a subsequent rising dome located in the central San Juans between Silverton and Lake City, which was created by magma bodies inflating within the shallow crust immediately prior to Oligocene volcanism. Any remaining surface was overwhelmed by volcanic deposits (discussed later in this chapter).

The Late-Eocene pre-volcanic topography of the San Juan Mountain region is conjectural, although geologists (Epis et al. 1976; Scott 1975) suggest that there was low relief with rounded hills and, in places, deeper valleys.

OLIGOCENE VOLCANOES (35–25 MA)

In the latest Eocene (35 Ma), magma forcefully pierced the crust north of what is now Saguache. During the ensuing 5 million years (Oligocene), volcanic centers formed northeast of the present-day San Juan Mountains, then farther south and west (Lipman and McIntosh, chapter 2, this volume). The volcanic eruptions built andesitic composite volcanoes (stratovolcanoes) similar to Washington's Mount St. Helens (plate 9); together, more than twenty of these composite volcanoes formed the San Juan Volcanic Field. The widespread volcanic deposits dumped upon the Eocene erosional surface consisted of andesitic explosion breccias, lava flows, and ash flows. Each volcanic center produced its own circular apron of debris and flows that often buried parts of adjacent volcanoes. The accumulative volume of these andesitic rocks was approximately 25,000 km^3 (6,000 mi.^{3}) (Lipman 2007). Thirty million years ago, the San Juan Mountains region might have resembled the landscape now seen in the Cascade Mountains of Oregon and Washington, with one significant difference: snowcaps and glaciers were probably absent. (The world had not yet entered an ice age, and the Antarctic ice cap was merely developing.) Many of the volcanoes likely rose several thousand meters above sea level, although the modern-day relief of the Rocky Mountains did not yet exist.

Around 30 Ma a new phase of eruptions began, with the development of huge calderas tens of kilometers in diameter (plate 10). These roughly cylindrical depressions of down-dropped crust developed catastrophically. This crustal collapse often carried with it the mass of one or more composite volcanoes and filled subcrustal voids created simultaneously from the vigorous venting of rhyolite ash that formed radiating ignimbrite sheets. The ejected ignimbrite sheets blanketed thousands of square kilometers, burying the bases of existing volcanoes and forming a widespread volcanic plateau (Steven, Hon, and Lanphere 1995). The original volcanic field might have occupied an area of more than 38,000 mi^2 (100,000 km^2) and an accumulative volume of greater than 12,000 mi^3 (50,000 km^3), including the previously erupted andesite (Lipman and McIntosh 2008). Thirty or so ash-flow eruptions occurred in the San Juan Mountains during a 2.5-million-year period, generating about 3,500 mi^3 (15,000 km^3) of ash (Lipman 2007). Between 30 and 26 Ma, the region probably displayed rolling terrain with partially buried older volcanoes, a poorly integrated drainage system, and scattered caldera depressions (plate 10). Those highland surfaces were part of a constructed landscape (Steven, Hon, and Lanphere 1995), in contrast to the erosional San Juan Peneplain proposed by Atwood and Mather (1932).

The Creede Caldera eruption, at 26.9 Ma, was one of the last in the San Juans. Shortly after the caldera formed, it was disrupted by a resurgent dome (Snowshoe Mountain) that created a circular moat with no outlet. As a consequence, a ring-shaped lake formed and was filled with sediment of the Creede Formation. The moat deposits contained fossil leaves, seeds, and stems, along with occasional impressions of flies, mosquitoes, and bees. The lake sediment was deposited over a period of 0.3–0.6 million years and not later than 26.26 Ma, as inferred from the age of Fisher Dacite lava flows overlying the lake deposits (Lanphere 2000). In places, the lake sediments exceed 3,900 ft (1,200 m) in thickness (Hulen 1992).

EROSION AND RIFT DEVELOPMENT (25–15 MA)

Steven, Hon, and Lanphere (1995) indicate that during the Late Oligocene and Early Miocene, the volcanic plateau was subjected to erosion alternating with eruption of trachybasalt lava flows and localized rhyolite lavas and tuffs (plate 11). These lava flow caps are mostly found today at 11,000 ft. (3,600 m) or higher and are exemplified by the Cannibal Plateau rim northeast of Lake City, Trout Mountain southeast of Creede, Fox Mountain south of South Fork, and Jarosa Peak north of Cumbres Pass (Steven, Hon, and Lanphere 1995; Steven et al. 1974). Steven, Hon, and Lanphere (1995) note that a paleocanyon with a relief of 2,000 ft (600 m) is partly preserved in the Rio Grande Valley between South Fork and Creede. The canyon was probably formed after Lake Creede was formed (26.3 Ma) and before trachybasalt lava (23–22 Ma) filled the Creede caldera moat (covering the lake sediment) and said canyon. The canyon was part of a buried landscape that consisted of resistant tablelands and valley benches composed of welded ignimbrites, alternating with slopes eroded from weaker ash-flow layers. The uplands were relatively low-relief plateaus. The canyon indicates the presence of an ancestral Upper Rio Grande.

Initial subsidence along the Rio Grande Rift, east of the volcanic plateau (where the San Luis Valley is today), began around 27–26 Ma, as inferred from the outpouring of the Hinsdale trachybasalts, the asymmetric eastward dips of the Summer Coon volcano (one of the trachybasalt sources), and the westward extension of the Rio Grande drainage (Brister and Gries 1994; Lipman 1976; Noblett and Loeffler 1987). As the rift slowly developed, the ancestral Rio Grande and other drainages fed sediment into a broad alluvial fan in the down-warped basin, south of what is now Del Norte (Brister and Gries 1994; Lipman, Steven, and Mehnert 1970).

From the initiation of rifting to the present, the San Luis Valley has been a catchment basin for sediment from the San Juan Mountains to the west and the

Sangre de Cristo Range to the east. The latter were uplifted simultaneously with the expanding rift. During the Late Oligocene and continuing through the Miocene, the San Luis Valley probably remained a closed drainage basin that resembled some of the basins now seen in the Basin and Range province of Nevada. San Luis Basin fill (mostly the Santa Fe Formation) reaches a maximum thickness of 21,000 ft (6,300 m) immediately northwest of Great Sand Dunes National Park (Kluth and Schaftenaar 1994). Did the Upper Rio Grande flow south along the rift valley, into what is now New Mexico, during the Late Oligocene and Miocene? We do not know. No evidence has been found to support this possibility.

MID- TO LATE-MIOCENE EROSION (15–5 MA)

The widespread eruption of the basaltic lava flows of the Hinsdale Formation ceased around 15.4 Ma (Steven, Hon, and Lanphere 1995). The flows followed the valleys and lowlands (plate 12); thus much of the eastern San Juans could be viewed as a broad lava-capped volcanic plateau displaying moderate relief (less than 3,300 ft [1,000 m]). The intervening highlands consisted of eroded ignimbrites and elevated resurgent domes, producing a bench-and-tableland topography (Steven, Hon, and Lanphere 1995). In the Creede region, this surface experienced moderate erosion (~1,000 ft [~300 m]) between 15 and 5 Ma (Rye et al. 2000; Steven, Hon, and Lanphere 1995). Today, over much of the eastern San Juan Mountains, this surface is seen perched above existing river valleys, exemplifying "hanging" topography. The isolation of these elevated surfaces started about 5 Ma and marks the beginning of the "canyon cycle of erosion" in the Southern Rocky Mountains (Lee 1923). These high surfaces were further modified by Pleistocene glacial ice fields (see the section Quaternary Glaciation later in this chapter).

LATE-MIOCENE AND PLIOCENE
UPLIFT AND EROSION (5–2.6 MA)

Further evidence of canyon-cycle incision comes from Rye and others (2000), who dated samples of alunite from a sulfide mineral deposit (25 Ma) just below the high topographic surface near Creede. They ascertained that the alunite formed between 4.8 and 3.1 Ma. Because alunite formation requires a water-saturated environment, they inferred that a high water table existed until after 5 Ma. Associated with the alunite is jarosite, which requires an aerated environment associated with the vadose zone above the water table. This jarosite was dated from vertical fractures as having an age range from 2.6 Ma (at the top of the fractures) to 0.9 Ma (at the bottom). Rye and others interpreted their results as indicating a post–5 Ma water-table drop, which was associated with deep incision and erosion from nearby streams.

Between 5 and 2.6 Ma, possibly beginning as early as 7 Ma, uplift and block faulting accelerated and accompanied localized eastward tilting of 2–7 degrees along the eastern margin of the San Juans (plate 13) (Epis et al. 1976; Steven, Hon, and Lanphere 1995). During this period, the northern San Luis Valley subsided further, and the Sangre de Cristo Mountains continued to rise, generating most of the approximately 30,000 ft (9,200 m) of vertical movement observed today along the Sangre de Cristo Fault, which defines the eastern edge of the San Luis Valley (Kluth and Schaftenaar 1994; Steven, Hon, and Lanphere 1995). The Servilleta Basalt (4.8–3.7 Ma) erupted along the margins of the rift, just north of the existing Colorado–New Mexico border, filling a portion of the San Luis Valley and creating a hydrologic dam to any drainage south into New Mexico.

So far, geologists have found no evidence (gravel deposits or paleochannel) that the Rio Grande flowed south into New Mexico prior to the Pliocene lava flows. Did the Rio Grande exit the San Luis Basin at any time during the last 20 Ma? If so, where? Based on geologic mapping and interpretation of well logs, the San Luis Basin supported a lake during the Late Pliocene and Early Pleistocene (Machette, Marchetti, and Thompson 2007). The lake, called Lake Alamosa, initially formed between 4 and 3 Ma and finally overflowed the lava dam at about 0.44 Ma, at which time the lake became permanently drained. The lake occupied more than 1,350 mi^2 (3,500 km^2) and exceeded 1,540 mi^2 (4,000 km^2) at its highest stands. Lake sediment deposition recorded a climatic record spanning more than 2 million years (K. Rogers et al. 1992).

At the same time these physical changes were occurring in the San Juans, the Earth was cooling episodically at a rate of approximately 1°C per million years (Lawrence, Liu, and Herbert 2006). For example, ocean temperatures off the eastern coast of South America decreased from 27°C (5 Ma) to 23°C (today). This cooling trend is revealed in figure 4.1 and shows an average cooling over the past 5 million years, in spite of the large oscillations driven by Milankovitch cycles, particularly the 41,000- and 100,000-year cycles. Milankovitch cycles are climate variations that result from solar-radiation fluctuation caused by eccentricity of the Earth's orbit (100-Ka cycle), fluctuation in the tilt of the Earth's axis (41-Ka cycle), and precession of the Earth's axis (23-Ka cycle) (search Milankovitch cycles on the web for more detail). This cooling was the precursor to the ice ages and glaciers that have since scarred the modern landscape.

During the Late-Miocene–Pliocene uplift that affected the Southern Rocky Mountains, the accompanying erosion and incision of rivers produced a radial drainage pattern flowing away from the higher elevations, as the spokes of a bicycle wheel radiate from its hub. The causes of the uplift are still debated (see Gonzales and Karlstrom, chapter 1, this volume). Ancestral rivers bearing the bulk of sediment load included the San Miguel, Dolores, Animas, San Juan, Conejos, Rio Grande, and Lake Fork of the Gunnison.

As the drainage systems developed, the Rio Grande Watershed expanded more than the others. As we follow the Continental Divide from north to south through

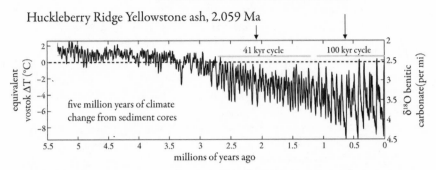

FIGURE 4.1 $\delta^{18}O$ isotopes in benthic foraminifera from deep-sea sediment cores and correlation with Vostock ice core equivalent temperature data. Downstrokes indicate cooling, while upstrokes signify warming trends. The onset of the "Pleistocene" glacial epoch is indicated between 2.7–2.6 Ma. The first major Northern Hemisphere ice sheet advance (Nebraskan) is indicated around 650 Ka. The seven major glacial stages correspond with plate 15 and can be loosely matched (last 350 Ka) with terrace development south of Durango, Colorado. The frequency of temperature oscillations closely matches the 100 and 41 Ka periods predicted by the Milankovitch Cycles. From Wikimedia Commons, based on Lisiecki and Raymo 2005.

the San Juans, the divide makes an abrupt 60-mi (100-km) jog to the west and then back to the east (plate 14). This westward transgression is a result of the dominance and aggressiveness of the Rio Grande Watershed, which "out-competed" the other watersheds. The surface of the Rio Grande Watershed is generally composed of thicker, softer, fractured volcanic rocks, whereas the surfaces of the Animas, San Miguel, and San Juan Watersheds are composed mostly of compact sedimentary rocks and hard plutonic and metamorphic rocks, which are relatively resistant to erosion. In addition, the San Luis Valley, into which the Rio Grande and its tributaries flowed, subsided to the east (along the Rio Grande rift), lowering the depositional base level and creating steeper stream gradients and more aggressive headward erosion toward the west. By the end of the Pliocene, the drainages were well-established and set the stage for glaciers that later occupied the same valleys.

QUATERNARY GLACIATION (2.6 MA–10 KA)

Glacial Events

The earliest evidence of global glacial conditions dates from 2.6 to 1.8 Ma (Kerr 2008). The Antarctic ice sheet initially developed around 34 Ma (Lear et al. 2008), the Greenland ice sheet developed in the Late Pliocene (Raymo 1994), and beginning around 2.6 Ma, glacial ice sheets periodically occupied portions of

Asia, Europe, and North America. Figure 4.1 suggests that cooling-cycle temperature oscillations increased in amplitude through time, and for each greater surge in cooling, the respective equilibrium-line altitude (ELA—the theoretical average freezing elevation above sea level and its intersection with the topography) became 1,000–2,000 ft (300–600 m) lower in the San Juan highlands. Today, only the highest San Juan summits lie above the ELA. Given sufficient terrain lying above the ELA, snow may become permanent and eventually accumulate into ice fields from which glaciers flow.

Detailed analyses by K. Rogers and others (1992) of sediment cores taken at Hansen Bluff, southeast of Alamosa in the San Luis Valley and dated 2.6–0.74 Ma (based on fossil pollen, invertebrates, and vertebrates), suggest that the first significant ice field in the San Juans developed around 800 Ka. This may represent the first time ice began to effectively alter the San Juan Mountains landscape through glacial erosion, deposition of ice-borne sediment, and meltwater action along rivers. Before then, much smaller glaciers might have been isolated on the higher peaks. Based on $\delta^{18}O$ values from marine sediment cores and CO_2 values from Antarctic ice cores, the most intense and widespread glaciations occurred since 650 Ka (plate 15). Seven of those glaciations correspond to the 100,000-year Milankovitch Cycle.

Glaciation in the San Juans

During the height of the last glacial expansion (22–20 Ka), the San Juan highlands supported a vast ice field (1,900 mi^2 [5,000 km^2]) dotted with exposed mountaintops called nunataks, which towered above the surrounding glaciers (plate 14). Hundreds of cirque basins developed beneath the glaciers and expanded laterally to shape the nunataks into glacial horns, such as Pigeon and Turret peaks in the Needle Mountains. At the peak of the last glacial advance, the Animas Glacier extended 56 mi (90 km) to its terminal moraine (the Animas City moraine on the north edge of Durango). The Rio Grande drainage was fed by a vast ice field. Prominent moraines are exposed about 10 mi (16 km) southwest of Creede (see Blair et al., chapter 16, POI 38).

Glacial Erosion

The San Juans are home to numerous peaks sculpted by glacial erosion, including thirteen that are higher than 14,000 ft (4,267 m). All are west and north of the Continental Divide, not along it. This apparent paradox results because the center of the San Juan Uplift (the Needle Mountains) is composed of relatively resistant rocks, such as gneiss and granite, and lies west of the Divide. The best-preserved features of glacial erosion are found in the western San Juans. As gravity pulled massive tongues of ice down existing river valleys, glacial loads of pulverized rock ground down bedrock and over-deepened valleys to form U-shaped canyons, such as the

Upper Uncompahgre and Upper Animas Gorges. East of the Continental Divide, where more fractured volcanic rocks dominate, these features are relatively subdued because they have been modified to a greater degree by postglacial weathering, mass wasting, and erosion.

Glacial Deposition

Debris carried by ice was deposited in ridges, such as end moraines at the glacier snouts and lateral moraines along valley sides. Thinner ground moraines, erratic boulders, and recessional moraines are also found within the glaciated valleys. As glaciers began to retreat, end moraines acted as dams for temporary lakes that filled with outwash and lake sediment. Such old lake beds underlie the present flat-floored valleys and can be seen just upstream from the latest end moraines in the Lower Animas, Uncompahgre, and Upper Rio Grande Valleys. The low-elevation glacial moraines seen today at Durango, Ridgway, Telluride, and Lake City are remnants from the last glacial maximum, which ended around 20 Ka. During the ensuing 20,000 years, San Juan glaciers retreated and eventually disappeared.

Glacial Cycles, Moraines, and Terraces

To date, the most extensive study of relationships among glacial cycles, moraines, and river terraces along the Animas River drainage has been conducted by Mary Gillam (1998). She interpreted the oldest high terraces (2.4 and 1.3 Ma) along the Animas River between Durango and Farmington as climate-related. Cool-wet periods are associated with glacial advance and outwash gravel deposition, and warm-dry periods are associated with river incision and terrace isolation during interglacials.

Some of the earlier evidence of glaciation in the San Juan Mountains comes from outwash alluvium on Florida Mesa, southeast of Durango, and contains dated (640 Ka) deposits of Lava Creek B ash from Yellowstone (Gillam 1998; Lanphere et al. 2002). This ignimbrite volcanic eruption occurred during glacial advance #7 on plate 15 and is recorded in isolated pockets around the Durango area (Gillam 1998).

Moraines provide direct evidence of glacial ice in the San Juan Mountains. The oldest are the Durango I moraines, at 360–345 Ka (#4 on plate 15; table 4.1) (Gillam 1998). Gillam (1998) also mapped three younger groups of moraines along the Animas River: the Durango (oldest), Spring Creek, and Animas City (youngest) (table 4.1). These moraines were also linked to distinct outwash terraces. In total, she identified thirty-four terrace platforms, the oldest (highest) of which likely dates to 2.6–2.2 Ma. Detailed studies are lacking for the eastern San Juans; thus comparative analysis would be conjectural.

Table 4.1. Correlation of glacial stages with Animas Valley glacial moraines and terraces. The 100,000-year cycles (see plate 15) are numbered for the purpose of this chapter.

Pleistocene stages (see plate 15)	Terraces (Gillam 1998)	Moraines	Age (Ka)	Comments
		Bakers Bridge (Pinedale 4?)	> 17.1	^{10}Be granite outcrop
	t7c, t7b	Animas City 2 & 3 (late & mid-Pinedale)	19.4	^{36}Cl Rio Grande, ^{10}Be terrace
Wisconsin Glacial (1)	t7a	Animas City 1 (early Pinedale)	22–60	regional correlation
	t6b	Spring Creek 2 (late Bull Lake)	60–100	thermal luminescence
Sangamon interglacial				
Illinoisan glacial (2)	t6a	Spring Creek 1 (early Bull Lake)	130–190	regional correlation
Yarmouth III interglacial				
Kansan (4) glacial (3)	t5e, t5d, t5c	Durango 2	250–275	river incision rate
Yarmouth II interglacial				
Kansan (3) glacial (4)	t5b, t5a	Durango 1	345–360	river incision rate
Yarmouth I interglacial				
Kansan (2) glacial (5)	t4e, t4d (?)		420–460	Antarctic CO_2 data, Lake Alamosa drains
Aftonian II interglacial				
Kansan (1) glacial (6)	t4d (?), t4c		510–540	Antarctic CO_2 data
Aftonian I interglacial				
Nebraskan glacial (7)	t4b, t4a		620–670	Lava Creek B ash
	t3b, t3a, t2i t2h, t2g, t2f, t2e, t2d, t2c, t2b, t2a	100,000-year cycles dominate 41,000-year cycles dominate		11 terraces likely associated with 26 or more global cooling cycles; see plate 15
	t1a		2.2–2.6 Ma	

Sources: Data from Benson et al. 2005; Coe et al. 2007; Gillam 1998; Guido, Ward, and Anderson 2007.

Rates of River Incision

River incision rates were determined from the dating of abandoned elevated terraces using amino-acid racemization, paleomagnetic polarity, and correlation with marine-oxygen-isotope data (Gillam 1998). The earliest incision rates considered for the Animas River, from terrace surfaces dated 2.6–2.2 Ma, were highly variable and gave a median rate of down cutting of 3 in (8 cm) per thousand years. Between 615 Ka and 130 Ka the Animas River lowered its channel approximately 4.9 in (12.5 cm) per thousand years, and, between 130 Ka and the present the incision rate was approximately 9.4 inches (24 cm) per thousand years. The data suggest that accelerated incision rates over the past 2 million years are affected by glacial cycles (Dethier 2001), although the possibility of local uplift, variations in bedrock geology, and changes in hydrology complicate the interpretation. However, glacial signals from CO_2 and $\delta^{18}O$ (figure 4.1) support the interpretation that increased incision rates during the Pleistocene were driven by temperature fluctuations that were increasing in amplitude and decreasing in frequency (for a discussion of the relationship between CO_2, $\delta^{18}O$, and temperature, see Wolff 2000). Since 700 Ka, the intensity and magnitude of glacial advances have abruptly increased with the dominance of the 100,000-year Milankovitch Cycle. It is generally accepted that during peak glacial advances and the waning stages of the glaciation, there is more sediment supply than the existing rivers can handle, thus promoting aggradation and formation of broad floodplains. During the intervening interglacials, the rivers periodically flood and effectively incise their channels, leaving former floodplains as high terraces.

The San Juan Mountains contain no evidence of any glacial till older than 650 Ka. Explanations include the possibilities that (1) older till was reworked and destroyed by younger glacial surges, (2) natural erosion and mass wasting eroded the deposits, or (3) early glacial cycles were limited in extent and magnitude. The data presented in figure 4.1 and the work of K. Rogers and others (1992) favor explanation 3 and suggest that the first extensive glacial ice field in the San Juan Mountains developed around 650 Ka. The incision-rate increase during the past several hundred thousand years is associated with over-deepening of valleys by aggressive valley glaciers and their greater sediment loads. During the interglacials, down cutting commenced with vigor because of abundant hard gravels available for eroding the bedrock channel, especially during intense short-duration flooding events. The relief from valley incision is greatest in the mountains and foothills and decreases when the rivers exit the range, while flatter gradients allow greater deposition. This can lead to terrace levels merging down-valley, as seen along the Animas River between Durango and Farmington.

The relief and gradients of the glacial valleys west of the Continental Divide are greater than those of corresponding valleys east of the Divide. This is caused in part by the resistant metamorphic and igneous rocks found west of the Divide, in contrast to the softer, fractured volcanic rocks that dominate the landscape east of the Divide.

HISTORY OF GLACIAL RETREAT

Eric Small (1995) mapped numerous features formed by stagnant ice—such as kames, kettles, and eskers—in the Upper Rio Grande Valley, upstream of the moraines near Hogback Mountain southwest of Creede. These landforms are generally rare in alpine glacial valleys and are commonly found along continental ice sheet margins. Small believes the low valley gradients and lower relief topography in the Rio Grande Watershed allowed for rapid widespread loss of the glacial ice when the ELA rose rapidly toward the end of the last ice advance (20 Ka). Kames, eskers, and kettles were likely more prevalent in the more subdued landscapes of the eastern San Juans because those tend to be associated with thinner ice that can become detached from the retreating valley glacier. The glacial valleys west of the Continental Divide have steeper gradients, are narrower, and support thicker ice in general, thereby allowing for a more gradual retreat of the ice terminus. Small (1995) noted the paucity of stagnant ice features in valleys west of the Divide. The lower valley gradients and broader valleys found in the Rio Grande drainage are attributed to the less resistant, fractured volcanic rock.

Two studies (Benson et al. 2005; Guido, Ward, and Anderson 2007) document the glacial retreat of the Rio Grande Valley Glacier and the Animas Valley Glacier, respectively. Using [10]Be cosmogenic dating of glacial outwash terrace gravels and select bedrock exposures, Guido's team determined that glacial-ice retreat in the Animas Valley began around 19.4 Ka and continued until 12.3 Ka, when the San Juan ice field became nonexistent. The presence of rock glaciers (discussed later in this chapter) in many of the high cirques suggests that some of the retreating glaciers became buried in their own detritus and colluvium from the oversteepened valley walls. The loss of the ice in the Animas Valley occurred over a 7,200-year period, with an average ice-terminus retreat rate of 50.8 ft/year (15.5 m/year). Benson and others (2005), using [36]Cl cosmogenic dating techniques, concluded that glacial retreat began between 20 and 19 Ka (ages recalibrated by Guido, Ward, and Anderson 2007) in the Rio Grande Valley, near Hogback Mountain southwest of Creede.

THE HOLOCENE LANDSCAPE (LAST 10 KA)

Rock Glaciers

One might reasonably assume that all glacial ice in the San Juans disappeared 12,000 years ago, but this inference may be in error. One of the more common landforms recognized by geologists working in the San Juan Mountains is the rock glacier. These landforms fill glacial cirques and glacial valleys and were once thought to be peculiar forms of talus or rock slides (Howe 1909). More recent work on these landforms in the Rocky Mountains indicates, however, that many of those occupying cirques may actually be relict glaciers buried in their own debris (Konrad et al. 1999; Outcalt and Benedict 1965).

White (1979) identified over 650 rock glaciers in the San Juan Mountains, mostly west of the Continental Divide. Six percent of these were classified as ice cored and are likely "fossil glaciers" of unknown age. No age dating has been done on the ice or the foreign matter in the ice. These ice-cored rock glaciers move centimeters per year, in contrast to active modern glaciers that move tens of meters per year. The Snowdon Peak rock glacier, southeast of Silverton, is considered to be an ice-cored rock glacier and was surveyed for a strain net on its surface in 1980 and resurveyed in 1982, 1985, 1986, and 1998. The results demonstrate a surface movement between 1.0 in and 1.8 in (2.5 and 4.5 cm) per year, with an average of 1.2 in/yr (3.1 cm/yr) (Stoddard and Blair 1999).

Rock glaciers compose part of the permafrost system and, along with landslides and rivers, have accounted for most of the natural movement of surface materials in the San Juans during the past few millennia. Rock glaciers can account for 60 percent of all mass transport in some glaciated valleys (Barsch 1977; Giardino and Vitek 1988). A survey conducted by Brenning, Grasser, and Friend (2007) in the San Juan Mountains determined that rock glaciers occupy approximately 29 mi^2 (75 km^2) of area, store the water equivalency of 0.12 to 0.18 mi^3 (0.5 to 0.76 km^3) (stored as ice), and indicate a postglacial denudation rate from 20 to 43 in/Ka (50 to 110 cm/Ka) in the talus-shed source area.

Rock glaciers in the San Juans contain a record of the waning years of glaciation. The ice likely contains a record of pollen and chemical isotopes that can assist us in understanding mid-latitude climate change during the past 10,000 to 20,000 years. This research has yet to be done, and, with present warming trends threatening high-altitude ice deposits, the opportunity is slowly diminishing.

Fires and Debris Fans

As climate changes, so do geomorphic processes. The fires in and around Mesa Verde National Park and the 2002 Missionary Ridge Fire near Durango have highlighted some of the consequences associated with massive forest burn-off, such as soil erosion, increased volume, and frequency of debris flows—including sudden deposition of sediment on debris fans and production of fire-spall detritus from exposed bedrock. Meyer and others (1995) present evidence from Yellowstone National Park indicating that as much as 30 percent of sediment composing some alluvial fans in the region was directly related to increased erosion as a result of fire events. These fires were also correlated with hyper-arid intervals that often lasted for decades. The 2002 Missionary Ridge Fire, which burned 73,000 acres (29,000 hectares) northeast of Durango, resulted in large pulses of debris added to existing fans. It appears that the uppermost sediments on many of the alluvial fans in the Animas Valley and adjacent drainages are also from historic, or Late-Holocene, fires associated with hyper-arid intervals (Bigio, Swetnam, and Baisan 2005; Cannon et al. 2003a; Cannon et al. 2003b; Frechette et al. 2003).

Landslides

The same gravitational force that tugs at rock glaciers also pulls on fractured rock and loose, debris-covered slopes to create complex slope failure, landslides, earthflows, and rockfall. These processes of mass wasting are relentless and are continuously modifying the terrain and confounding home owners who build in threatened areas. One well-known slide is the Slumgullion Earth Flow, near Lake City. Over its 4.3-mi (7-km) length, movement continues episodically, especially during the spring snowmelt. In the upper part, movements of 1.6 to 20 ft (0.5 to 6 m) per year have been recorded (Coe et al. 2000). In July 1991, Dave English, a horse packer, witnessed a 10.5-million-yd^3 (8-million-m^3) debris avalanche sweep down northeast-facing slopes near Pole Mountain, above West Lost Trail Creek (4.3 mi [7 km] northwest of Rio Grande Reservoir), and traverse 0.6 mi [1 km] in 25 to 30 seconds (W. Rogers et al. 1992). On July 5, 1998, a 36,600-yd^3 (28,000-m^3) block of Dakota Sandstone toppled from the rim of Missionary Ridge just north of Durango, producing a rock slide that is still clearly visible and that continues to move, thanks to a spring issuing from the Dakota-Morrison contact. More of the same will occur in the future wherever cliffs, canyons, and mountainsides exist.

Human Impact

Humans are now the greatest agent of change on the planet (Zalasiewicz et al. 2008). They account for more transport of debris on land than all geologic processes combined (Hooke 2000). The impact of human activity, both directly and indirectly, on the San Juan landscape is inescapable, although systematic documentation is lacking. In the last 150 years, the invasion of gold seekers, farmers and ranchers, developers, recreational enthusiasts, and second-home owners has created roads, towns, dams, mines, and mine waste. (More than 1,500 mine workings have been documented in San Juan County.) Mining activity in the late 1800s and early 1900s accounted for surges of sediment load down major drainages, such as the Animas River, where tailings supplied sediment at 50 to 4,700 times pre-mining rates in some places (Vincent and Elliott 2007). In the Upper Animas Valley, between Howardsville and the old town of Eureka, sediment surges have increased channel braiding and added heavy metals to floodplain deposits (Blair, Yager, and Church 2002; Church, Fey, and Unruh 2007).

Collectively, human impacts will last millennia and longer. Research is needed to document and evaluate human impacts, so that future land planners and policy makers can make informed decisions that will benefit the natural environment and human society.

REFERENCES

Atwood, W. W., and K. F. Mather. 1932. *Physiography and Quaternary Geology of the San Juan Mountains, Colorado*. Professional Paper 166. Washington, DC: US Geological Survey.

Baars, D. L. 1992. *The American Alps*. Albuquerque: University of New Mexico Press.

Barsch, D. 1977. Nature and Importance of Mass-Wasting by Rock Glaciers in Alpine Permafrost Environments. *Earth Surface Processes* 2:231–245.

Benson, L., R. Madole, G. Landis, and J. Gosse. 2005. New Data for Late Pleistocene Pinedale Glaciation from Southwestern Colorado. *Quaternary Science Reviews* 24:49–65.

Bigio, E., T. Swetnam, and C. Baisan. 2005. The Integration of Tree-Ring and Alluvial Fan Records of Fire History at the Missionary Ridge Fire near Durango, Colorado. *Geological Society of America Abstracts with Programs* 37(7):111.

Blair, R. W., Jr., D. B. Yager, and S. E. Church. 2002. *Surficial Geologic Maps along the Riparian Zone of the Animas River and Its Headwater Tributaries, Silverton to Durango, Colorado, with Upper Animas River Watershed Gradient Profiles*. US Geological Survey Digital Data Series DDS–71, on CD-ROM.

Brenning, A., M. Grasser, and D. A. Friend. 2007. Statistical Estimation and Generalized Additive Modeling of Rock-Glacier Distribution in the San Juan Mountains, Colorado, United States. *Journal of Geophysics Research* 112:FO2S15; doi: 10.1029/2006JF000528.

Brister, B. S., and R. R. Gries. 1994. Depth and Geometry of the Northern Rio Grande Rift in the San Luis Basin, South-Central Colorado. In G. R. Keller and S. M. Cather, eds., *Basins of the Rio Grande Rift—Structure, Stratigraphy, and Tectonic Setting*. Special Paper 291. Boulder: Geological Society of America, 39–58.

Cannon, S. H., J. A. Michael, J. E. Gartner, and J. A. Gleason. 2003a. *Assessment of Potential Debris-Flow Peak Discharges from Basins Burned by the 2002 Missionary Ridge Fire, Colorado*. Open File Report 03–332. Denver: US Geological Survey.

Cannon, S. H., W. Romme, R. Wu, and M. B. Thurston. 2003b. The Geologic Impact of Wildfires in the San Juan Mountains, Southwestern Colorado. *Geological Society of America Abstracts with Programs* 35(6):9.

Cather, S. M. 2004. Laramide Orogeny in Central and Northern New Mexico and Southern Colorado. In G. H. Mack and K. A. Giles, eds., *The Geology of New Mexico: A Geologic History*. Special Publication 11. Socorro: New Mexico Geological Society, 203–248.

Church, S. E., D. L. Fey, and D. M. Unruh. 2007. Trace Elements and Lead Isotopes in Modern Streambed and Terrace Sediment—Determination of Current and Premining Geochemical Baselines. In S. E. Church, P. von Guerard, and S. E. Finger, eds., *Integrated Investigations of Environmental Effects of Historical Mining in the Animas River Watershed, San Juan County, Colorado*. Professional Paper 1651. Reston, VA: US Geological Survey, 569–642.

Coe, J. A. (ed.), E. R. Bigio, R. W. Blair Jr., M. Burke, S. H. Cannon, V. G. deWolfe, J. Ey, J. E. Gartner, M. L. Gillam, N. D. Knowlton, P. M. Santi, and W. H. Schulz. 2007. *Mass*

Wasting Following the 2002 Missionary Ridge Fire near Durango, Colorado: A Field Trip Guidebook. Open File Report 2007–1289. Denver: US Geological Survey.

Coe, J. A., , J. W. Godt, W. L. Ellis, W. Z. Savage, J. E. Savage, P. S. Powers, D. J. Varnes, and P. Tachker. 2000. *Preliminary Interpretation of Seasonal Movement of the Slumgullion Landslide as Determined from GPS Observations, July 1998–July 1999.* Open File Report 00-102. Denver: US Geological Survey.

Dethier, D. P. 2001. Pleistocene Incision Rates in the Western United States Calibrated using Lava Creek B Tephra. *Geology* 29(9):783–786.

Epis, R. C., G. R. Scott, R. B. Taylor, and C. E. Chapin. 1976. Cenozoic Volcanic, Tectonic and Geomorphic Features of Central Colorado In R. C. Epis and R. J. Weimer, eds., *Professional Contributions of the Colorado School of Mines, Studies in Colorado Field Geology* 8 (November):323–338.

Frechette, J. D., D. A. Gonzales, R. Kenny, and J. R. Thompson. 2003. Evidence for a Connection between Wildfires, Erosion and Landscape Development over the Past 3600 Years in Southwestern Colorado. *Geological Society of America Abstracts with Programs* 35(6):36.

Giardino, J. R., and J. D. Vitek. 1988. The Significance of Rock Glaciers in the Glacial-Periglacial Landscape Continuum. *Journal of Quaternary Science* 3:97–103.

Gillam, M. L. 1998. Late Cenozoic Geology and Soils of the Lower Animas River Valley, Colorado and New Mexico. PhD dissertation, University of Colorado, Boulder.

Guido, A. S., D. J. Ward, and R. S. Anderson. 2007. Pacing the Post-Last Glacial Maximum Demise of the Animas Valley Glacier and the San Juan Mountain Ice Cap, Colorado. *Geology* 35(8):739–742.

Hooke, R. LeB. 2000. On the History of Humans as Geomorphic Agents. *Geology* 28: 843–846.

Howe, E. 1909. *Landslides in the San Juan Mountains, Including a Consideration of Their Causes and Classifications.* Professional Paper 67. Washington, DC: US Geological Survey.

Hulen, J. B. 1992. *Field Geologic Summaries for Core Holes CCM-2 (Airport 1–6) and CCM-1 (Hosselkus 1–10).* Creede Scientific Drilling Project Report ESL-920111-TR.

Kerr, R. A. 2008. Geology: A Time War over the Period We Live In. *Science* 319(5862): 402–403.

Kluth, C. F., and C. H. Schaftenaar. 1994. Depth and Geometry of the Northern Rio Grande Rift in the San Luis Basin, South-Central Colorado. In G. R. Keller and S. M. Cather, eds., *Special Paper 291.* Boulder: Geological Society of America, 27–37.

Konrad, S. K., N. F. Humphrey, E. J. Steig, D. H. Clark, N. Potter, and W. T. Pfeffer. 1999. Rock Glacier Dynamics and Paleoclimatic Implications. *Geology* 27(12):1131–1134.

Lanphere, M. A. 2000. Duration of Sedimentation of the Creede Formation from 40Ar/39Ar Ages. In P. M. Bethke and R. L. Hay, eds., *Ancient Lake Creede—Its Volcano-Tectonic Setting, History of Sedimentation, and Relation to Mineralization in the Creede Mining District.* Special Paper 346. Boulder: Geological Society of America, 71–76.

Lanphere, M. A., D. E. Champion, R. L. Christiansen, G. A. Izett, and J. D. Obradovich. 2002. Revised Ages for Tuffs of the Yellowstone Plateau Volcanic Field: Assignment of the Huckleberry Ridge Tuff to a New Geomagnetic Polarity Event. *Geological Society of America Bulletin* 114(5):559–568.

Larsen, E. S., and W. Cross. 1956. *Geology and Petrology of the San Juan Region, Southwestern Colorado.* Professional Paper 258. Washington, DC: US Geological Survey.

Lawrence, K. T., Z. Liu, and T. D. Herbert. 2006. Evolution of the Eastern Tropical Pacific through Plio-Pleistocene Glaciation. *Science* 312(5770):79–83.

Lear, C. H., T. R. Bailey, P. N. Pearson, H. K. Coxall, and Y. Rosenthal. 2008. Cooling and Ice Growth across the Eocene-Oligocene Transition. *Geology* 36(3):251–254.

Lee, W. T. 1923. Contributions to the Geography of the United States, 1922—Peneplains of the Front Range and Rocky Mountain National Park, Colorado. *US Geological Survey Bulletin* 730.

Lipman, P. W. 1976. Geologic Map of the Del Norte Area, Eastern San Juan Mountains, Colorado. Miscellaneous Investigations Map I-952. Reston, VA: US Geological Survey.

———. 2006. Geologic Map of the Central San Juan Caldera Cluster, Southwestern Colorado [1:50,000]. Geologic Investigations Series I-2799. Reston, VA: US Geological Survey.

———. 2007. Incremental Assembly and Prolonged Consolidation of Cordilleran Magma Chambers: Evidence from the Southern Rocky Mountain Volcanic Field. *Geosphere* 3:42–70; doi: 10.1130/GES00061.1.

Lipman, P. W., and W. C. McIntosh. 2008. Eruptive and Noneruptive Calderas, Northeastern San Juan Mountains, Colorado: Where Did the Ignimbrites Come From? *Geological Society of America Bulletin* 120(7/8):771–795; doi: 10.1130/B26330.1.

Lipman, P. W., T. A. Steven, and H. H. Mehnert. 1970. Volcanic History of the San Juan Mountains, Colorado, as Indicated by Potassium-Argon Dating. *Geological Society of America Bulletin* 81:2329–2352.

Lisiecki, L. E., and M. E. Raymo. 2005. A Pliocene-Pleistocene Stack of 57 Globally Distributed Benthic δ^{18}O Records. *Paleoceanography* 20:PA1003; doi: 10.1029/2004PA001071.

Machette, M. N., D. W. Marchetti, and R. A. Thompson. 2007. Ancient Lake Alamosa and the Pliocene to Middle Pleistocene Evolution of the Rio Grande. In M. N. Machette, M.-M. Coates, and M. L. Johnson, eds., *2007 Rocky Mountain Section Friends of the Pleistocene Field Trip—Quaternary Geology of the San Luis Basin of Colorado and New Mexico, September 7–9, 2007.* Open File Report 2007–1193. Denver: US Geological Survey, 157–167.

Meyer, G. A., S. G. Wells, and A.J.T. Jull. 1995. Fire and Alluvial Chronology in Yellowstone National Park: Climatic and Intrinsic Controls on Holocene Geomorphic Processes. *Geological Society of America Bulletin* 107(10):1211–1230.

Noblett, J. B., and B. M. Loeffler. 1987. The Geology of Summer Coon Volcano near Del Norte, Colorado. In S. S. Beus, ed., *Centennial Field Guide*, vol. 2. Boulder: Rocky Mountain Section of the Geological Society of America, 349–352.

Outcalt, S. I., and J. B. Benedict. 1965. Photo-Interpretation of Two Types of Rock Glacier in the Colorado Front Range. *Journal of Glaciology* 5:849–856.

Raymo, M. E. 1994. The Initiation of Northern Hemisphere Glaciation. *Annual Review of Earth and Planetary Sciences* 22:353–383.

Rogers, K. L., E. E. Larson, G. Smith, D. Katzman, G. R. Smith, T. Cerling, Y. Wang, R. G. Baker, K. C. Lohmann, C. A. Repenning, P. Patterson, and G. Mackie. 1992. Pliocene and Pleistocene Geologic and Climatic Evolution in the San Luis Valley of South-Central Colorado. *Palaeogeography, Palaeoclimatology, Palaeoecology* 94:55–86.

Rogers, W. P., D. English, R. L. Schuster, and R. M. Kirkham. 1992. Large Rock Slide/Debris Avalanche in the San Juan Mountains, Southwestern Colorado, USA, July 1991. *Landslide News* [international newsletter] (August):22–24.

Rye, R. O., P. M. Bethke, M. A. Lanphere, and T. A. Steven. 2000. Neogene Geomorphic and Climatic Evolution of the Central San Juan Mountains, Colorado: K/Ar Age and Stable Isotope Data on Supergene Alunite and Jarosite from the Creede Mining District. In P. M. Bethke and R. L. Hay, eds., *Ancient Lake Creede: Its Volcano-Tectonic Setting, History of Sedimentation, and Relation to Mineralization in the Creede Mining District.* Special Paper 346. Boulder: Geological Society of America, 95–103.

Scott, G. R. 1975. Cenozoic Surfaces and Deposits in the Southern Rocky Mountains. In B. F. Curtis, ed., *Cenozoic History of the Southern Rocky Mountains.* Memoir 144. Boulder: Geological Society of America, 227–248.

Small, E. E. 1995. Hypsometric Forcing of Stagnant Ice Margins: Pleistocene Valley Glaciers, San Juan Mountains, Colorado. *Geomorphology* 14:109–121.

Steven, T. A., K. Hon, and M. A Lanphere. 1995. *Neogene Geomorphic Evolution of the Central San Juan Mountains near Creede, Colorado.* Miscellaneous Investigations Series, Map I-2504. Reston, VA: US Geological Survey.

Steven, T. A., P. W. Lipman, W. J. Hail Jr., F. Barker, and R. G. Luedke. 1974. Geologic Map of the Durango Quadrangle, Southwestern Colorado. Miscellaneous Investigations Series Map 1-764, scale 1:250,000. Reston, VA: US Geological Survey.

Stoddard, C., and R. Blair. 1999. Movement of the Snowdon Rock Glacier, San Juan Mountains, Colorado. *Geological Society of America Abstracts with Programs* 31(7): A269–A270.

Suppe, J. 1985. *Principles of Structural Geology.* Englewood Cliffs, NJ: Prentice-Hall.

Van Houten, F. B. 1957. Appraisal of Ridgway and Gunnison "Tillites," Southwestern Colorado. *Bulletin of the Geological Society of America* 68:383–388.

Vincent, K. R., and J. G. Elliott. 2007. Response of the Upper Animas River Downstream from Eureka to Discharge of Mill Tailings. In S. E. Church, P. von Guerard, and S. E. Finger, eds., *Integrated Investigations of Environmental Effects of Historical Mining in the Animas River Watershed, San Juan County, Colorado*, vol. 2. Professional Paper 1651. Reston, VA: US Geological Survey, 889–941.

White, G. P. 1979. Rock Glacier Morphometry, San Juan Mountains, Colorado, Part 2. *Geological Society of America Bulletin* 90:924–952.

Wolff, E. W. 2000. History of the Atmosphere from Ice Cores. In C. Boutron, ed., *European Research Course on Atmospheres,* vol. 4: *From Weather Forecasting to Exploring the Solar System.* Les Lilis, France: EDP Sciences, 147–177.

Zalasiewicz, J., M. Williams, A. Smith, T. L. Barry, A. L. Coe, P. R. Brown, P. Brencheley, D. Cantrill, A. Gale, P. Gibbard, F. J. Gregory, M. W. Hounslow, A. C. Kerr, P. Pearson, R. Knox, J. Powell, C. Waters, J. Marshall, M. Oates, P. Rawson, and P. Stone. 2008. Are We Now Living in the Anthropocene? *GSA Today* 18(2):4–8.

The Hydrogeology of the San Juan Mountains

Jonathan Saul Caine and Anna B. Wilson

K NOWLEDGE OF THE OCCURRENCE, STORAGE, AND FLOW of groundwater in mountainous regions is limited by the lack of integrated data from wells, streams, springs, and climate. In his comprehensive treatment of the hydrogeology of the San Luis Valley, Huntley (1979) hypothesized that the underlying, fractured volcanic bedrock of the San Juan Mountains has relatively high bulk permeability and a regional-scale water table with a low hydraulic gradient. Other (some more recent) studies of fractured crystalline bedrock in mountainous terrain indicate that these rock units can act as aquifers (Kahn et al. 2008; Manning and Caine 2007; Robinson 1978; Stober and Bucher 2005). The body of recent work also suggests that the conception that fractured crystalline bedrock is of such low permeability that it constitutes a "no-flow zone" may be inappropriate. In addition to establishing a new baseline, the data presented here are used to test Huntley's (1979) hypotheses that suggest that the San Juan Mountains may be underlain by a substantial groundwater system.

With the advent of computers and digital databases, many types of publicly available data can be used to test hypotheses and provide new insights into mountain hydrogeology at the regional scale in the San Juan Mountains. Plate 16 illustrates processes that suggest several fundamental questions arising from our lack of knowledge of mountain hydrogeology. These questions include: What are the dynamic interrelationships among the tectonics of mountain building, climate, and groundwater, and what are the time scales over which associated processes operate? How does extreme topographic relief allow for groundwater recharge along steep surfaces rather than simply causing precipitation to run off? How does extreme relief translate into hydraulic gradients that drive groundwater flow? Can extreme gradients drive large volumes of meteoric water deep into the Earth's upper crust? Once in the subsurface, what are the residence times of these waters? Finally, how does complex geology, commonly associated with mountainous terrain, influence these processes and control potentially heterogeneous and tortuous flow pathways?

This chapter presents a synthesis of hydrogeological data, in a reconnaissance style, at the regional scale for the San Juan Mountains. Analyses of these data shed some light on the questions posed earlier for the San Juan Mountains and on mountain hydrogeologic processes in general. These analyses are based on public digital data from geologic and topographic maps, precipitation networks, stream gauges, groundwater wells, and springs. These data can be integrated using the hydrologic cycle expressed as a mass balance between inputs and outputs. The data types noted earlier form the basic set of measurements used to explore, characterize, and quantify elements of the hydrologic cycle. This exploration at a variety of scales yields insight into the relationships among the physical geological framework, climatological and hydrological budgets, and the hydraulic properties of the major aquifers in the San Juan Mountains. Each of these factors has been broken down and investigated separately and then integrated at the end of the chapter, using a conceptual model.

Although the San Juan Mountains contain extensive precious- and base-metal deposits that have led to natural and mining-related groundwater contamination, this topic is not addressed here. Interested readers should refer to the extensive body of US Geological Survey work in Gray et al. (1994), Plumlee et al. (1995), Wirt et al. (1999), Johnson and Yager (2006), Johnson et al. (2007), and Church, von Guerard, and Finger (2007). Huntley (1979) also provided a large database for regional hydro-geochemistry of the San Juan Mountains (SJM).

GENERAL METHODS

In this chapter, SJM watersheds are used to explore and quantify relationships among topography, geology, climate, and hydrology. Watersheds are topographically, geologically, and hydrologically self-contained land units in which topographic divides and gradients cause water to flow and be stored nearly exclusively within their three-dimensional confines (plate 16). Using a Geographic Information System (GIS), datasets were compiled for the major SJM watersheds and integrated into the maps and figures presented in this chapter. Contouring of data, such as precipitation or topography, was done using standard methods. The comprehensive datasets and results of various calculations were compiled and distilled using computer spreadsheets. These results were used to annotate elements of the maps and to construct summary tables. Other methods are explained in individual sections, where appropriate. Not all data types are available for each watershed. Thus, results from various watershed calculations and associated estimates of, for example, quantities of water or aquifer properties are only representative and may not adequately characterize the entire San Juan Mountains and all of the processes that occur within them. Because different data types are reported from various data sources using both English (inches, feet, and gallons) and SI (centimeters, meters, and liters) units, conversions and non-intuitive units are kept to a minimum; both English and SI units are used herein. Conversion factors are provided in table 5.1.

GEOLOGIC AND TOPOGRAPHIC AQUIFER FRAMEWORK

The interior of the San Juan Mountains is composed of a complex of volcanic centers that formed 27–35 Ma. These centers are marked by a set of nested calderas, from which the San Juan Volcanic Field emanated (plate 17; Lipman et al. 1973; Steven et al. 1974). The rocks that make up the volcanic field—one of the largest in the world—include sub-horizontally layered andesitic to silicic lavas, pyroclastic debris flows, and mudflows. Stocks, sills, and dikes intrude all of these rocks. Precambrian metamorphic rocks and Paleozoic and Mesozoic sedimentary rocks flank the central deposits of Tertiary volcanic rocks (plate 17). Surficial deposits, dominantly from glacial and fluvial processes, also exist and do not form major, regional-scale aquifers.

Each of the major rock types described here is part of a geologically heterogeneous aquifer system. The system is made up of individual rock layers, or units, which have varying thicknesses, three-dimensional geometries, and hydraulic properties, such as porosity and permeability. Geologic and topographic maps and geologic well-log data, used together, can help define the physical configuration and internal structure of the units, and water-well hydraulic data provide information on subsurface hydraulic properties. Although the units and some of their properties within the aquifer system can be defined and measured, there are also heterogeneities that operate at different scales. For example, the extent to which parts of the aquifer system are controlled by faults and fractures (scales of centimeters to hundreds of meters) versus intergranular permeability (scales of microns to millimeters), or some combination of these, is poorly understood.

The most significant topographic relief in the San Juan Mountains is found primarily where there are Tertiary extrusive rocks from the San Juan Volcanic Field (compare plates 17 and 18). Other areas of significant relief surrounding the volcanic field are underlain by Precambrian, Paleozoic, and Mesozoic rocks. The 7,800 mi^2 (20,200 km^2) SJM region, as defined for this study, is based on topography, geology, and watersheds (plate 18). About 80 percent of the terrain is above 10,000 ft (3,048 m) in elevation. Geology and topography control the shape, size,

Table 5.1. Unit conversion factors

English to SI

Multiply	By	To obtain
Length		
inch (in)	2.54	centimeter (cm)
foot (ft)	0.3048	meter (m)
mile (mi)	1.609	kilometer (km)
Area		
acre	0.004047	square kilometer (km^2)
square mile (mi^2)	2.590	square kilometer (km^2)
Volume		
gallon (gal)	3.785	liter (L)
gallon (gal)	0.003785	cubic meter (m^3)
cubic inch (in^3)	16.39	cubic centimeter (cm^3)
cubic foot (ft^3)	0.02832	cubic meter (m^3)
Flow Rate		
cubic foot per second (ft^3/s)	0.02832	cubic meter per second (m^3/s)
gallon per minute (gal/min)	0.06309	liter per second (L/s)

SI to English

Multiply	By	To obtain
Length		
centimeter (cm)	0.3937	inch (in.)
meter (m)	3.281	foot (ft)
kilometer (km)	0.6214	mile (mi)
Area		
square kilometer (km^2)	247.1	acre
square kilometer (km^2)	0.3861	square mile (mi^2)
Volume		
liter (L)	0.2642	gallon (gal)
cubic meter (m^3)	264.2	gallon (gal)
cubic centimeter (cm^3)	0.06102	cubic inch (in^3)
cubic meter (m^3)	35.31	cubic foot (ft^3)
Flow Rate		
cubic meter per second (m^3/s)	35.31	cubic foot per second (ft^3/s)
liter per second (L/s)	15.85	gallon per minute (gal/min)

Temperature in Celsius (°C) can be converted to Fahrenheit (°F) as follows: °F = (1.8 × °C) + 32
Temperature in Fahrenheit (°F) can be converted to Celsius (°C) as follows: °C = (°F − 32) / 1.8

and relief of the major SJM watersheds. Watershed relief ranges from a maximum of 7,270 ft (2,217 m) to a minimum of 2,103 ft (641 m) and has a mean of 4,511 ft (1,375 m) (plate 18; table 5.2). The extreme relief and rough topography at the scale of the major watersheds result in topographic, and potentially local hydraulic, gradients greater than 0.01 and possibly up to 0.1 (table 5.2).

. . . .

Table 5.2. Physical characteristics of major San Juan Mountain watersheds

Watershed and Gauge Name	Dam	Geology	Gauge Location		Stream Metrics						
			LAT (d m s)	LON (d m s)	WSA (km^2)	Orient (d)	Length (m)	HWE (m)	ME (m)	REL (m)	Grad
Animas Du	no	**PC,PZ,T**	37 16 45	107 52 47	1792	203	77242	3475	2011	1493	0.019
Dolores Do	no	PZ,**MZ**	37 28 21	108 29 49	1305	234	61066	3753	2133	1638	0.027
Lake Fork GW	no	T	38 17 56	107 13 46	865	027	56146	3730	2378	1344	0.024
Piedra Ar	no	MZ	37 05 18	107 23 50	1629	208	63990	3508	1889	1634	0.026
Rio Grande 30mi*	yes	T	37 43 29	107 15 18	422	102	26732	3715	2914	880	0.033
Rio Grande DN*	yes	T	37 41 22	106 27 38	3419	096	96794	3715	2438	1283	0.013
San Miguel PV	no	**MZ**,T	38 02 33	108 07 54	803	294	33818	2784	2194	641	0.019
San Miguel Ur	no	**MZ**	38 21 26	108 42 44	3882	302	95088	2784	1584	1219	0.013
Uncompahgre De*	yes	MZ	38 44 31	108 04 49	2888	338	99882	3718	1522	2217	0.022
Uncompahgre Ou	no	T	38 02 36	107 40 57	199	340	14513	3718	2377	1402	0.097
Minimum					199		14513	2784	1522	641	0.013
Maximum					3882		99882	3753	2914	2217	0.097
Mean					1721		62527	3486	2144	1375	0.029

Notes and abbreviations: * indicates dammed stream, Du = Durango, Do = Dolores, GW = Gateway, Ar = Arboles, 30mi = 30 Mile Bridge, DN = Del Norte, PV = Placerville, Ur = Uravan, De = Delta, Ou = Ouray. The Lake Fork is part of the Upper Gunnison, and the Piedra is part of the Upper San Juan Watershed. Gauges have 25 years of record, except for the Rio Grande 30mi (24 years), San Miguel Ur (22 years), and Uncompahgre Ou (7 years). PC = Precambrian crystalline rocks, PZ = Paleozoic sedimentary rocks, MZ = Mesozoic sedimentary rocks, T = Tertiary volcanic rocks. (Dominant rock type is bold-faced.) LAT = latitude, LON = longitude, km = kilometers, m = meters, d = degrees, m = minute, s = seconds, WSA = watershed area. Orient refers to the average downstream direction of the main stream and its drainage, HWE = headwater elevation, ME = mouth elevation, REL = relief, Grad = gradient.

Source: Stream metrics calculated by J. Caine (2010).

WATERSHEDS, TOPOGRAPHY, AND PRECIPITATION

Geology, climate, and erosion largely control the location and relief of watershed topographic divides, which are starting points for surface water flow, subsurface groundwater divides, and associated hydraulic gradients (plate 16). However, groundwater divides in the subsurface are unlikely to be vertical and could substantially change in orientation with depth. Plate 18 shows the major river drainages and watershed divides draped over topography. Seven major, representative watersheds form a regional-scale (southwest quadrant of Colorado), east-west-trending, oblong, radial drainage pattern emanating from the highest peaks of the San Juan Mountains. The locations of the major watershed divides were taken from US Geological Survey hydrologic-unit map data (e.g., Seaber, Kapinos, and Knapp 1987). Watershed areas, calculated using GIS, range from a maximum of 1,499 mi^2 (3,882 km^2) to a minimum of 77 mi^2 (199 km^2) (table 5.2). Several of the major watersheds, including the Dolores, San Miguel, Uncompahgre, and Piedra, extend outside the SJM region (plate 18). To facilitate basic hydrologic calculations representative of the mountain environments, sub-watersheds for the Rio Grande, San Miguel, and Uncompahgre were defined using stream-gauge locations that capture drainage exclusively from the mountains (plate 18; table 5.2). An interesting feature of SJM watershed divides is that they include the Continental Divide, which shows a major re-entrant, or jog, into the San Juan Mountains formed by incision of the Rio Grande headwaters into the volcanic field (plate 18). This is one of the largest such re-entrants along the Continental Divide in North America.

Plate 19 shows a contour map of average annual precipitation for the San Juan Mountains, from thirty years of rain and snow data from PRISM (Parameter-Elevation Regressions on Independent Slopes Model) (PRISM 2007). The maximum precipitation ranges from about 65 in (165 cm) per year in the regions of the highest peaks to about 7.5 in (20 cm) per year in the surrounding lowlands (plate 19). Watershed annual mean precipitation was also calculated from GIS data, using the mean value and area of each contour interval within each watershed summed over each watershed's total area (table 5.3). This results in watershed precipitation estimates ranging from a high of 37 in (94 cm) to a low of 17.7 in (45 cm) per year (plate 19; table 5.3). The primary form of precipitation is snow. A more detailed treatment of SJM climate is presented in Keen (1996).

Comparison of plates 18 and 19 shows clear relationships among watershed orientation, elevations, and precipitation. Orographic effect makes the San Juan Mountains a regional capture zone for precipitation. Further examination of elevation, precipitation, and geology (plates 17, 18, and 19) leads to the interpretation that there are correlations among uplift of the volcanic field, the relative ease of erodibility of the volcanic rocks, and orographic effect, which together likely caused the re-entrant into the Continental Divide—possibly during the original inflation of the volcanic field in the Tertiary and (or) during rebound from

Table 5.3. Water budget estimates for major San Juan Mountain watersheds

Watershed Name	PPT (cm)	TSF (m³)	BF (m³)	TSF (cm³)	WSA (cm²)	ATSF (cm)	ABF (cm)	AOF (cm)	ET+GO (cm)	% TSF of PPT	% BF of PPT	% BF of TSF
Animas Du	89	8×10^8	1×10^8	8×10^{14}	2×10^{13}	43	8	35	46	49	9	19
Dolores Do	75	4×10^8	4×10^7	4×10^{14}	1×10^{13}	32	3	29	43	43	4	8
Lake Fork GW	70	2×10^8	3×10^7	2×10^{14}	9×10^{12}	25	4	21	45	35	5	15
Piedra Ar	72	4×10^8	5×10^7	4×10^{14}	2×10^{13}	25	3	22	47	35	4	12
Rio Grande 30mi*	94	2×10^8	3×10^6	2×10^{14}	4×10^{12}	45	1	45	49	48	1	1
Rio Grande DN*	68	8×10^8	1×10^8	8×10^{14}	3×10^{13}	24	3	21	44	35	5	14
San Miguel PV	73	2×10^8	4×10^7	2×10^{14}	8×10^{12}	29	5	24	44	40	7	18
San Miguel Ur	51	4×10^8	4×10^7	4×10^{14}	4×10^{13}	10	1	8	42	19	2	12
Uncompahgre De*	45	3×10^8	8×10^7	3×10^{14}	3×10^{13}	10	3	8	35	23	6	27
Uncompahgre Ou	82	1×10^8	2×10^7	1×10^{14}	2×10^{12}	50	8	42	32	60	10	16
Minimum	45	1×10^8	3×10^6	1×10^{14}	2×10^{12}	10	1	8	32	19	1	1
Maximum	94	8×10^8	1×10^8	8×10^{14}	4×10^{13}	50	8	45	49	60	10	27
Mean	72	3×10^8	4×10^7	3×10^{14}	1×10^{13}	29	4	25	43	39	5	14

[Input] − [Output] = [PPT] − [ET + TSF + GO] = 0; TSF = BF + OF; [ET + GO]: [ET + GO] = [PPT − TSF]
Notes and abbreviations: * indicates dammed stream; Du = Durango, Do = Dolores, GW = Gateway, Ar = Arboles, 30mi = 30 Mile Bridge, DN = Del Norte, PV = Placerville, Ur = Uravan, De = Delta, Ou = Ouray, PPT = precipitation, TSF = total stream flow (A as prefix = areally distributed), BF = base flow, WSA = watershed area, ET = evapotranspiration, GO = groundwater outflow, OF = overland flow, cm = centimeters, m = meters, yr = year. All data are average annual values.

Pleistocene alpine glaciation (Barton, Steven, and Hayba 2000; Steven, Hon, and Lanphere 1995).

WATER BUDGETS, STREAM FLOW, AND GROUNDWATER

A water budget is an estimate of the hydrologic inputs and outputs within a watershed (Freeze and Cherry 1979; Healy et al. 2007). It is a fundamental expression of the mass balance for components of the hydrologic cycle within a watershed (plate 16), where:

Input – Output +/– Storage = 0
IP – OP +/– S = 0

For simplicity and because of a lack of data, changes in storage are ignored, thus:

Precipitation – Evapotranspiration – Total Stream Flow – Groundwater Outflow = 0,
PPT – ET – TSF – GO = 0

Total Stream Flow = Base Flow + Overland Flow.
TSF = BF + OF

These terms are defined in detail in subsequent paragraphs, and some are also explained in the glossary. Although the analyses of the hydrologic-cycle budgets presented here do not apply the most sophisticated methods (e.g., Claassen, Reddy, and Halm 1986; Healy et al. 2007; Tallaksen 1995), they are useful for providing basic insight into the SJM hydrologic cycle, with a focus on groundwater and aquifer hydraulic properties at the watershed scale.

Another important consideration when evaluating or quantifying components of the hydrologic cycle in a region is human influence on the hydrologic system. Within the study area are dams on the Rio Grande and the Uncompahgre River. Diversions such as irrigation ditches are also not accounted for in this dataset, potentially limiting the analysis. For streams dammed above the gauging stations, the base-flow component of each hydrograph, computed with the methods used here, should reflect a minimum stream-flow contribution that is only an approximation of natural flow. Overland flow (OF) in a dammed watershed is controlled by a number of factors, particularly dam release, and it is not a good indicator of natural hydrologic conditions.

Precipitation (PPT) is the fundamental hydrologic input into a watershed. When estimating a water budget, we assume that, within a watershed, almost all of the water input can be measured as rain or snow. Major outputs of water include evapotranspiration (ET), groundwater outflow (GO), and total stream flow (TSF). These components are expressed as units of volume per unit time, or volume per unit time per unit area. For example, when comparing watersheds of different sizes, it is useful to determine the amount of precipitation, or total stream flow, distributed over a watershed's entire area. This results in volumes per unit area, or simply depth. These depths can be thought of as infinitesimally narrow columns of water that are representative of the entire watershed and are thus useful for comparing hydrologic components within a single watershed or from one watershed to another (table 5.3).

Evapotranspiration, about which little is known in mountainous terrain, is a key output of the hydrologic cycle. It is part of a set of complex processes that reflect vegetative type and density, slope, aspect, wind speed and duration, and solar insolation. Evapotranspiration is difficult to measure directly and is estimated using a number of methods involving lysimeters, evaporation pans, meteorological stations, and hydrograph analyses (Healy et al. 2007). Estimates of evapotranspiration in mountainous terrain range from 10 to 35 percent of precipitation (Claassen, Reddy, and Halm 1986; Kahn et al. 2008; Liu, Williams, and Caine 2004).

Groundwater outflow is water that leaves the watershed in the subsurface, within the groundwater-flow system. It cannot be measured directly. Both evapotranspiration and groundwater outflow are estimated here as a combined component by subtracting total stream flow from precipitation (plate 21; table 5.3). In undammed streams in the San Juan Mountains, estimates of $ET + GO$ account for 40 to 65 percent of precipitation, with a mean of 56 percent (plates 20 and 21). After $ET + GO$ and OF, the residual water may be available for infiltration into the groundwater-flow system.

Mechanisms of water recharge are poorly understood, particularly in mountainous terrain (Manning and Caine 2007; Manning, Solomon, and Sheldon 2003). However, snowmelt is likely the primary source of SJM groundwater recharge because the melting of a snowpack is usually sufficiently slow to allow adequate time for infiltration. High to extreme relief may impede surface waters from infiltrating into the groundwater-flow system because of their rapid travel times out of such a watershed.

Total stream flow, sometimes called discharge (Q), can be measured at the mouth of an individual watershed at a stream gauge (plates 16 and 18; table 5.3). Such measurements are a fundamental way to quantify the components of outputs, such as base flow and overland flow, from a watershed (plate 16; table 5.3). Total stream flow has been measured by the US Geological Survey at stream gauging stations at a variety of locations within SJM watersheds (plate 18). In some cases, there are nearly 100 years of record.

Plots of total stream flow versus time are called hydrographs (e.g., plate 20). These represent the integrated flow of groundwater (base flow) and surface-water (overland flow) components of total stream flow moving through a watershed. Hydrographs are used here to estimate water budgets for each SJM watershed (table 5.3). Average daily total stream flow for SJM streams ranges from a few tenths of a cubic meter to a few hundred cubic meters per second (m^3/s). This translates into hundreds of millions of cubic meters of water flowing through each of the major streams each year (table 5.3). For perspective, the annual flow in almost any one of the major SJM streams is, on average, equivalent to nearly 50 percent of the current total annual domestic water use by the city of Denver, Colorado (http://quickfacts. census.gov; http://www.greenprintdenver.org).

Base flow (*BF*) is the component of stream flow that comes directly from groundwater discharge into a stream channel (plates 16 and 20). Using stream hydrographs, base flow for each watershed was estimated by averaging the lowest one-day flow for each year over the period of record (plates 20 and 21; table 5.3). Average annual base flows obtained from nearly twenty-five years of record (generally 1975 through 1999) range from 3.5 to 162.5 ft^3/s (0.1 to 4.6 m^3/s) for each of the major SJM streams (table 5.3). Although these flows are a relatively small proportion of total stream flow (14 percent on average), they equate to tens of millions of cubic meters of water discharging from SJM aquifers each year, a portion of which comes exclusively from bedrock aquifers (table 5.3).

Base flow is also a reflection of hydraulic gradient, recharge, and precipitation over time. Although the base-flow analyses presented do not directly reveal bulk aquifer hydraulic properties, base flow, as a percentage of average annual precipitation distributed over the area of each watershed, is an indirect indicator of these properties. For the undammed streams that drain the SJM region (plates 20 and 21; table 5.3), base flow ranges from 4 to 10 percent of precipitation. In a general way, mountain watersheds underlain by high-permeability bedrock aquifers may produce larger base-flow percentages of precipitation when compared with watersheds underlain by low-permeability aquifers because of the higher flux (volume of water per unit time) of groundwater that can move through rocks of higher permeability. Although few data, such as from aquifer tests, are available to adequately evaluate this assertion for the San Juan Mountains, it is consistent both with what is observed and with the hypotheses of Huntley (1979).

Overland flow is surface water that does not infiltrate into the groundwater-flow system, particularly when the ground is saturated, and ends up directly in a stream channel. An estimate of overland flow is obtained by subtracting base flow from total stream flow (table 5.3). Overland flow in the San Juan Mountains can be extreme because of the high-relief topography. This causes streams to be "flashy," meaning average overland flows can change into torrents with flows in the hundreds of m^3/s in very short time spans (plate 20). However, high-gradient overland flow does little to support groundwater because this component of stream flow quickly leaves the watershed.

Plate 21 shows the magnitude and distribution of each hydrologic-cycle component for each of the SJM watersheds analyzed. Note that base flow per unit area in undammed watersheds appears independent of watershed size, precipitation, gradient, and aspect (table 5.3). For example, the Animas and Uncompahgre Watersheds have comparable precipitation, base flow per unit area, and gradient, while the latter watershed is much smaller in area than the former (plates 20 and 21; table 5.3).

Aspect plays a potential role in controlling recharge. For example, snow on north-facing slopes may have more time to infiltrate into the groundwater-flow system rather than rapidly running off, as commonly happens on southerly facing slopes. However, aspect does not appear to be a major control of base flow in SJM watersheds. In the San Juan Mountains, the Animas Watershed has a generally south-facing aspect and the Uncompahgre a generally north-facing aspect (plate 18), while the Dolores and Piedra face south to southwest, and the Lake Fork faces north and is smaller (plate 18); yet all five watersheds have comparable precipitation and base flow per unit area, suggesting that aspect is not a major control of base flow at the regional scale. In addition, the Lake Fork has the largest base flow per unit area of the latter three watersheds and is dominantly underlain by volcanic aquifers. In contrast, the Dolores, Piedra, and San Miguel Watersheds are each underlain by both Mesozoic sedimentary and Tertiary volcanic aquifer units, and each has relatively low base flow per unit area when compared with the watersheds predominantly underlain by volcanics (plate 17; table 5.3).

Evapotranspiration cannot be ruled out as a major factor in controlling the distribution of the hydrologic-cycle components shown in plates 20 and 21. However, the mean elevations and range of elevations of each of the SJM watersheds are similar. Because evapotranspiration is somewhat dependent on elevation (the upper reaches of alpine watersheds tend to have low densities of vegetation and thus low evapotranspiration), neither evapotranspiration nor elevation appears to be a major control of base flow per unit area. Thus the hydrologic-cycle component estimates for the regional scale of the San Juan Mountains indicate that geology may be the most important influence on base flow per unit area, which is consistent with the relatively high bulk permeability of the Tertiary volcanic-rock aquifers.

ESTIMATES OF SJM AQUIFER HYDRAULIC PROPERTIES

Few details are known about the bedrock aquifers of the San Juan Mountains. Identification of aquifers usually involves delineation of the lateral extent, thickness, and representative hydraulic properties of rock or sediment bodies that can store and transmit groundwater. Although few wells have been drilled specifically to test and delineate bulk aquifer hydraulic properties in the San Juan Mountains, a wealth of domestic water-supply well data is available from the Colorado Office of the State Engineer (COSE). Huntley (1979) also estimated hydraulic properties from spring data, estimated values for local hydraulic conductivity, measured intergranular hydraulic conductivities in a laboratory, and constructed a numerical groundwater-flow model. These and other data, along with previous model results,

Table 5.4. San Juan Mountains groundwater well data and statistics

Basic Depth Data (in feet)

	Depth	Screen Thickness	Median Screen Depth	Water Level	Yield for All Data (gpm)
Minimum	1	2	7	1	1
Maximum	3,136	1,050	1,305	2,020	1,500
Mean	143	53	151	62	18
Count	4,562	2,681	2,681	3,895	4,285

All Well-Yield Data and Yields Sorted by Top Screen Depth (in feet)

% gpm	All Depths	Depth ≥ 0, ≤ 25	Depth ≥ 26, ≤ 50	Depth ≥ 51, ≤ 75	Depth ≥ 76, ≥ 100	Depth ≥ 101, ≤ 200	Depth ≤ 201, ≤ 500	Depth ≥ 501
% ≤ 5	19	14	10	9	13	25	38	42
% ≤ 10	39	31	25	26	37	49	60	56
% ≥ 10 & ≤ 30	74	76	82	84	80	70	56	42
% ≥ 50	6	8	7	6	6	4	5	9
% ≥ 100	2	2	2	3	1	1	1	2

Notes and abbreviations: Data determined from reported groundwater well permits from the Colorado Office of the State Engineer. See glossary for definitions of well screen and associated terms. All depth data in FBG = feet below ground surface; gpm = gallons per minute. Count refers to the number of data points for each type of data.

have been integrated herein to report simple estimates of regional-scale hydraulic properties of SJM aquifers and to estimate groundwater residence times in each of the major watersheds.

Plate 22 shows the locations of 19,031 domestic wells. Nearly 4,600 wells within the SJM study area have been segregated from the entire dataset and their records tabulated. Pertinent data reported to the COSE by water-well drillers as part of the permitting process include location, total depth, well-screen interval, and well yield (table 5.4). The reliability of these data depends on the well drillers' ability to accurately measure and report these data.

Most groundwater wells are in the mountain valleys, along roads and streams (plate 22). This indicates that many of the wells were drilled into relatively high-permeability surficial deposits, providing little specific information about the hydraulic properties of the volcanic-bedrock aquifer system. Although some well reports include geological logs that could be used to delineate the depth to bedrock, those have not been used in this reconnaissance study. However, several insights have been gained regarding surficial deposits versus bedrock-geology hydraulics by using well-yield data sorted by the depth of the top of the well screen.

Figure 5.1 and table 5.4 show the basic statistics for wells drilled in the San Juan Mountains. For 4,562 wells, the mean depth is 143 ft (43.6 m). Both well-screen thickness and well-screen median depth can be indicative of the depth range for groundwater production, or where most of the water enters a well. The mean screen thickness is 53 ft (16.2 m), and average median screen depth is 151 ft (46 m). Of 4,285 wells, reported yields are a minimum of 1, maximum of 1,500, and mean of 18 gallons per minute (gpm, table 5.4). For comparison, 1,028 reported well yields in fractured, Precambrian metamorphic basement rocks in the Colorado Front Range indicate a minimum of 0.06 and maximum of 50 gpm (Caine and Tomusiak 2003).

Few data exist on the thickness of SJM surficial deposits that could make for high-permeability aquifers. Johnson and Yager (2006) drilled three research wells in the Upper Animas Watershed in which depths to bedrock range from 23 to 64 ft (7 to 19.5 m). Thirteen research wells drilled in an alpine environment similar to the San Juan Mountains, in the Colorado Front Range, indicated depths to bedrock from 0 feet near ridges to 91 ft (27.7 m) in the valley bottoms (Caine et al. 2006). Figure 5.1b shows the well-yield data sorted by depth intervals of the tops of well screens, with a few interesting results. Although the 50th (median) and 75th percentiles for yield at top-screen depths are equal, indicating the data are possibly skewed to these values, the yields at top-screen depths from 0 to 100 ft (30.5 m) are 15 gpm. For top-screen depths deeper than 100 feet, the 50th percentile yields decrease, are not equal to the 75th percentiles, and level out at 8 gpm. If one assumes an average depth to bedrock is about 100 feet or less, which is consistent with the few data available, the depth-sorted yield data may indicate that unconsolidated surface deposits have a median yield of about 15 gpm and bedrock a median yield of about 8 gpm (figure 5.1). This again assumes the data are representative and that the possibility of skewed data from top-screen depths shallower than 100 feet can be reconciled by inherent natural hydraulic properties of the aquifer rather than by deficiencies in the measurement and reporting of the data (drillers have been known to record 15 gpm as a minimum value and not determine the upper limit).

Although well yield is not simply related to permeability (k) or hydraulic conductivity (K), the few hydraulic conductivity data available for SJM volcanic bedrock indicate values from approximately 10^{-6} to 10^{-9} m/s, inclusive of fracture and intergranular K, and are consistent with the high reported yields (Huntley 1979; Johnson and Yager 2006). Thus reported yields for depths below 100 feet to greater than 500 feet for reported bulk K values are also consistent with a high-permeability, crystalline bedrock aquifer system, given the various assumptions inherent in interpreting this type of data.

Finally, Huntley (1979) used groundwater-flow-model results to estimate groundwater-flow velocities that range from about 6.6 to 1,716 ft (2 to 523 m) per year in San Juan Mountains volcanic rocks. However, these velocities cannot be

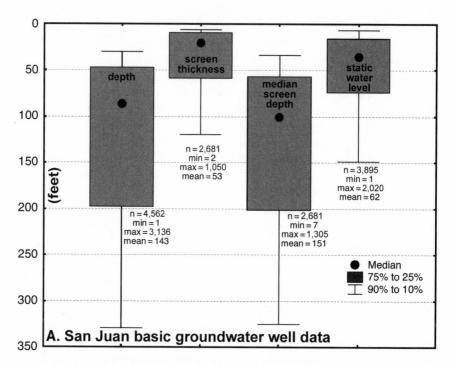

A. San Juan basic groundwater well data

depth
n = 4,562
min = 1
max = 3,136
mean = 143

screen thickness
n = 2,681
min = 2
max = 1,050
mean = 53

median screen depth
n = 2,681
min = 7
max = 1,305
mean = 151

static water level
n = 3,895
min = 1
max = 2,020
mean = 62

● Median
■ 75% to 25%
⊥ 90% to 10%

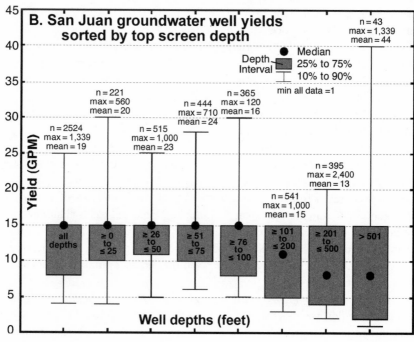

B. San Juan groundwater well yields sorted by top screen depth

n = 43
max = 1,339
mean = 44

● Median
Depth Interval
■ 25% to 75%
⊥ 10% to 90%
min all data = 1

n = 221
max = 560
mean = 20

n = 444
max = 710
mean = 24

n = 365
max = 120
mean = 16

n = 2524
max = 1,339
mean = 19

n = 515
max = 1,000
mean = 23

n = 395
max = 2,400
mean = 13

n = 541
max = 1,000
mean = 15

Yield (GPM)

all depths

≥ 0 to ≤ 25

≥ 26 to ≤ 50

≥ 51 to ≤ 75

≥ 76 to ≤ 100

≥ 101 to ≤ 200

≥ 201 to ≤ 500

> 501

Well depths (feet)

FIGURE 5.1 Box plots showing percentile (10th and 90th marked by whiskers on the vertical lines, 25th and 75th marked by the tops and bottoms of the gray boxes, 50th [median] marked by the black dots) distributions and other basic statistics for San Juan Mountain–region domestic water-well data. All data reported by water-well drillers to the Colorado Office of the State Engineer. A. Total well depth, screen thickness, median screen depth, and static water levels. B. Reported well yield, sorted by depth of the top of the well screen. Each depth interval is shown within each gray box.

easily used to estimate regional-scale groundwater residence times for each major watershed. More recent studies in the Upper Animas Watershed using isotopic groundwater age determinations (e.g., Manning and Solomon 2005) resulted in apparent residence times (the duration of time water remains in the subsurface) that range from a few years to a few decades (Johnson et al. 2007). Deep groundwaters are likely to reside in the regional flow system for hundreds to tens of thousands of years; however, few data are available to evaluate this assertion for the San Juan Mountains. Huntley (1979) also hypothesized that there is a very high degree of permeability anisotropy in the sub-horizontally layered volcanic rocks and that k is nearly constant with depth in the volcanic bedrock (horizontal k can be up to 1,000 times greater than vertical k). Thus the layered nature of volcanic-rock units that have different bulk permeabilities causes a large amount of lateral groundwater flow.

HOT SPRINGS OF THE SAN JUAN MOUNTAINS

The presence of a significant number of hot springs in the San Juan Mountains is indicative of a deeper groundwater-flow system. The available spring data may also represent how the deeper system interacts with the shallower, cooler groundwater-flow system. Plate 23 shows the locations, temperature (T) distributions, concentrations of total dissolved solids (TDS), and flow volumes for sixty-two hot springs in the San Juan Mountains. Although the US Geological Survey and Britt and Hornbeck (1996) datasets may not provide a complete list of all the known springs within the San Juan Mountains, they represent a sample of what may exist.

Forty-one percent of the sixty-two springs in the San Juan Mountains have temperatures greater than 86°F (30°C) (plate 23). The hotter springs are distributed throughout the San Juan Mountains, are found in all lithologies and at elevations from about 6,594 to 10,105 ft (2,010 to 3,080 m), and are weakly correlated with elevation (plate 23). Many of these hotter springs have high flow rates, occur close to warm and cool springs, and are poorly correlated with total dissolved solid concentrations (plate 23). The distribution and occurrence of SJM hot springs are likely indicative of fracture-and-fault-controlled (discrete) flow systems superimposed upon continuum flow systems controlled by intergranular permeability. In general, deeper groundwaters are hotter because temperature increases with depth, and fractures and faults provide efficient flow paths for hotter waters to ascend from depth. The hot springs also clearly indicate elevated SJM regional heat flow associated with the geologically recent emplacement of the San Juan Volcanic Field, consistent with possibly deeper circulation and flow of groundwater (Decker 2001).

Hot-spring flow rates range from 0.5 to 800 gpm, with a mean flow of 48 gpm, and 63 percent of hot springs have flows greater than 10 gpm (plate 23). These high flow rates are consistent with a bedrock aquifer system that has relatively high bulk permeability.

CONCEPTUAL MODEL FOR GEOLOGICAL CONTROLS OF GROUNDWATER

Plate 24 is a depiction of the fundamental geological elements that may control the occurrence, storage, and flow of groundwater in the volcanic, crystalline bedrock of the San Juan Mountains. Surficial deposits occur in valley bottoms and thin along their flanks. These deposits generally have higher bulk permeability (~15 gpm) than the underlying volcanic bedrock (~8 gpm), inferred from reported well-yield data (figure 5.1; table 5.4). Surface deposits may be locally significant shallow aquifers that support some degree of base flow to stream channels important to various plants, animals, and biogeochemical processes.

The dominant SJM aquifer system is composed of a heterogeneous array of sub-horizontally layered volcanic rocks. Some have very large lateral extents (tens of kilometers), many orders of magnitude greater than their vertical extents (tens of meters). The lateral continuity of layering with vertical heterogeneity of rock type, the variable response to fracturing, and possibly the effects of hydrothermal alteration together are responsible for both vertical anisotropy and unconfined to confined conditions within the SJM aquifer system, as discussed by Huntley (1979). Total thickness of the volcanic field ranges from tens of meters along the outer flanks to several kilometers in regions of nested calderas (Huntley 1979; Lipman 2000).

Welded and non-welded ash-flow tuffs, lava flows, and lahars are variably weathered, hydrothermally altered, and fractured, with decimeter- to micron-scale intergranular pore spaces (plate 24). Data at a variety of scales clearly indicate that both fractures and intergranular pores play an important role in the porosity and permeability of individual aquifer units. Igneous intrusive stocks, dikes, sills, and associated metallic, mineral-rich veins cut the layered volcanic rocks. At the time of emplacement, the intrusions and their heat are likely to be responsible for deposition of a variety of sulfide minerals, both in discrete veins and in large-scale zones of disseminated hydrothermal alteration (Bove et al. 2007). More chaotic volcanic deposits, such as caldera collapse and debris deposits, also disrupt the layered volcanics. Finally, all of these geological features are cut by a variety of fault zones, many of them extensional and steeply inclined discontinuities (Yager and Bove 2002) that may act as conduits and (or) barriers to groundwater flow (plate 24; Caine, Evans, and Forster 1996).

As orographically enhanced precipitation falls and the seasonal snowpack grows, water molecules begin their journey from the atmosphere into the watershed and potentially into the groundwater-flow system. As the snowpack melts, water that is not lost back to the atmosphere through sublimation or evapotranspiration or as outflow from the San Juan Mountains through overland flow, groundwater outflow, or other mechanisms such as human consumption infiltrates below the water table and into the groundwater-flow system (plates 16 and 24).

Although the topography of the San Juan Mountains has near-extreme relief and is very rough and uncertainty exists about static water levels away from stream valleys, the permeability of much of the volcanic bedrock is likely high enough to allow for rapid flow-through and a regionally low-angle water table, as proposed by Huntley (1979). The presence of both hot and cool high-elevation springs (plate 23) and of perennial, near-headwater streams (plate 18) suggests that the regional water table reaches well into the highest elevations of the San Juan Mountains. The evidence for hydraulic properties and the occurrence of groundwater also indicates that the volcanic-bedrock aquifer is well connected, albeit heterogeneously, to great depth. This also suggests that the aquifer can be treated as a continuum at the regional scale (Manning and Caine 2007). However, at the scale of a single well or field of wells, there may be a high degree of heterogeneity that could cause complex pathways for contaminant transport.

The large lateral extents and thicknesses of the volcanic aquifer system, existing base-flow data, and water-budget estimates are consistent with a large flux of water flowing through the subsurface of the SJM each year. Base-flow discharge to all of the major streams that were analyzed accounts for an average of over 550 million m^3, or about 150 billion gallons, of water per year (table 5.3). This equates to approximately 3.5 years of water use by the city of Denver, Colorado, according to population and per capita water-use estimates from the year 2000 (http://quickfacts.census.gov; http://www.greenprintdenver.org). Surface and groundwater outflows from the San Juan Mountains probably recharge the surrounding San Luis, San Juan, San Miguel, Uncompahgre, and Gunnison Basins. Another portion of groundwater may also recharge the Earth's uppermost crust and be transported far from its origin in the high San Juan Mountains. The flow of this water could be driven by the great hydraulic head formed by the combined processes of tectonically driven topographic uplift and orographic capture of precipitation over recent and geologic time scales.

ACKNOWLEDGMENTS

This work was written in memory of Laurie Wirt, who deeply loved the San Juan Mountains and expressed her love through conscientious efforts to understand their environment and aqueous geochemistry. She was supported by the Mineral Resources Program of the US Geological Survey. Discussions with Dana Bove, Stan Church, Andy Manning, Alisa Mast, Ray Johnson, Laurie Wirt, and Douglas Yager helped initiate this work. Reviews by Cathy Ager, Rob Blair, Kip Bossong, George Bracksieck, and Shaul Hurwitz are greatly appreciated and improved the manuscript.

REFERENCES

Barton, P. B., T. A. Steven, and D. O. Hayba. 2000. Hydrologic Budget of the Late Oligocene Lake Creede and the Evolution of the Upper Rio Grande Drainage System. In P. M. Bethke and R. L. Hay, eds., *Ancient Lake Creede: Its Volcano-Tectonic Setting, History of Sedimentation, and Relation to Mineralization in the Creede Mining District.* Special Paper 346. Boulder: Geological Society of America, 105–126.

Bove, D. J., M. A. Mast, J. B. Dalton, W. G. Wright, and D. B. Yager. 2007. Major Styles of Mineralization and Hydrothermal Alteration and Related Solid- and Aqueous-Phase Geochemical Signatures. In S. E. Church, P. von Guerard, and S. E. Finger, eds., *Integrated Investigations of Environmental Effects of Historical Mining in the Animas River Watershed, San Juan County, Colorado.* Professional Paper 1651. Reston, VA: US Geological Survey, chapter E5.

Britt, T. L., and J. M. Hornbeck. 1996. Energy Resources. In R. Blair, ed., *The Western San Juan Mountains: Their Geology, Ecology, and Human History.* Niwot: University Press of Colorado, 96–112.

Caine, J. S., J. P. Evans, and C. B. Forster. 1996. Fault-Zone Architecture and Permeability Structure. *Geology* 24:1025–1028.

Caine, J. S., A. H. Manning, P. L. Verplanck, D. J. Bove, K. G. Kahn, and S. Ge. 2006. *Well Construction Information, Lithologic Logs, Water Level Data, and Overview of Research in Handcart Gulch, Colorado: An Alpine Watershed Affected by Metalliferous Hydrothermal Alteration.* Open File Report 2006–1189. Reston, VA: US Geological Survey.

Caine, J. S., and S.R.A. Tomusiak. 2003. Brittle Structures and Their Role in Controlling Porosity and Permeability in a Complex Precambrian Crystalline-Rock Aquifer System in the Colorado Rocky Mountain Front Range. *Geological Society of America Bulletin* 115:1410–1424.

Church, S. E., P. von Guerard, and S. E. Finger, eds. 2007. *Integrated Investigations of Environmental Effects of Historical Mining in the Animas River Watershed, San Juan County, Colorado.* Professional Paper 1651. Reston, VA: US Geological Survey.

Claassen, H. C., M. M. Reddy, and D. R. Halm. 1986. Use of the Chloride Ion in Determining Hydrologic-Basin Water Budgets—A Three-Year Case Study in the San Juan Mountains, Colorado, U.S.A. *Journal of Hydrology* 85:49–71.

Decker, E. R. 2001. Thermal Data for Creede Caldera Moat Project Coreholes. In P. M. Bethke, ed., *Preliminary Scientific Results of the Creede Caldera Continental Scientific Drilling Program.* Open File Report 94–260-O. Denver: US Geological Survey.

Freeze, R. A., and J. A. Cherry. 1979. *Groundwater.* New York: Prentice-Hall.

Gray, J. E., M. F. Coolbaugh, G. S. Plumlee, and W. W. Atkinson. 1994. Environmental Geology of the Summitville Mine, Colorado. *Economic Geology, Special Issue on Volcanic Centers as Exploration Targets* 89:2006–2014.

Healy, R. W., T. C. Winter, J. W. LaBaugh, and O. L. Franke. 2007. Water Budgets: Foundations for Effective Water Resources and Environmental Management. Circular 1308. Boulder: US Geological Survey.

Huntley, D. 1979. Ground-Water Recharge to the Aquifers of Northern San Luis Valley, Colorado. *Geological Society of America Bulletin* 90(8):1196–1281.

Johnson, R. H., L. Wirt, A. H. Manning, K. J. Leib, D. L. Fey, and D. B. Yager. 2007. *Geochemistry of Surface and Ground Water in Cement Creek from Gladstone to Georgia Gulch and in Prospect Gulch, San Juan County, Colorado.* Open File Report 2007–1004. Washington, DC: US Geological Survey.

Johnson, R. H., and D. B. Yager. 2006. *Completion Reports, Core Logs, and Hydrogeologic Data from Wells and Piezometers in Prospect Gulch, San Juan County, Colorado.* Open File Report 2006–1030. Denver: US Geological Survey.

Kahn, K. G., S. Ge, J. S. Caine, and A. H. Manning. 2008. Characterization of the Shallow Groundwater System in an Alpine Watershed: Handcart Gulch, Colorado. *Journal of Hydrogeology* 16:103–121.

Keen, R. A. 1996. Weather and Climate. In R. Blair, ed., *The Western San Juan Mountains: Their Geology, Ecology, and Human History.* Niwot: University Press of Colorado, 113–126.

Lipman, P. W. 2000. Central San Juan Caldera Cluster: Regional Volcanic Framework. In P. M. Bethke and R. L. Hay, eds., *Ancient Lake Creede: Its Volcano-Tectonic Setting, History of Sedimentation, and Relation to Mineralization in the Creede Mining District.* Special Paper 346. Boulder: Geological Society of America, 9–69.

Lipman, P. W., T. A. Steven, R. G. Luedke, and W. S. Burbank. 1973. Revised Volcanic History of the San Juan, Uncompahgre, Silverton and Lake City Calderas in the Western San Juan Mountains, Colorado. *U.S. Geological Survey Journal of Research* 1:627–642.

Liu, F., M. W. Williams, and N. Caine. 2004. Source Waters and Flow Paths in an Alpine Catchment, Colorado Front Range, United States. *Water Resources Research* 40: W09401; doi: 10.1029/2004WR003076.

Manning, A. H., and J. S. Caine. 2007. Groundwater Noble Gas, Age, and Temperature Signatures in an Alpine Watershed: Valuable Tools in Conceptual Model Development. *Water Resources Research* 43:W04404; doi: 10.1029/2006WR005349.

Manning, A. H., and D. K. Solomon. 2005. An Integrated Environmental Tracer Approach to Characterizing Groundwater Circulation in Mountain Block. *Water Resources Research* 41:W12412; doi: 10.1029/2005WR004178.

Manning, A. H., D. K. Solomon, and A. L. Sheldon. 2003. Applications of a Total Dissolved Gas Pressure Probe in Groundwater Studies. *Ground Water* 41:440–448.

Plumlee, G. S., R. K. Streufert, K. S. Smith, S. M. Smith, A. Wallace, M. Toth, J. T. Nash, R. Robinson, and W. H. Ficklin. 1995. *Geology-Based Map of Potential Metal-Mine Drainage Hazards in Colorado.* Open File Report 95–26. Denver: US Geological Survey.

PRISM. 2007. Parameter-Elevation Regressions on Independent Slopes Model. Available at www.wcc.nrcs.usda.gov/climate/prism.html.

Robinson, C. S. 1978. Hydrology of Fractured Crystalline Rocks, Henderson Mine, Colorado. *Mining Engineering* 30:1185–1194.

Seaber, P. R., F. P. Kapinos, and G. L. Knapp. 1987. *Hydrologic Unit Maps.* Water-Supply Paper 2294. Denver: US Geological Survey.

Steven, T. A., K. Hon, and M. A. Lanphere. 1995. Neogene Geomorphic Evolution of the Central San Juan Mountains near Creede, Colorado. Miscellaneous Investigations Series Map I-2504. Reston, VA: US Geological Survey.

Steven, T. A., P. W. Lipman, W. J. Hail Jr., F. Barker, and R. G. Luedke. 1974. Geologic Map of the Durango Quadrangle, Southwestern Colorado. Miscellaneous Investigations Series Map I-764. Reston, VA: US Geological Survey

Stober, I., and K. Bucher. 2005. The Upper Continental Crust, an Aquifer and Its Fluid: Hydraulic and Chemical Data from 4 km Depth in Fractured Crystalline Basement Rocks at the KTB Test Site. *Geofluids* 5:8–19.

Tallaksen, L. M. 1995. A Review of Baseflow Recession Analysis. *Journal of Hydrology* 165:349–370.

Tweto, O. 1979. Geologic Map of Colorado. Reston, VA: US Geological Survey Special Map.

Wirt, L., K. J. Leib, D. J. Bove, M. A. Mast, J. B. Evans, and G. P. Meeker. 1999. *Determination of Chemical-Constituent Loads during Base-Flow and Storm-Runoff Conditions near Historical Mines in Prospect Gulch, Upper Animas River Watershed, Southwestern Colorado.* Open File Report 99–159. Denver: US Geological Survey.

Yager, D. B., and D. J. Bove, compilers. 2002. Generalized Geologic Map of Part of the Upper Animas River Watershed and Vicinity, Silverton, Colorado. Miscellaneous Field Studies Map MF-2377. Reston, VA: US Geological Survey.

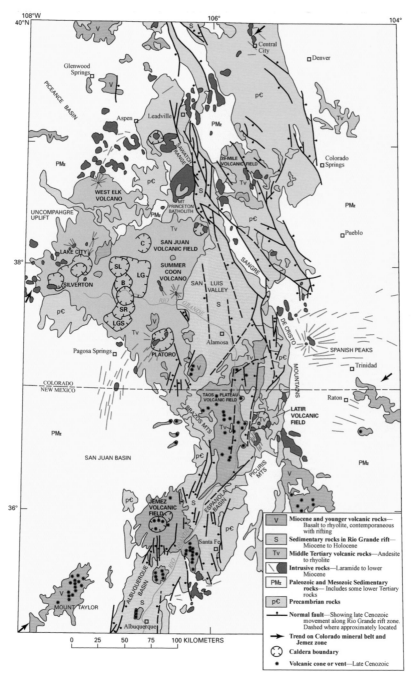

PLATE 1 Generalized geologic map of the San Juan Volcanic Field (from Lipman 2006). The exposed Precambrian rocks in southwestern Colorado define the Laramide San Juan uplift.

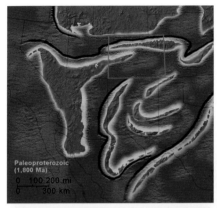

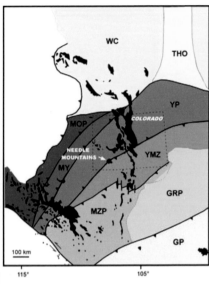

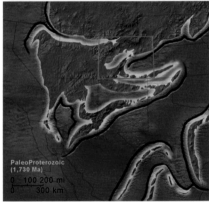

PLATE 2 Paleogeographic reconstructions of the southwestern United States from 1,800 to 1,680 Ma with present-day Colorado outlined in red (Images from Dr. Ron Blakey, Northern Arizona University, Flagstaff, Arizona; http://cpgeosystems.com/index.html). During this period of time, the edge of Laurentia grew through additions of different crustal blocks. The top right figure shows Proterozoic exposures (black) in present-day North America with the boundaries and ages of crustal provinces in billions of years: **WC** (Wyoming Craton), > 2.5; **THO** (Trans-Hudson), 2.0 to 1.8; **MOP** (Mohave Province), 2.0 to 1.8; **MY** (Mohave-Yavapia Transition Zone); **YP** (Yavapai Province), 1.8 to 1.7; **YMZ** (Yavapai-Mazatzal Transition Zone); **MZP** (Mazatzal Province), 1.7 to 1.6; **GRP** (Granite-Rhyolite Province), ~1.4; **GP** (Grenville Province), 1.3 to 1.0. Uncolored areas indicate crustal domains that are < 1 billion years old. Figure modified after Karlstrom et al. (2004).

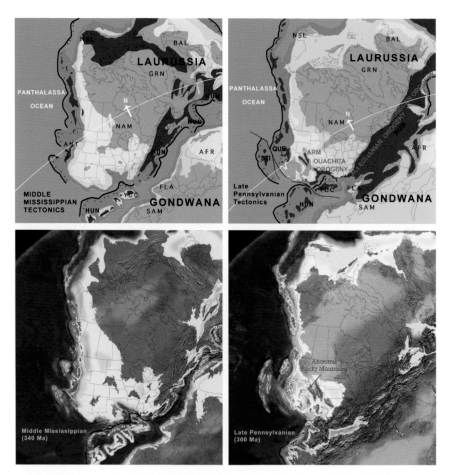

PLATE 3 Plate tectonic reconstructions of North America from 340 to 300 Ma, showing the development the formation the supercontinent Pangaea by the collision of Laurussia (Laurasia) and Gondwana along the eastern margin of North America. This collision led to the uplift of crustal blocks in Colorado to form the Ancestral Rocky Mountains. The white line on the upper images shows the approximate position of the equator with the arrow indicating geographic north. (Images are from Dr. Ron Blakey, Northern Arizona University, Flagstaff, Arizona; http://cpgeosystems.com/index.html.)

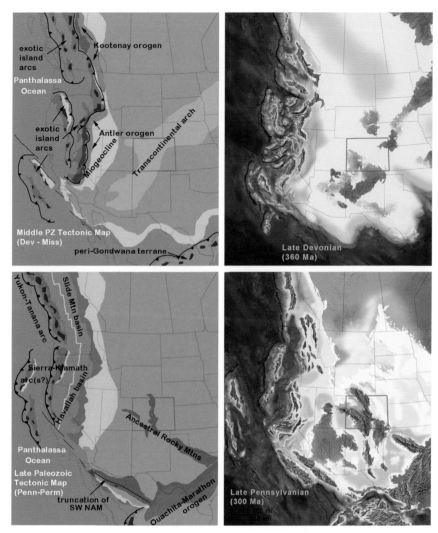

PLATE 4 Paleogeographic reconstructions of the southwestern United States at 360 to 300 Ma. An extensive seaway developed in this region from 500 to 300 Ma. Uplift of the Ancestral Rocky Mountains causes the retreat of this seaway, and development of fluvial systems off of the highlands. Present-day Colorado is outline in red. Images are from Dr. Ron Blakey, Northern Arizona University, Flagstaff, Arizona; http://cpgeosystems.com/index.html).

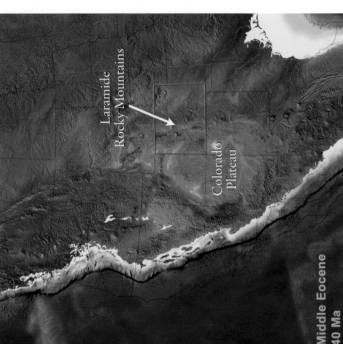

PLATE 5 Paleogeographic reconstructions of southwestern United States from 75 to 40 Ma. Over this period of time, an extensive intracontinental seaway that covered much of what is now the western United States during the Cretaceous period, retreats as the region is compressed and uplifted to form the Laramide Rocky Mountains. Images provided are from Dr. Ron Blakey, Northern Arizona University, Flagstaff, Arizona; http://cpgeosystems.com/index.html).

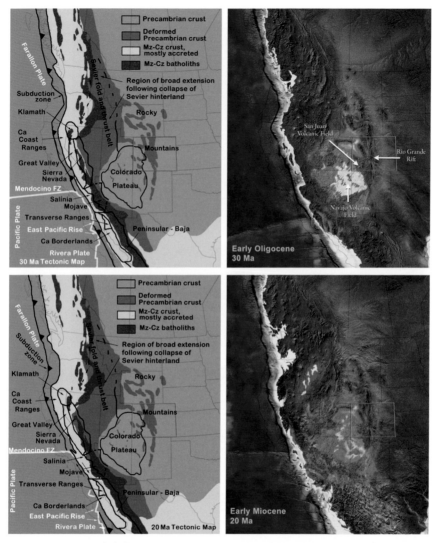

PLATE 6 The upper left figure shows the Paleogeographic reconstruction of North America and the southwestern United States at ~ 30 Ma and ~20 Ma. This was period of time was marked by widespread magmatism and some of the greatest and most extensive volcanic eruptions in the geologic record. Huge calderas formed above large bodies of magma to form the San Juan volcanic field. Diatremes of the Navajo volcanic field also formed at this time, and the Rio Grande Rift was initiated. Images are from Dr. Ron Blakey, Northern Arizona University, Flagstaff, Arizona; http://cpgeosystems.com/index.html).

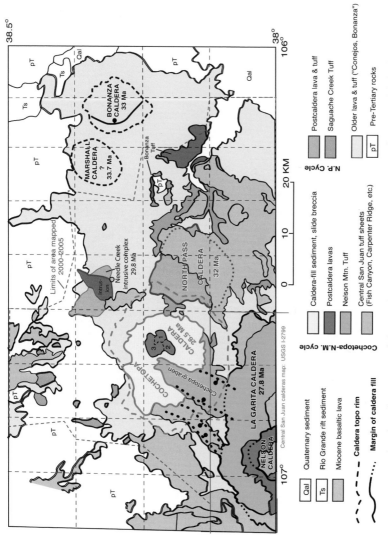

PLATE 7 Generalized geologic map of the northeastern San Juan region. Based on unpublished mapping (P. W. Lipman 2001–2006); compilation of volcanic geology is incomplete in areas beyond mapped limits.

38.5°

106°

38°

N.P. Cycle

Postcaldera lava & tuff

Saguache Creek Tuff

Older lava & tuff ("Conejos, Bonanza")

pT   Pre-Tertiary rocks

Cochetopa-N.M. cycle

Qal   Quaternary sediment

Ts   Rio Grande rift sediment

Miocene basaltic lava

– – –   Caldera topo rim

········   Margin of caldera fill

Caldera-fill sediment, slide breccia

Postcaldera lavas

Nelson Mtn. Tuff

Central San Juan tuff sheets
(Fish Canyon, Carpenter Ridge, etc.)

107°   Central San Juan calderas map: USGS I-2799

20 KM

0   10   20

BONANZA
CALDERA
33 Ma

MARSHALL
CALDERA
?
33.7 Ma

Bonanza
Tuff

pT

Ts

Qal

Limits of area mapped
2000-2005

Needle Creek
Intrusive complex
29.8 Ma

Intrus-
ion

NORTH PASS
CALDERA
32 Ma

COCHETOPA
CALDERA
26.5 Ma

Cochetopa graben

3
2

LA GARITA CALDERA
27.8 Ma

NELSON
CALDERA

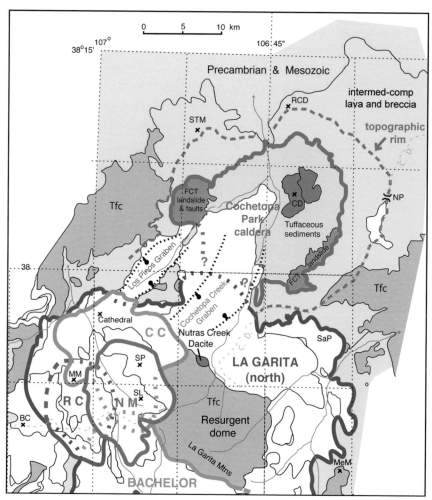

PLATE 8 Generalized geologic map of San Luis complex and Cochetopa Park caldera. Modified from Lipman 2006, figure 5C. Calderas of the San Luis complex: CC = Cebolla Creek; NM = Nelson Mountain; RC = Rat Creek. Tfc = Fish Canyon Tuff. Geographic names: BC = Baldy Cinco; CD = Cochetopa Dome; MeM = Mesa Mountain; MM = Mineral Mountain; NP = North Pass; RCD = Razor Creek Dome; SaP = Saguache Park; SL = San Luis Peak; SP = Stewart Peak; STM = Sawtooth Mountain.

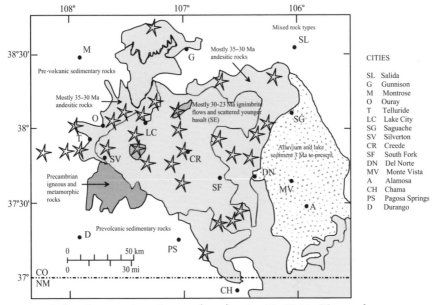

PLATE 9 Inferred composite volcanoes (stars) active 35–30 Ma. These volcanoes are the source for the andesitic rocks (gray) representing the Conejos and San Juan volcanics later buried by ignimbrites flows (beige). Map modified from Lipman 2006; Stevens, Hon, and Lanphere 1995.

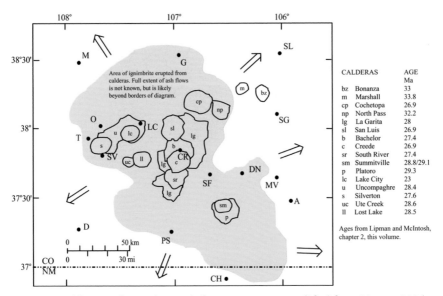

PLATE 10 Calderas and ignimbrite ash flows 33–23 Ma. Modified from Lipman 2006; Steven, Hon, and Lanphere 1995.

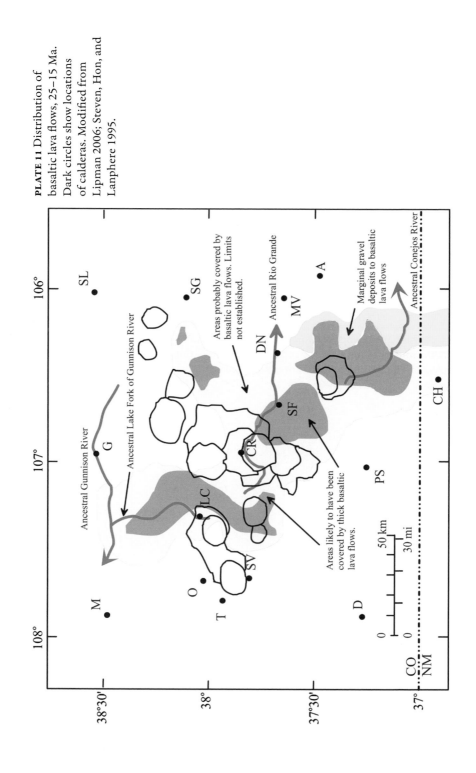

PLATE 11 Distribution of basaltic lava flows, 25–15 Ma. Dark circles show locations of calderas. Modified from Lipman 2006; Steven, Hon, and Lanphere 1995.

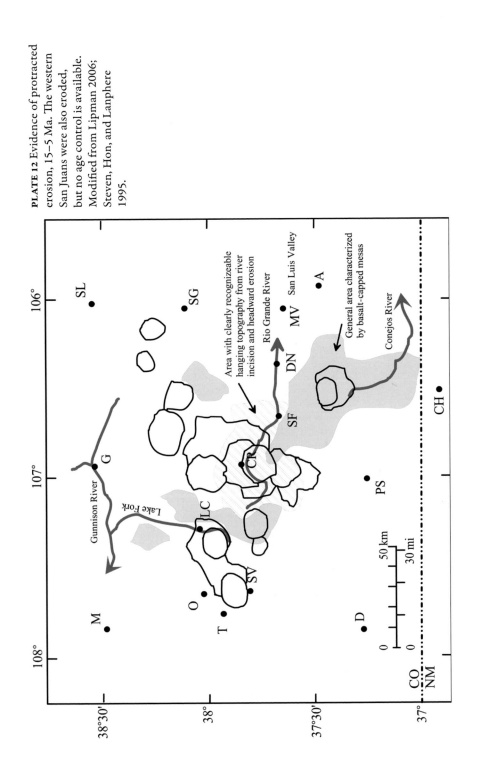

PLATE 12 Evidence of protracted erosion, 15–5 Ma. The western San Juans were also eroded, but no age control is available. Modified from Lipman 2006; Steven, Hon, and Lanphere 1995.

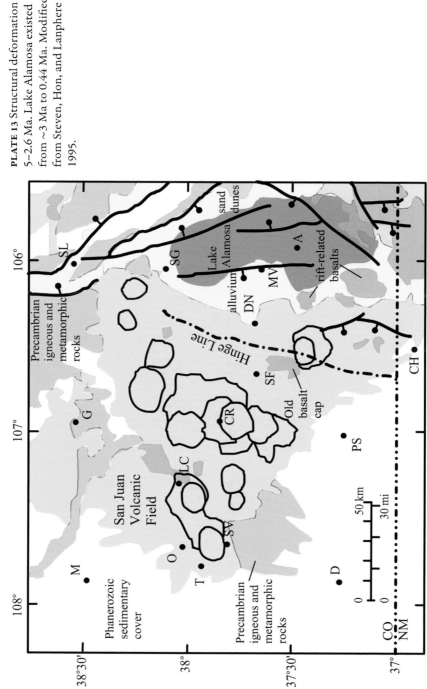

PLATE 13 Structural deformation 5–2.6 Ma. Lake Alamosa existed from ~3 Ma to 0.44 Ma. Modified from Steven, Hon, and Lanphere 1995.

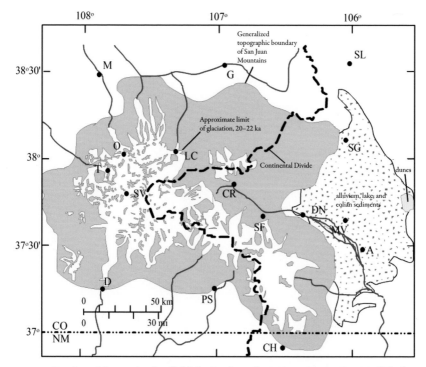

PLATE 14 San Juan Mountains ice field during last glaciations 20–22 Ka. Modified from Atwood and Mather 1932; Guido, Ward, and Anderson 2007; Machette, Marchetti, and Thompson 2007.

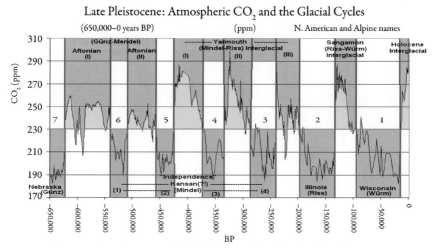

PLATE 15 Ice-core data for atmospheric CO_2 related to the glacial cycles: en.wikipedia.org/wiki/Image: Atmospheric_CO2_with_glaicers_Cycles.gif. Wikimedia Commons as presented by Tom Ruen, 2005.

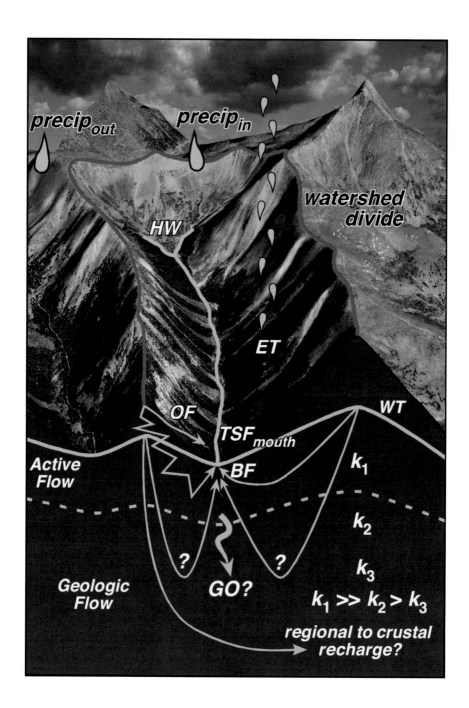

PLATE 16 Illustration of typical mountainous terrain, showing components of the hydrologic cycle as it pertains to the occurrence, storage, and flow of mountain surface and groundwaters. Precipitation that falls within the watershed divides (PPT_{in}) is included in water-budget calculations for individual watersheds. PPT_{out} falls outside the watershed divides and is not accounted for in water-budget calculations. Total stream flow (TSF) from the headwaters (HW) to the mouth (TSF_{mouth}) of the main stream is composed of base flow (BF) and overland flow (OF). Water leaves the watershed primarily by evapotranspiration (ET) and groundwater outflow (GO). Groundwater flow occurs perpendicular to the cross-section, out of the watershed and down gradient, toward the reader. The water table (WT) largely mimics topography, and smooth groundwater flow lines illustrate continuous flow, whereas jagged lines illustrate tortuous flow controlled by discrete features, such as fractures and faults. An active flow zone is indicated where most groundwater is locally driven by extreme to high, mountainous hydraulic gradients over human time scales. Flow that may be driven regionally over geologic time scales, possibly recharging the Earth's upper crust, is also illustrated.

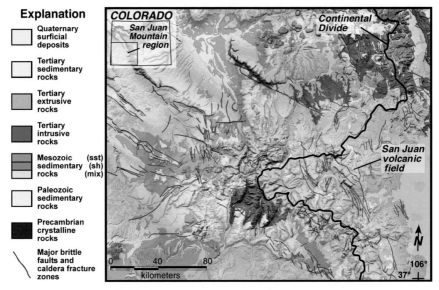

PLATE 17 Simplified geologic map of the San Juan Mountain region draped over a 30-meter digital-elevation model, showing relationships among geology, topography, and geomorphology (modified from Tweto 1979). Faults are indicated by dark gray lines; sst = sandstone; sh = shale; mix = mixed clastic sedimentary rocks.

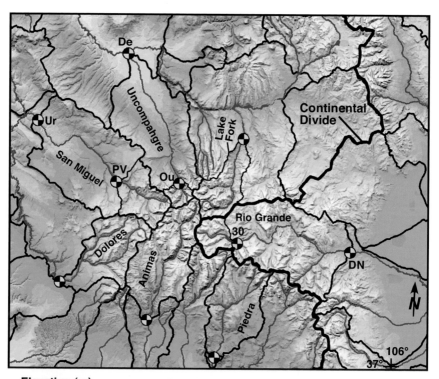

Elevation (m)

- 3,300 to 4,300
- 3,000 to 3,300
- 2,700 to 3,000
- 2,400 to 2,700
- 2,100 to 2,400
- 1,800 to 2,100
- 1,500 to 1,800
- 1,200 to 1,500

- major stream
- watershed divide
- stream gauge

0 ___ 40
kilometers

PLATE 18 Shaded relief topographic map of the San Juan Mountain region, showing the major streams, watershed divides, and stream gauges at the watershed mouths. Abbreviations for watersheds that have sub-basins: De = Uncompahgre at Delta; Ou = Uncompahgre at Ouray; PV = San Miguel at Placerville; Ur = San Miguel at Uravan; 30 = Rio Grande at 30 Mile Bridge; and DN = Rio Grande at Del Norte.

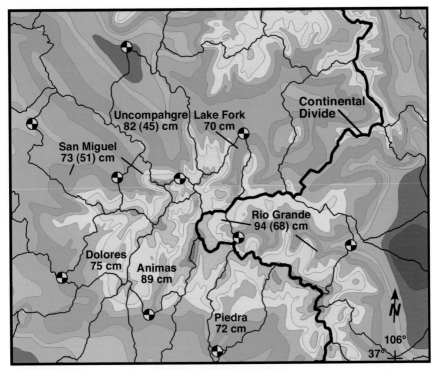

Annual Average Precipitation (cm)

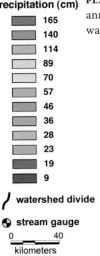

- 165
- 140
- 114
- 89
- 70
- 57
- 46
- 36
- 28
- 23
- 19
- 9

/ watershed divide

◑ stream gauge

0 ————— 40
kilometers

PLATE 19 Map of the San Juan Mountain region, showing average annual precipitation over thirty years, from 1975 to 2005, for each watershed and sub-watershed. Modified from PRISM 2007.

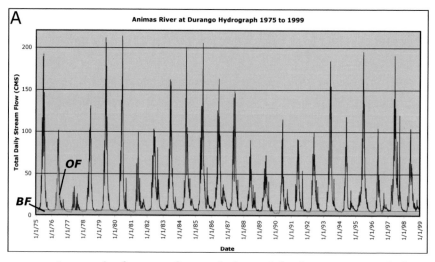

PLATE 20 An example of a twenty-five-year hydrograph for the Animas watershed. The average twenty-five-year base flow (*BF*) component of flow is illustrated by the turquoise box intersecting the lowest flows, or troughs, for each yearly flow cycle. One year of overland flow (*OF*) is illustrated for the 1976–1977 water year as the turquoise-filled curve above *BF*. Total stream flow (*TSF*) for the twenty-five-year period is the total area under the dark-blue jagged curve.

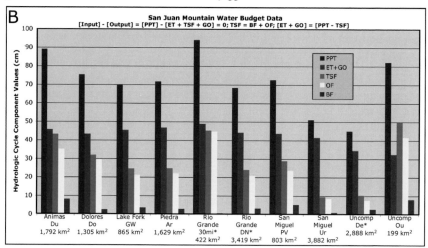

PLATE 21 Water budgets for major components of the hydrologic cycle, expressed as water distributed over the area of each watershed or as columns of some depth, with dimensions of length, for the San Juan Mountain region. Each budget was derived from precipitation and hydrograph data for each watershed. The hydrologic equation is shown with other variables, including precipitation (*PPT*), evapotranspiration (*ET*), and groundwater outflow (*GO*). Abbreviations as for tables 5.2 and 5.3.

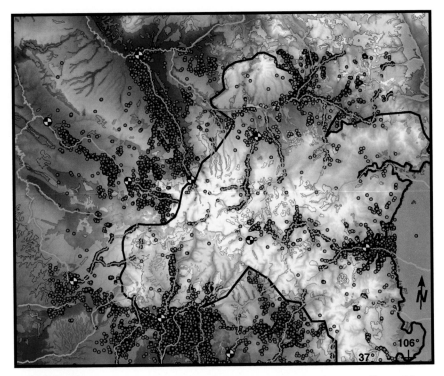

Explanation

(boundary of
SJMs

● well outside
SJMS

● well within
SJMs

● stream gauge

∫ major stream

0 ___ 40
kilometers

PLATE 22 Location map of domestic groundwater wells outside
(purple) and within (green) the San Juan Mountains (black line),
overlain on geologic and digital elevation maps.

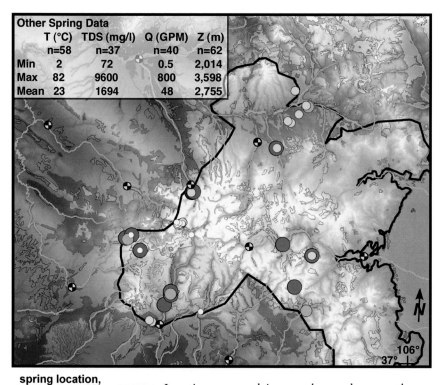

Other Spring Data				
	T (°C)	TDS (mg/l)	Q (GPM)	Z (m)
	n=58	n=37	n=40	n=62
Min	2	72	0.5	2,014
Max	82	9600	800	3,598
Mean	23	1694	48	2,755

spring location, temperature (°C), percent (n = 62)

(boundary of SJMs

○ 0 to 6 °C, 22%

○ 6 to 18 °C, 29%

◉ 18 to 30 °C, 8%

● > 30 °C, 41%

⊕ stream gauge

∫ major stream

0 ▭▭ 40
kilometers

PLATE 23 Location map overlain on geology and topography, data table, and distribution of natural springs in the San Juan Mountain region. The size and color of spring location markers are proportional to spring-water temperature. The percentage of the total for each temperature range is shown in the legend. Basic statistics are shown for reported temperature (T), concentration of total dissolved solids (TDS), discharge or flow (Q), and elevation (Z). Data from US Geological Survey.

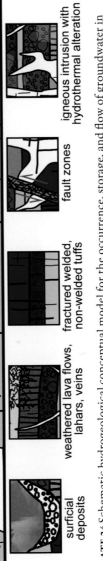

Snowpack

Mine

Groundwater Flow Lines

surficial
deposits

weathered lava flows,
lahars, veins

fractured welded,
non-welded tuffs

fault zones

igneous intrusion with
hydrothermal alteration

PLATE 24 Schematic hydrogeological conceptual model for the occurrence, storage, and flow of groundwater in the San Juan volcanic aquifer system. Groundwater flow lines likely take numerous pathways—some tortuous, some continuous, and some through local hydrothermal alteration zones, where water may become enriched in metallic-mineral-deposit contaminants.

PLATE 25 Upper Sunlight Lake, an alpine lake located at 12,545 ft elevation in the Weminuche Wilderness. Photo by Blake Meneken.

PLATE 26 Basin fen in Prospect Basin near Telluride, Colorado, showing floating mat surrounding open water. Vegetation is *Menyanthes trifoliata* in the pond and *Carex utriculata* and *Pedicularis groenlandica* in the foreground.

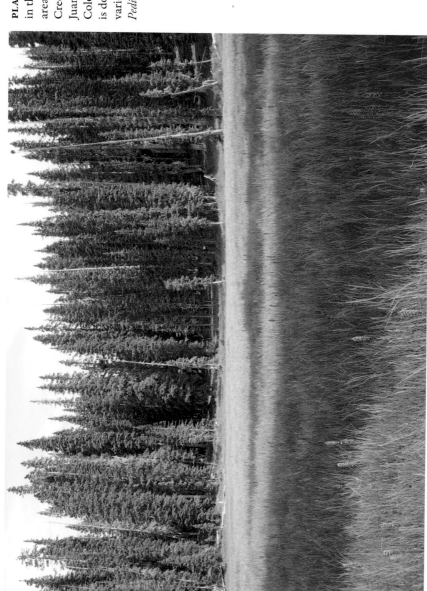

PLATE 27 Sloping fen in the North Mosca area near Williams Creek Reservoir, San Juan National Forest, Colorado. Vegetation is dominated by a variety of sedges and *Pedicularis groenlandica*.

PLATE 28 *Amanita muscaria*, a mushroom mycorrhizal with spruce. Note the white gills, the annulus (ring) around the stipe (stalk), the volva (cup) at the base of the stalk, and the red cap with white warts. Image © Eric Isselée / Shutterstock.

PLATE 29 *Boletus edulis*. Note the net-like pattern at the top of the stipe, the bulbous stipe, the mahogany-colored cap, and the pores instead of gills. Photograph © Birute Vijeikiene / Shutterstock.

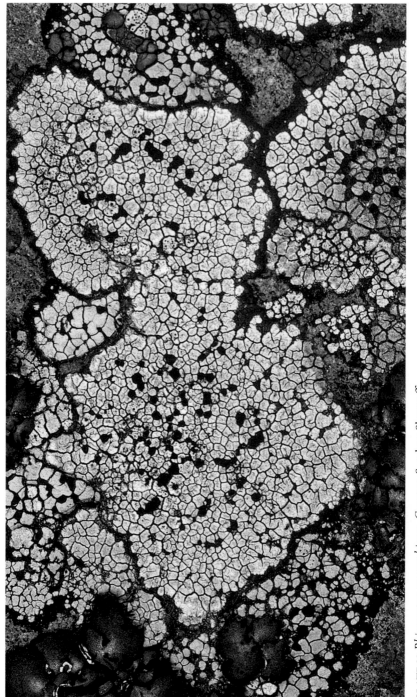

PLATE 30 *Rhizocarpon geographicum. Courtesy,* Stephen Sharnoff.

PLATE 31 An example of an old-growth ponderosa pine stand in the eastern San Juan Mountains that retains historical stand structure characteristics. Note the openness of the trees, presence of snags, and herbaceous understory. This stand has never been logged or grazed. Photo by Peter M. Brown.

PLATE 32 This ponderosa pine stand is located in the Kenny Flats area south of Pagosa Springs. It has been grazed and logged and has experienced fire suppression. Before 1880, Kenny Flats had an average fire interval of six years, with large landscape-scale fires occurring, on average, every thirteen years. Note the high density of trees and lack of understory vegetation. Photo by Sara Brinton.

PLATE 33 Typical aspen grove

PLATE 34 Trail Ridge, Pagosa Ranger District. White fir has taken over the warm-dry mixed-conifer understory to the exclusion of ponderosa-pine regeneration. Lack of fire has allowed more white-fir regeneration to survive than would have been the case under the historical fire regime. Photo by Sara Brinton.

PLATE 35 (a) bighorn sheep, (b) bull elk, (c) puma (mountain lion), (d) Canada lynx, (e) Rio Grande cutthroat (USFWS photograph), (f) Colorado River cutthroat (*courtesy,* Gary Vos, USFS). All photos from Colorado Division of Wildlife except as noted.

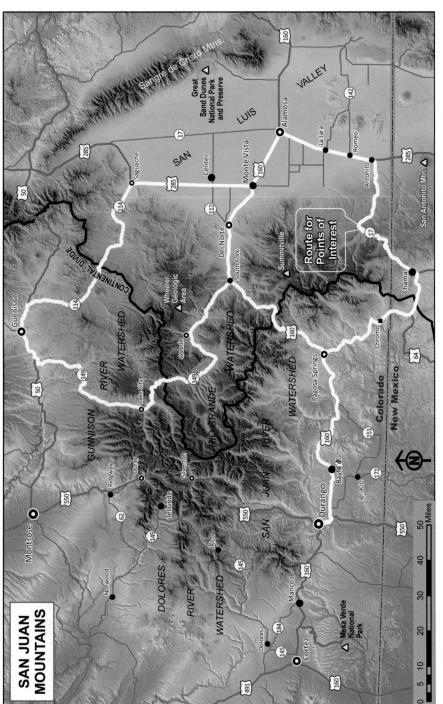

PLATE 36 The San Juan Mountains.

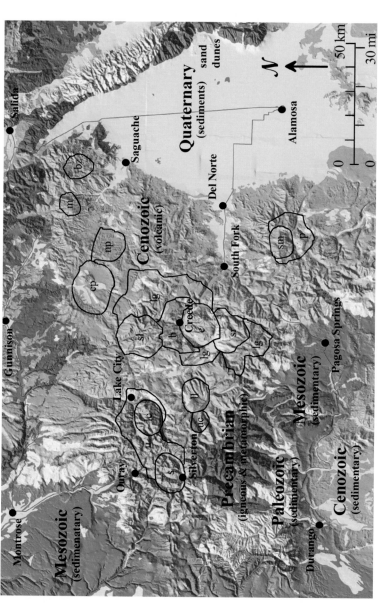

PLATE 37 General geologic map of the San Juan Mountains. Caldera abbreviations: bz = Bonanza; cp = Cochetopa; np = North Pass; lg = La Garita; sl = San Luis; b = Bachelor; c = Creede; sr = South River; sm = Summitville; p = Platoro; lc = Lake City; u = Uncompaghre; s = Silverton; uc = Ute Creek; ll = Lost Lakes. Red line depicts the linkage of several prominent highways.

Chronostratigraphic units (ages in millions of years, Ma)	STRATIGRAPHIC CHART FOR THE SAN JUAN MOUNTAINS
	WESTERN / EASTERN

Chronostratigraphic units (ages in millions of years, Ma)

ERA / PERIOD	EPOCH	Age (Ma)
CENOZOIC ERA — PALEOGENE / NEOGENE — QUATERNARY	HOLOCENE	(.01)
	PLEISTOCENE	(2.6)
	PLIOCENE	(5.3)
TERTIARY	MIOCENE	(23)
	OLIGOCENE	(33.9)
	EOCENE	(55.8)
	PALEOCENE	(65.5)
MESOZOIC ERA — CRETACEOUS		
		(145.5)
JURASSIC		(199.6)
TRIASSIC		(251)
PALEOZOIC ERA — PERMIAN		(299)
PENNSYLVANIAN		(318)
MISSISSIPPIAN		(359)
DEVONIAN		(416)
SILURIAN		(444)
ORDOVICIAN		(488)
CAMBRIAN		(542)
PRECAMBRIAN — PROTEROZOIC		(2,500)
ARCHEAN		

STRATIGRAPHIC CHART FOR THE SAN JUAN MOUNTAINS

WESTERN

- STREAM DEPOSITS
- GLACIAL AND STREAM DEPOSITS
- BASALTS (15–26 Ma) Hinsdale Fm.
- IGNIMBRITES (26–30 Ma) multiple ash-flow sheets
- ANDESITES (30–35 Ma) San Juan Fm. (west), Conejos Fm. (east)
- NACIMIENTO, SAN JOSE FMS, TELLURIDE CONG, BLANCO BASIN FM.
- ANIMAS FORMATION
- KIRTLAND SHALE
- FRUITLAND FORMATION
- PICTURED CLIFFS SANDSTONE
- LEWIS SHALE
- MESA VERDE GROUP: CLIFF HOUSE SS / MENEFEE FM / POINT LOOKOUT SS
- MANCOS SHALE
- DAKOTA SANDSTONE
- BURRO CANYON FORMATION
- MORRISON FORMATION: BRUSHY BASIN MEM / SALT WASH MEM
- JUNCTION CREEK SANDSTONE
- WANAKAH FORMATION
- ENTRADA SANDSTONE
- DOLORES FORMATION (CHINLE)
- WINGATE SANDSTONE ?
- CUTLER FORMATION
- HERMOSA FM: HONAKER TRAIL MEM / PARADOX MEM / PINKERTON TRAIL MEM
- MOLAS FORMATION
- LEADVILLE LIMESTONE
- ELBERT FORMATION
- OURAY LIMESTONE
- IGNACIO FORMATION
- TRIMBLE Gr, ELECTRA LAKE Gb (1,460), EOLUS Gr (1,460) AND RELATED ROCKS
- igneous intrusives – granite, gabbro
- UNCOMPAHGRE FM (quartzite)
- VALLECITO FM (metaconglomerate, < 1,690 Ma)
- TENMILE (1,720), BAKERS BRIDGE (1,720), TWILIGHT (1,800), IRVING FM (1,800 Ma)

EASTERN

- SAN LUIS VALLEY
- Lake sed (0.4–3.5)
- Basin fill (27–0 Ma)

Ignimbrite unit (ashflow)   Age (Ma)
(Lipman and McIntosh, chap 2)

Ignimbrite unit (ashflow)	Age (Ma)
Sunshine Peak	23.0
Snowshoe Mtn	26.9
Nelson Mtn	26.9
Cebolla Creek	26.9
Rat Creek	26.9
Wason Park	27.4
Blue Creek	27.45
Carpenter Ridge	27.5
Crystal Lake	27.6
Fish Canyon	28.0
Chiquito Peak	28.35
Sapinero Mesa	28.35
Dillon Mesa	28.4
Blue Mesa	28.5
Ute Ridge	28.6
Masonic Park	28.7
South Fork	28.7
Ra Jadero	28.8
Ojito Creek	29.1
La Jara Canyon	29.1
Black Mtn	29.4
Saguache Creek	32.2
Bonanza	33.0

For the eastern San Juans the sedimentary section is buried beneath the Tertiary volcanic pile above. Only around the perimeter are some of the sedimentary units exposed, i.e., south of Gunnison and north of Chama.

Stippled areas indicate the geologic record is missing.

These Precambrian units are exposed in Cochetopa Canyon southeast of Gunnison.

COCHETOPA GRANITE (1,650 Ma)
COCHETOPA GNEISS (1,720)

* Modified after Blair 1996, *Western San Juan Mountains*, p. 2; ages from International Stratigraphic Chart 2008.

Long-Term Temperature Trends in the San Juan Mountains

Imtiaz Rangwala and James R. Miller

TEMPERATURES IN THE SAN JUAN MOUNTAINS are highly variable in both time and space. During a single day, temperatures in the high mountains can vary from well below freezing to above 15°C (60°F). Large extremes can also occur in the lower-elevation, more arid regions of southwestern Colorado. At a single instant in time, temperatures can range from well below freezing at the region's highest elevations to well above freezing at the lower elevations. There are also significant seasonal changes, with extremes over 38°C (100°F) in summer and as low as −46°C (−50°F) in winter. Because of the complex terrain, and because most observation stations are biased according to where people live—most commonly in the valleys—it is often difficult to effectively examine the mountain climate.

Globally, surface-air temperatures have risen by about 0.7°C (1.3°F) during the last century. At least a portion of this increase is attributed to human activities, primarily from burning fossil fuels, which increase the amount of carbon dioxide in the atmosphere. Carbon dioxide is one of several greenhouse gases that leads to increasing surface-air temperatures by absorbing some of the infrared radiation (from the Earth's surface) that would otherwise escape into space. The increased temperature can trigger other physical processes that can further amplify warming at the Earth's surface, producing a positive feedback on temperature. Observations from many of the planet's high-elevation regions during the second half of the twentieth century suggested that those have warmed at a greater rate than the global average. Moreover, surface temperatures at the majority of high-elevation stations across the globe are increasing faster than the free-air temperatures (temperature of air in an atmospheric column) at the same elevation. However, the nature of the specific reported changes in mountain systems is often different from region to region.

In a prequel to this book, *The Western San Juan Mountains*, Richard Keen presents a summary of the general climate of the San Juan Mountains. He found no long-term trends in temperature or precipitation during the twentieth century (Keen 1996). The climate record in his analysis ends in 1994. In this chapter, we revisit temperature trends in the San Juans by considering an extensive network of observations that extends the record through 2009. A more detailed analysis of these trends can be found in Rangwala and Miller (2010). Here we address whether any long-term temperature trends in the San Juan Mountains have emerged, considering that Keen (1996) found no such trends. Increases in surface temperatures in these mountains may significantly affect hydrological and ecological systems. Impacts on stream-flow amounts and timing will extend to the large downstream regions covered by the Colorado River and the Rio Grande.

To examine temperature trends in the instrumental record, we used observations from three sources—station data from National Weather Service (NWS) and Snow Telemetry (SNOTEL) stations (Julander, Curtis, and Beard 2007) and gridded data from PRISM (Parameter-Elevation Regressions on Independent Slopes Model) (Daly et al. 2008). The location of the observation stations is shown in figure 6.1. SNOTEL stations provide an independent record of temperature trends, but for a shorter period: 1984–2005. These stations also capture the climate at higher elevations in the San Juans (average elevation = 10,000 ft [3,048 m]) and are, on average, 2,500 ft (762 m) higher than the NWS stations. When we compared the mean seasonal temperature trends between the NWS and SNOTEL stations, we found very high correlations (Rangwala and Miller 2010; figure 6.2). Additional information on these datasets is provided in the appendix to this chapter.

LONG-TERM TRENDS

Figure 6.2 shows the annual mean surface-air temperature anomalies (differences from the 30-year average, 1960–1990) in the San Juans during the twentieth century. In general, the figure shows a cooler period from 1910–1930, followed by a warmer period between 1935 and 1960, a persistent cooling trend between 1960 and 1975, and a rapid warming trend between 1993 and 2005. The rapid warming during the latter decade has retreated somewhat since 2005. The year 2008, which was very snowy, was the first to have below-normal temperatures since 1993. The twentieth-century warming trend did not result from a gradual long-term warming process but rather from temperature increases that occurred after 1993. This is not inconsistent with Keen's climate summary, which found no trend in temperatures

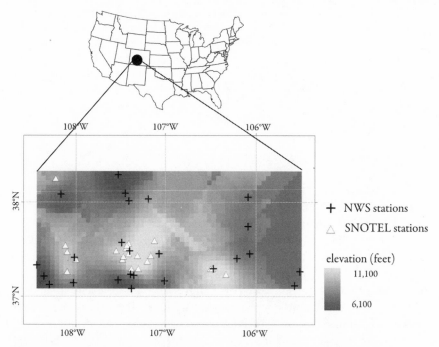

FIGURE 6.1 Spatial distribution of the National Weather Service (NWS) and SNOTEL stations. Most of the observation stations are in these twelve counties: Archuleta, Conejos, Dolores, Hinsdale, La Plata, Mineral, Montezuma, Ouray, Rio Grande, Saguache, San Juan, and San Miguel.

in the San Juans prior to 1994. The large temperature increases since 1994 were also observed for all of Colorado (Ray, Barsugli, and Averyt 2008, figure 2.5). Among the NWS stations, 1934 and 2003 were nearly tied as the warmest year on record, at 1.7°C (3°F) above normal. PRISM temperatures indicated that 2003 was slightly warmer than 1934. SNOTEL stations reported 2005 as the warmest year in their record.

The daily mean temperature is based on the average of the daily minimum (T_{min}) and maximum (T_{max}) temperatures. To better understand long-term warming in the San Juans, we examine changes in each of these variables separately. Figure 6.3 shows that the trend in maximum temperature is similar to that for the mean temperature, with a cooler period between 1910 and 1930, a warmer period between 1935 and 1960, a persistent cooling trend between 1960 and 1975, and rapid warming between 1993 and 2005. Since 2005, there has been a decreasing trend in T_{max}; T_{max} for 2008 and 2009 was below normal for the period since 1993. Conversely, between 1910 and 1994 there was a long-term gradual warming trend in minimum temperatures of about 0.7°C (1.26°F) (figure 6.3b). The dashed line

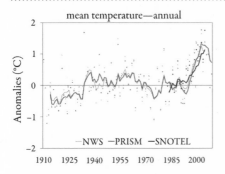

FIGURE 6.2 Anomalies in mean annual surface-air temperature (°C) in the San Juan Mountain region between 1910 and 2009, relative to the 1960–1990 base period. The curves show five-year running averages. The dots are the actual temperature anomalies.

in figures 6.3a and b is a linear regression for the 1910–1994 period, which further illustrates a long-term warming signal in T_{min} and no such trend in T_{max}; T_{min} also shows a lower inter-annual variability relative to T_{max}. Between 1993 and 2005, T_{min} shows a more rapid and continuous increase. Unlike T_{max}, T_{min} anomalies after 1993 remain above normal until the present, even for the snowy 2008. There is a stronger warming signal in T_{min}, both in the long- and near-term historical record. This is consistent with other studies of late-twentieth-century climate change in different mountain regions around the world, which have found that minimum temperatures have been increasing roughly twice as fast as maximum temperatures. Based on the linear regression fitted for the entire 100-year period (1910–2009), maximum temperatures have been increasing at a rate of 0.06°C (0.11°F) and minimum temperatures at a rate of 0.12°C (0.22°F) per decade. Between 1993 and 2005, both maximum and minimum temperatures increased rapidly, at similar rates of 1.3°C (2.3°F) per decade.

What contributed to the cooling trend in mean temperatures between 1960 and 1975? Can it be attributed primarily to the significant decreasing trend in maximum temperatures between 1960 and 1990? A possible explanation could be related to the "solar-dimming" phenomenon during this period. One of the direct effects of particulate pollution in the atmosphere is a reduction in incoming solar radiation at the surface, causing a decrease in daytime surface heating and hence affecting maximum temperatures more than minimum temperatures. This effect is thought to be responsible in part for global-scale cooling between 1950 and 1980, although it was restricted mostly to the Northern Hemisphere. Estimates of atmospheric sulfate burden for the twentieth century indicate an increasing trend between 1960 and 1980 for the San Juans. This is consistent with the largest decrease in maximum temperatures during this period, as shown in figure 6.3a. Moreover, the sudden increases in maximum temperatures around 1935 might have resulted in part from reductions in atmospheric load of mineral dust from the western United States rangeland, such as the Colorado Plateau, driven by imposed restrictions on grazing. Large and sustained dust loadings of the atmosphere will reduce incoming solar radiation at the surface and thereby lower daytime maximum temperatures.

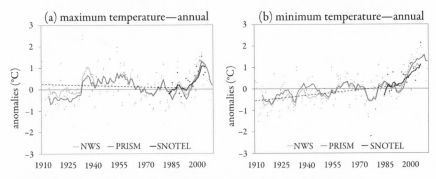

FIGURE 6.3 Same as figure 6.2 but for the (a) maximum and (b) minimum temperatures. The dashed line is a linear regression between 1910 and 1994.

SEASONAL TRENDS

We will now explore possible seasonal differences in these long-term trends in annual temperatures. Figure 6.4 shows that prior to the mid-1990s there was no warming trend for maximum temperatures in any season. In fact, in all seasons except winter, the maximum temperatures decreased, with the cooling particularly pronounced in the fall. Although maximum temperatures were warmer in winter during the last decade, there have been similar warmer decades in the past (e.g., the 1950s). Figure 6.5 shows that the minimum temperature gradually increased in all seasons during the twentieth century. Relative to the long-term variability of temperatures during that century, the increases in maximum and minimum temperatures during the spring and summer between 1993 and 2005 are unprecedented, accounting for much of the sudden and rapid warming in the San Juans over the last two decades. During this same period, there were large positive T_{min} anomalies during the winter and relatively smaller anomalies during the fall.

Figures 6.4 and 6.5 also show that the inter-annual and inter-decadal variability in temperature anomalies is much greater in winter than in summer. This large variability could result in part from variations in winter precipitation and therefore snow cover, which is influenced by synoptic-scale Pacific climate variability resulting from the El Niño Southern Oscillation and the Pacific Decadal Oscillation (PDO). El Niño and the warm phase (+) of PDO are generally observed to be associated with cold, wet winters in the San Juans. There is also a greater inter-annual variability associated with T_{max} relative to T_{min} for all seasons; T_{max} is more strongly influenced by factors such as snow cover and soil moisture, which are affected by precipitation variability. Our preliminary analysis of the precipitation record suggests significant negative correlations between T_{max} and precipitation, particularly during summer.

The spring of 1934 maintains the record for the highest average daytime temperature on record, at 4°C (7.2°F) above normal. Interestingly, that spring was

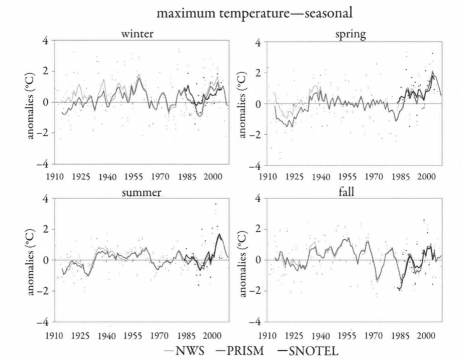

FIGURE 6.4 Same as figure 6.2 but for T_{max} during each season.

juxtaposed between years with colder-than-normal springs. For summer, some of the highest recorded daytime extremes have occurred only recently (1994, 2000, 2002, and 2003). The summer of 2002 experienced by far the highest daytime temperatures on record: 3.5°C (6.3°F) above normal. That year was one of the driest on record in Colorado, even when we include tree-ring records that go back several hundred years.

RECENT TEMPERATURE TRENDS

The temperature trends in the San Juans between 1990 and 2005 were unprecedented during the last century. The large temperature increases during this period were found in all seasons at both the NWS and SNOTEL sites, with the mean temperature at both sites increasing by about 1°C (1.8 °F) per decade (figure 6.6). Although the increases in maximum and minimum temperatures were similar at the NWS and SNOTEL sites, they were larger at NWS sites during the winter (1.5°C [2.7°F]/decade) and larger at SNOTEL sites during the summer (1.5°C/decade). The elevation of the SNOTEL sites is higher than that of the NWS sites; therefore, the former experience the bulk of snowmelt later in the year. For example, there is

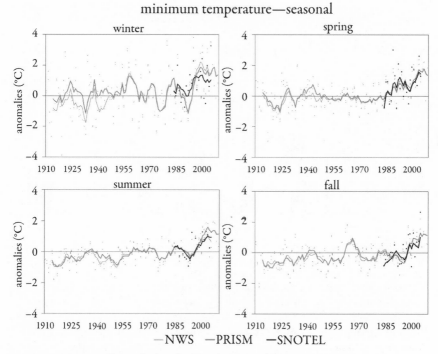

minimum temperature—seasonal

—NWS —PRISM —SNOTEL

FIGURE 6.5 Same as figure 6.2 but for T_{min} during each season.

likely to be negligible snow cover at the elevations associated with NWS stations (below 10,000 ft [3,048 m]) in early summer, whereas there is still substantial snow cover at the elevations associated with SNOTEL sites (above 10,000 ft). Therefore, temperature changes driven by the snow-albedo feedback mechanism would be important earlier in the summer at higher elevations. This positive feedback on temperature occurs because warmer temperatures melt snow, thus decreasing the surface reflectivity (albedo) and allowing more solar radiation to be absorbed, which increases the surface temperature even more. Enhanced snow-albedo effects in mountainous regions have been reported across the globe.

The effect of climate change on snowmelt and the latter's feedback on the climate are further complicated by deposition of desert dust on mountain snow-pack, particularly during the spring. Dust on snow has been observed to reduce the reflectivity (albedo) of snow and thereby to considerably increase the absorption of incoming solar radiation. This increased absorption, which occurs even if the dust is buried as much as 12 in (30 cm) below the snow surface, leads to earlier and faster snowmelt. The episodic dust storms from the Colorado Plateau are well recognized in the San Juans, but their impacts on mountain snowpack have only recently been studied. The Center for Snow and Avalanche Studies in Silverton is

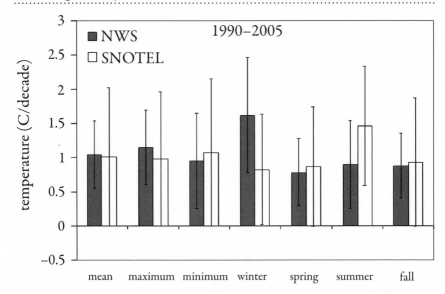

FIGURE 6.6 Linear regression in annual (mean, maximum, and minimum) and seasonal (winter, spring, summer, and fall) surface-air temperatures (°C/decade) during the 1990–2005 period in the San Juan Mountain region, from both NWS and SNOTEL observations. Error bars show one standard deviation. © Regents of the University of Colorado.

one of the entities involved in this study. The impacts of these episodic dust events during snowmelt season are expected to influence surface-temperature trends from year to year, primarily through the snow-albedo feedback mechanism. Future intensification of drought in dust-source regions could increase the frequency and intensity of dust-on-snow events, magnifying the effects of climate change on the snow hydrology of the San Juans.

The rate of winter warming at NWS sites has been about twice that of the warming during the spring, summer, and fall. However, significant variability is seen in the warming rate during the winter. At the SNOTEL sites, the winter warming has been about half that at the NWS sites. Maximum and minimum temperatures have increased by about the same amount during spring and summer, but the minimum temperatures have increased more in winter and fall. Warming during spring and summer is more strongly affected by factors that influence incoming solar radiation, such as changes in snow cover as a result of changes in precipitation and melting processes during spring and changes in cloud cover and soil moisture during summer. On the other hand, the warming during winter and fall appears to be caused by an increase in infrared heating of the surface.

We also analyzed temperature trends between 1990 and 2005 on a monthly basis and found that January, May, and July experienced the largest warming rates

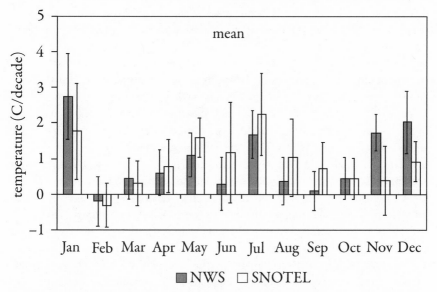

FIGURE 6.7 Linear regression for each month in daily mean temperature (°C/decade), standard deviation. © Regents of the University of Colorado.

overall (figure 6.7). The four months that showed the largest warming rates at the NWS sites, in descending order, were January, December, November, and July; at the SNOTEL sites those were July, January, May, and June. In general, winter months have warmed at a higher rate at the NWS sites, while summer months warmed more at the SNOTEL sites. At the NWS sites, T_{min} increased more than T_{max} in January; this was reversed from May through July. Greater increases in T_{max} also occurred at the SNOTEL sites during the summer. We found a conspicuous cooling trend in February at both the SNOTEL and NWS sites.

SUMMARY

Our analysis suggests that the San Juan Mountains warmed by about 1°C (1.8°F) between 1910 and 2009 and that almost all of this warming occurred between 1994 and 2005. However, we also found a gradual long-term warming trend in nighttime (minimum) temperatures between 1910 and 1994, which is somewhat offset by a slight decrease in daytime (maximum) temperatures. The general warming pattern in the San Juans during the twentieth century is similar to the pattern observed both in Colorado and globally. Although the late-twentieth-century warming trend observed globally began later in the San Juans, it occurred much more rapidly, with large increases in both maximum and minimum temperatures, between 1994 and 2005. Temperatures have decreased somewhat since then. Will the increased rate of warming in the San Juans continue as a result of some of the positive feedbacks mentioned earlier in this chapter, or will it become more consistent with the global rate of change?

The minimum temperature trends show a more continuous and robust warming signal since 1994. Minimum temperature is less strongly affected by inter-annual precipitation variability than is the maximum temperature. However, the recent drier years have experienced much greater positive extremes in maximum temperatures than was the case in the past. These daytime extreme temperatures in water-stressed years may have significant ecological impacts.

The temperature trends in the San Juans have been consistent with those shown by other recent studies: that the mountain region of the interior southwestern United States warmed at one of the highest rates in the contiguous United States in the early twenty-first century. This warming may be accompanied by drying trends in the associated river basins and modifications of climate type at higher elevations. The recent warming trend may be associated, in part, with regional circulation and precipitation changes. Further research is required to understand the causal mechanisms and to determine the relative contributions of natural variability and anthropogenic forcing to these recent temperature increases. Analysis of changes in precipitation, snow cover, cloud cover, and specific humidity during this period will provide useful insights to help predict whether these recent warming trends will continue during the coming decades. Climate models project an additional warming of 2 to 4°C (3.5 to 7°F) by the mid-twenty-first century in the region. Warming of this magnitude will have significant implications for water resources. Without significant changes in precipitation in the Colorado River and Rio Grande Basins, there will be more severe water-supply shortages than in the past.

REFERENCES

Daly, C., M. Halbleib, J. J. Smith, W. P. Gibson, M. K. Dogget, G. H. Taylor, J. Curtis, and P. A. Pasteris. 2008. Physiographically-Sensitive Mapping of Temperature and Precipitation across the Conterminous United States. *International Journal of Climatology* 28:2031–2064.

Julander, R. P., J. Curtis, and A. Beard. 2007. The SNOTEL Temperature Dataset. *Mountain Views Newsletter,* available at http://www.fs.fed.us/psw/cirmount/.

Keen, R. A. 1996. Weather and Climate. In R. Blair, ed., *The Western San Juan Mountains: Their Geology, Ecology, and Human History*, 113–126. Niwot: University Press of Colorado.

Rangwala, I., and J. R. Miller. 2010. Twentieth Century Temperature Trends in Colorado's San Juan Mountains. *Arctic, Antarctic and Alpine Research* 42(1):89–97.

Ray, A. J., J. J. Barsugli, and K. B. Averyt. 2008. The Observed Record of Colorado Climate. In *Climate Change in Colorado*. Report for the Colorado Water Conservation Board, 5–15. Boulder: University of Colorado.

APPENDIX

We used observations from twenty-five NWS stations (see table 6.1 for the station list) from 1950 to 2005. A homogenized dataset was available for the period prior to 1950 at only six NWS stations in our study region—Del Norte, Telluride, Manassa, Saguache, Montrose, and Hermit. Therefore, we used these six stations to estimate temperature trends between 1910 and 1949. Our other source of observations since the mid-1980s was provided by the twenty-three higher-elevation SNOTEL stations (see table 6.2 for the station list) managed by the National Resources Conservation Service. Since these stations have been used primarily to measure snow-water equivalents, we performed quality control by examining the daily record for erroneous values and omitting those in our calculations. Furthermore, we used PRISM data corresponding to the twenty-five NWS station locations to extend our analysis to 2009. PRISM is a statistical technique used to generate climate maps from point observations. PRISM data from a grid corresponding to a weather station should closely capture that station's observed record. Moreover, the PRISM process tends to remove inhomogeneities in the climate record, because it tends to conform the climate trend at a particular location to those of its neighboring locations.

Table 6.1. NWS stations list

Name	Elevation	Duration
Alamosa WSO AP (050130)	7,540	1950–2005
Center 4 SSW (051458)	7,670	1950–2005
Cortez (051885)	6,180	1950–2005
Del Norte (052184)	7,880	1895–2005
Durango (052432)	6,600	1950–1990
Fort Lewis (053016)	7,600	1950–2005
Hermit 7 ESE (053951)	9,000	1895–2005
Ignacio IN (054250)	6,460	1950–1993
Lake City (054734)	8,880	1950–2005
Lemon Dam (054934)	8,090	1982–2005
Manassa (055322)	7,710	1895–2005
Mesa Verde National Park (055531)	7,070	1950–2005
Monte Vista (055706)	7,660	1950–2005
Montrose (055722)	5,785	1895–2005
Norwood (056012)	7,020	1950–2005
Ouray (056203)	7,840	1950–2005

continued on next page

Table 6.1—*continued*

Name	Elevation	Duration
Pagosa Spring (056258)	7,100	1950–2005
Rico (057017)	8,850	1959–2001
Ridgway (057020)	7,000	1982–2005
Saguache (057337)	7,690	1895–2005
Silverton (057656)	9,330	1950–2005
Telluride (058204)	8,770	1895–2005
Vallecito Dam (058582)	7,650	1950–2005
Wolf Creek Pass 1 E (059181)	10,640	1958–2001
Yellow Jacket 2 W (059275)	6,860	1962–2002

Note: USHCN v2 homogenized dataset was used for the six shaded stations.

Table 6.2. SNOTEL stations list.

No.	Name	Elevation (ft.)	Duration
1	Beartown	11,600	1984–2005
2	Cascade	8,880	1987–2005
3	Cascade #2	8,920	1991–2005
4	Columbine Pass	9,400	1987–2005
5	Columbus Basin	10,785	1995–2005
6	El Diente Peak	10,000	1987–2005
7	Idarado	9,800	1987–2005
8	Lily Pond	11,000	1984–2005
9	Lizard Head Pass	10,200	1986–2005
10	Lone Cone	9,600	1987–2005
11	Mancos	10,000	1997–2005
12	Middle Creek	11,250	1984–2005
13	Mineral Creek	10,040	1987–2005
14	Molas Lake	10,500	1987–2005
15	Red Mountain Pass	11,200	1986–2005
16	Scotch Creek	9,100	1987–2005
17	Slumgullion	11,440	1984–2005
18	Spud Mountain	10,660	1987–2005
19	Stump Lakes	11,200	1987–2005
20	Upper Rio Grande	9,420	1987–2005
21	Upper San Juan	10,200	1984–2005
22	Vallecito	10,880	1987–2005
23	Wolf Creek Summit	11,000	1987–2005

Part 2
Biological Communities of the San Juan Mountains

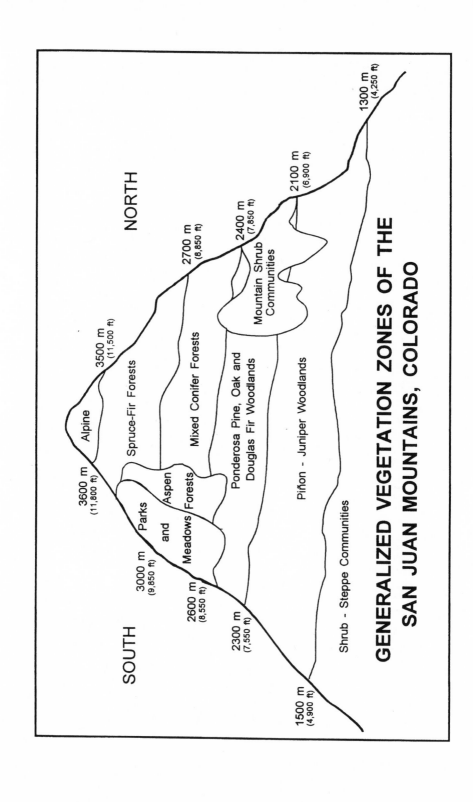

GENERALIZED VEGETATION ZONES OF THE SAN JUAN MOUNTAINS, COLORADO

NORTH

SOUTH

1300 m (4,250 ft)

2100 m (6,900 ft)

2400 m (7,850 ft)

2700 m (8,850 ft)

3500 m (11,500 ft)

3600 m (11,800 ft)

3000 m (9,850 ft)

2600 m (8,550 ft)

2300 m (7,550 ft)

1500 m (4,900 ft)

Alpine

Spruce-Fir Forests

Mixed Conifer Forests

Ponderosa Pine, Oak and Douglas Fir Woodlands

Mountain Shrub Communities

Piñon - Juniper Woodlands

Shrub - Steppe Communities

Parks and Meadows

Aspen Forests

CHAPTER SEVEN

Mountain Lakes and Reservoirs

Koren Nydick

E COLOGICAL PROCESSES AND SPECIES OCCURRENCES within most ecosystems in the semiarid western United States are limited, in some way or another, by moisture. Aquatic ecosystems are obvious exceptions. *The Western San Juan Mountains* focused on flowing (or *lotic*) water—rivers, streams, and their associated riparian wetlands. In this volume, we devote a chapter to "still" (or *lentic*) water and itsecosystems, specifically lakes and reservoirs, and a second chapter to fen wetlands. This chapter investigates the physical, chemical, and biological nature of San Juan lakes and reservoirs and discuss human impacts on these ecosystems.

Mountain lakes and reservoirs are valued for the many "ecosystem services" they provide to society. Our enchantment with mountain lakes was epitomized in John Denver's song "Rocky Mountain High" with the lyrics, "His sight has turned inside himself to try and understand the serenity of a clear blue mountain lake." Visitors surveyed in Rocky Mountain National Park in northern Colorado selected water as critical to their enjoyment of the outdoors, ranking water second only to mountain vistas (Taylor et al. 1996). Lakes and reservoirs are favorite recreation destinations for hiking, fishing, and boating (although many fisheries in mountain lakes are stocked artificially, as discussed later in this chapter). Lakes and reservoirs play other vital, often hidden roles. They provide drinking water and irrigation and are sources of electricity and power generation. They also absorb rainfall and runoff from land, help prevent floods, store pollutants that would otherwise travel downstream, provide habitat for wildlife, and contribute to biodiversity.

There are approximately 630 water bodies in the San Juan Mountains, excluding La Garita Mountains and the South San Juans. These include remote alpine lakes perched high in glacial cirques (plate 25) to subalpine lakes nestled in spruce-fir forests to a variety of human-made reservoirs. Many natural lakes in the San Juan Mountains were formed following glacial activity that carved out depressions or dammed valleys with glacial moraines. A few, such as Little Molas Lake, are a result of karst topography, where water has dissolved limestone bedrock. Alpine and subalpine lakes are generally chemically dilute and nutrient-poor, and they support less biological productivity than lower-elevation lakes and reservoirs. Organisms native to these ecosystems have evolved to be efficient at taking up low concentrations of nutrients, especially nitrogen (N) and phosphorus (P).

Many lakes remain relatively "pristine," but others are impacted by development, recreation, air pollution, fish stocking, and historic mining activities. Reservoirs, such as Vallecito and McPhee, are obvious in their human origin, but other, less apparent hydrological alterations have formed or augmented lakes. For example, Emerald Lake, at over 10,000 ft (3,048 m) near present-day Vallecito Reservoir, was dammed by settlers who used the water for irrigation and later blew up the dam during a drought (Warlick 2003). Lake Hope, located between Silverton and Telluride at 11,870 ft (3,618 m) elevation, stores water for the Ames Power Plant and is used to maintain constant water levels for Trout Lake. Very little research has investigated the impacts of these kinds of human activities in the San Juan Mountains.

COMPLEXITY AND VARIABILITY

While the following sections explore the physical, chemical, and biological characteristics of mountain lakes and reservoirs, these components are inextricably linked and influence each other in complex ways. External factors, such as climate, watershed characteristics, and atmospheric deposition, determine chemical inputs into lakes. The water chemistry influences the biological productivity and types of species that live in the lake. In turn, internal factors, such as lake morphology and food-web composition, influence the way these chemical inputs are processed, incorporated into biota, and transformed within the water body. Species composition and biological productivity regulate excretion and sedimentation, which release nutrients back into the water column.

In addition, the physical, chemical, and biological nature of lakes is extremely variable. Lakes vary physically in terms of light levels, temperature, and water currents and differ chemically in terms of nutrients, major ions, and contaminants. They contrast biologically in terms of both structure (including the food web, biomass of organisms, and types of species) and function (processes such as photosynthesis, nitrogen fixation, and decomposition). There is a great deal of spatial heterogeneity in all of these components, which produces variability both within and among lakes. Physical, chemical, and biological aspects of lakes also vary considerably over time, some quickly and others slowly.

During the summer of 2007, the Mountain Studies Institute (MSI) conducted a synoptic survey investigating twenty-eight lakes and reservoirs in the San Juan Mountain region (table 7.1). The purpose of this survey was to investigate mercury concentrations in zooplankton, but many other parameters were measured as well. These data, as well as the results of other investigations, are discussed later in this chapter. In many instances, studies that address a specific question have not been completed in the San Juans; in this case, I include findings from other mountain regions.

WATERSHEDS AND LAKE CHEMISTRY

In many respects, a lake is a reflection of its watershed. In Rocky Mountain watersheds, vegetation, soil fertility, steepness of slopes, and the amount and duration of snowpack are important controls on water chemistry (Brown, Wurtsbaugh, and Nydick 2008; Campbell et al. 2000; Clow and Sueker 2000). Steep, rocky alpine terrain has a lower capacity to retain and modify geologic and atmospheric chemical inputs than do more fertile, lower-elevation ecosystems (Baron et al. 2000). Therefore, alpine lakes are often more sensitive to impacts of air pollution than lower-elevation lakes. As vegetation and soil development increase below treeline, however, contributions of organic matter rise, and the retention and modification of these inputs increase. For example, alpine streams and lakes generally have higher concentrations of the ion nitrate (NO_3^-) than those at lower elevations, but the opposite is true for concentrations of dissolved organic nitrogen and carbon (DON and DOC) (Hood, Williams, and Caine 2003).

Snowpack insulates soil from subfreezing temperatures and allows microbial activity to continue during winter. In the spring, melting snow flushes dissolved constituents (some of which are products of microbial processing) out of soils and groundwater into lakes and streams. In years with deeper snowpacks, more nutrients and other chemicals are transported into lakes during spring snowmelt than in years with shallow snowpacks (Williams et al. 1996). The nutrients flushed into mountain lakes with the snowmelt pulse are critical to support the spring and summer food webs in these nutrient-poor ecosystems. Additional factors that influence water chemistry include the ratio of watershed area to lake area, the amount of flushing or mixing of the lake, and the landscape position of the lake in relation to wetlands and upstream lakes.

Geology is yet another critical influence on water quality, and the diversity of rock types in the San Juan Mountains produces a great deal of chemical variability. Resistant rocks, such as granite, gneiss, and quartzite and well-cemented volcanic tuffs, weather slowly, resulting in lakes that are relatively dilute, unproductive, and

Table 7.1. Minimum, maximum, and average values for environmental variables measured for twenty-eight lakes and reservoirs in the San Juan Mountain region.

	Unit	Minimum	Maximum	Average
Sample date		7/23/2007	9/02/2007	8/07/2007
Elevation	feet	6,683	13,100	11,177
Max depth	meters	3.0	53.3	15.4
Secchi depth (SD)	meters	1.4	16	5.7
Specific conductivity	$\mu S/cm^2$	8.6	1,137	125
Stratification		yes = 15	partial = 4	no = 9
pH		5.6	8.4	7.2
SO4	$\mu g/L$	375	648	39
Cl	$\mu g/L$	96	16	1.2
NO_3-N	$\mu g/L$	0	306	73
PO_4-P	$\mu g/L$	1	9	2
NH_4-N	$\mu g/L$	3	41	11
TDN—total dissolved N	$\mu g/L$	67	446	192
DON—dissolved organic N	$\mu g/L$	32	405	109
TN—total N	$\mu g/L$	67	474	217
TDP—total dissolved P	$\mu g/L$	1	10	3
TP—total P	$\mu g/L$	1	15	6
DOC—dissolved organic carbon	mg/L	1.5	6.8	2.7
Chlorophyll a (epilimnion)	$\mu g/L$	0.3	6.4	2.0
Chlorophyll a (metalimnion)	$\mu g/L$	0.3	19.0	4.1
Chlorophyll a (hypolimnion)	$\mu g/L$	0.7	96.1	10.5
Methyl Hg in zooplankton (> 300 μm)	$\mu g/L$	7.0	185	56
Total Hg in zooplankton (> 300 μm)	$\mu g/L$	36	806	154
Trophic State Index (SD)		20	56	38
Trophic State Index (Chlorophyll a)		19	49	34
Trophic State Index (TP)		0.9	43	27
DIN:TP—dissolved inorganic N to TP		0.3	209	30

Source: Data from Nydick 2010, unpublished data.

poorly buffered against acidification. Sedimentary rocks weather more easily than many other rock types and contribute more particulate and dissolved materials to lakes. Water bodies in sedimentary lithologies, especially limestone or calcite, have higher acid-neutralizing capacity and are sometimes more productive than lakes situated in metamorphic or igneous parent rock. Weathering of some igneous rocks in the San Juan Mountain mineralized belt produces metal-rich, acidic water from the dissolution of pyrite (Bove et al. 2007). Some lakes in the San Juan mineralized belt are termed "dead" because they support little life other than microbes. Some of these lakes, such as Silver Lake (near Silverton) and Kite Lake, are impacted by mine tailings as well.

While watershed properties influence the chemistry of water entering a lake, biological activity in the lake can alter its chemical composition. Lakes absorb and attenuate the snowmelt pulse of solutes flushed into them (Kling et al. 2000; Wurtsbaugh et al. 2005). Some materials are processed biologically, and products are deposited on lake bottoms or transported out of the lakes and downstream over time. In general, biological activity in lakes acts to reduce amounts of inorganic solutes, increase organic components, and buffer acidity (i.e., produce alkalinity and raise pH) (Brown, Wurtsbaugh, and Nydick 2008).

The 2007 MSI lake survey demonstrated high variability in water chemistry among lakes and reservoirs (table 7.1). Miramonte Reservoir was by far the least dilute water body sampled and had the highest specific conductivity and sulfate and chloride concentrations. It also had the highest levels of DOC and DON and very low water clarity. Interestingly, nitrate, phosphate, and chlorophyll a (index of algal biomass, discussed later) were not very high. On the other end of the spectrum were lakes, such as Snowdon Lake, that were very dilute, with specific conductivity as low as 9 $\mu S/cm^2$.

The US Forest Service and US Geological Survey monitor the inorganic water chemistry of high-elevation lakes in the Weminuche, La Garita, and South San Juan Wilderness Areas during the summer. This monitoring program is aimed at detecting impacts of changing air quality (namely N and sulfur deposition and acid rain) on sensitive lakes in Class I Federal Areas, which have special protection status under the Clean Air Act. In Rocky Mountain National Park, atmospheric N deposition from coal-burning power plants, motorized vehicles, and animal operations is implicated in changing the water chemistry (namely increasing nitrate, NO_3^-) and algal composition of alpine lakes on the east side of the Continental Divide (Baron et al. 2000). Indeed, atmospheric N deposition is thought to have caused N enrichment and eutrophication (increased algal biomass) of lakes near industrialized areas in the Northern Hemisphere (Bergström and Jansson 2006). Such changes have not been observed in the San Juan Mountains, although one may argue that this question has been inadequately addressed. About fifteen years of data exist for some of the approximately twenty-two wilderness lakes sampled every summer in the San Juans, but temporal trends in the data have not been explored.

While changes over time have yet to be determined, Musselman and Slauson (2004) investigated spatial differences in water chemistry using data collected from high-elevation lakes across Colorado in 1995. Although there was much variability within each region and overlap among regions, nitrate concentrations in San Juan lakes were found to be lower on average than in lakes in Colorado's Front Range. Similarly, in a 2006 survey, nitrate in alpine and subalpine lakes of the San Juan Mountains and the Elk Mountains, to the north, was found to be lower on average than in lakes in the Front Range (Elser at al. 2008). Average nitrate concentrations in alpine and subalpine lakes measured by MSI in 2007 were slightly higher than those found by Elser and others; high nitrate concentrations were present in several of the highest lakes, including Snowdon Lake, discussed earlier (Nydick 2010). Acid-neutralizing capacities of alpine and subalpine lakes in the San Juans are generally very low, with the exception of those located on limestone. Musselman and Slauson (2004) found that 73 percent of the wilderness lakes were sensitive to acidification, with acid-neutralizing capacity less than 200 µeq/L.

PHYSICAL CHARACTERISTICS AND STRATIFIED LAKES

Light, temperature, and wind mixing are key lake properties, and their interaction influences a water body's chemistry and biology. The water column of mountain lakes is often very clear because there are low amounts of light-absorbing, dissolved organic substances and suspended materials. As nutrient input and erosion from the watershed above the lake increase, water clarity decreases and light cannot penetrate as far.

The amount of light penetrating into the water column is an important control on temperature and photosynthesis. Photosynthesis by algae and attached plants converts sunlight and carbon dioxide into the food that supports much of the food web and also produces much of the dissolved oxygen in the water. Available light levels determine the depth of the euphotic zone, the portion of the lake where algae and plants can grow. The bottom of the euphotic zone occurs at a depth where the amount of light available is only about 1 percent of the light that hits the surface of the lake. The euphotic zone is measured by lowering a light meter through the water column, but it can also be estimated by lowering a black-and-white disk, called a Secchi Disk, until it disappears from view. The euphotic-zone depth is two to three times the "Secchi Depth," which is also a useful measurement of water clarity. The 2007 MSI lake study measured Secchi Depths ranging from 4.6 ft (1.4 m) in Woods Lake (a montane reservoir at 9,423 ft [2,762 m] altitude, near Fall Creek in San Miguel County) to 52.5 ft (16.0 m) in Hope Lake (an alpine lake at 11,867 ft [3,617 m] altitude, near Ophir in San Miguel County), with an average of 18.7 ft (5.7 m) (Nydick 2010). Some very clear lakes, such as Lake Tahoe in the Sierra Nevadas, have Secchi Depths of 130+ ft (40+ m). Alpine lakes below glaciers (tarns) can have reduced water clarity as a result of "rock flour" produced by the glaciers. In the San Juans, no aboveground glaciers have existed for thousands of years, but rock dust from underground rock glaciers and talus fields may reduce water clarity in some lakes. Snowdon Lake is an example. It has very dilute chemistry and low algal productivity, but its Secchi Depth was only 11.8 ft (3.6 m). Its aqua color results from the reflection of light off of the fine rock particles.

Many lakes in temperate environments (such as in the San Juans) tend to stratify, which means they separate into distinct depth layers based on the density of the water. Density differences are driven mostly by changes in temperature, but the amount of dissolved chemicals can also play a role. In winter, the water at the bottom of the lake is about 4°C (39.2°F), the temperature at which water is the densest. Water above the bottom layer will be colder until it reaches 0°C (32°F) and forms an ice layer at the surface. In spring, when the ice melts and surface water warms to 4°C, very little wind is required to completely mix the lake. This mixing is referred to as *spring turnover*. As the surface water continues to warm, it becomes less dense than the colder water below. Eventually, the density difference becomes too great for winds to mix the lake completely. The *epilimnion*, a well-mixed warm layer, forms on top, while the *hypolimnion*, a colder, denser layer, occurs at the bottom. In between is the *metalimnion*, a zone where temperature decreases rapidly with depth.

The deeper lakes in the San Juan Mountains become stratified in the summer, while the shallower lakes often do not. The MSI lake survey recorded fifteen, four, and nine lakes that were stratified, partially stratified, and unstratified, respectively. The average maximum depth of stratified lakes was 74.5 ft (22.7 m), compared with 19 ft (5.8 m) for unstratified lakes.

Highland Mary Lake, a deep alpine lake in San Juan County, is an example of a stratified lake (figure 7.1). Summer water temperature in the top 13 ft (4 m) is 12 to 13°C (53.6 to 55.4°F). Below 13 ft, temperature cools rapidly until it reaches 4°C (39.2°F) at 26 ft (8 m). In the epilimnion, the concentration of dissolved oxygen is high (> 10 mg/L) because of wind mixing and photosynthesis. Interestingly, dissolved oxygen increases significantly in the metalimnion and actually becomes super-saturated. This phenomenon is likely caused by an increase in algal biomass and photosynthesis, which produces oxygen at mid-depths and is termed a *deep chlorophyll maximum*. Indeed, the concentration of chlorophyll *a* (a photosynthetic pigment used as an index of algal biomass) in the metalimnion of Highland Mary Lake was about twice that in both the epilimnion and the hypolimnion. Near the bottom of the lake, dissolved oxygen concentration decreases to 4 ppm (4 mg/L). The Secchi Depth of this lake was 22 ft (6.7 m), which suggests that the euphotic zone ends somewhere between 42.6 and 65.6 ft (13 and 20 m). Below the euphotic zone, dissolved oxygen decreases, as it is used by decomposition and respiration and not replenished by photosynthesis. Anoxic (i.e., no oxygen) conditions in hypolimnetic waters promote chemically reducing reactions, including the release of phosphorus and heavy metals stored in the sediments. Anoxic conditions also promote methylation of mercury by bacteria. Fish require well-oxygenated water and cannot live in anoxic bottom waters of stratified lakes or reservoirs.

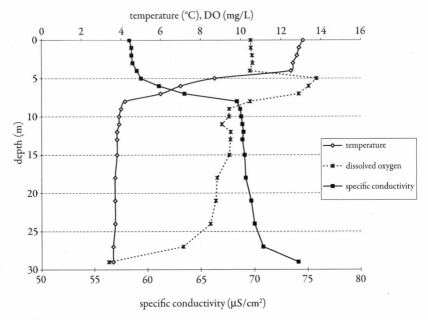

temperature (°C), DO (mg/L)

specific conductivity (μS/cm²)

FIGURE 7.1 Depth profile of temperature, dissolved oxygen (DO), and specific conductivity for Highland Mary Lake on July 20, 2007.

Wildfire and subsequent soil erosion can have large impacts on water bodies. The Missionary Ridge Fire in June and July 2002 and the Bear Creek Fire in August 2003 burned large areas of forest above Vallecito Reservoir. Large quantities of sediment were carried by runoff into the reservoir, which was at very low levels as a result of sustained drought. In August 2003, thousands of fish were killed. Dissolved oxygen levels were very low because of the huge input of extra organic matter, and they remained at reduced levels through 2006. Iron, manganese, and ammonia concentrations increased after the fire, sometimes to levels above toxicity standards for human consumption of fish (Wright 2008).

BIOLOGICAL PRODUCTIVITY AND FOOD WEBS

Since mountain lakes are typically relatively dilute and nutrient-poor, they tend to have low biological productivity, a *trophic state*. They have relatively simple food webs but can also have high diversity among short-lived species at the base of the food webs—that is, primary producers (algae), decomposers (bacteria and fungi), and primary consumers (zooplankton and other invertebrates). For example, hundreds of species of phytoplankton algae may coexist in a lake during the course of a year because each has a competitive niche related to water temperature, light levels, grazing avoidance, and present nutrients, such as nitrogen, phosphorus, and silica. Conditions fluctuate greatly, affecting all of these niches.

Chlorophyll *a* of phytoplankton (the algae that live in the water column), as well as the Secchi Depth and the concentration of total phosphorus (TP), can be used as indexes of *trophic state* (Carlson and Simpson 1996). Waters with high water clarity, low productivity, and low nutrient concentrations are termed *oligotrophic* (chlorophyll *a* < 1 µg/L). High-productivity lakes are called *eutrophic* (chlorophyll *a* > 7.3 µg/L), and those in between are *mesotrophic*.

Chlorophyll *a* in the epilimnion of lakes sampled by the MSI in 2007 ranged from 0.3 to 6.8 µg/L, with an average of 2.0 µg/L (table 7.1). Six of twenty-eight lakes had prominent deep chlorophyll maxima. As expected, most lakes had low nutrients and productivity. Using the Trophic State Indexes (Carlson and Simpson 1996), four lakes were classified as ultra-oligotrophic, six as oligotrophic, fourteen as oligo-mesotrophic, and three as mesotrophic. Elevation was not a good predictor of epilimnetic chlorophyll *a*, however, as some alpine lakes, such as Fuller and Clear, had chlorophyll *a* greater than 4 µg/L. In contrast, Trout Lake, a 9,710-ft (2,960-m)-high reservoir, had only 0.9 µg/L of chlorophyll *a*. Trophic state, therefore, varied somewhat unpredictably with elevation, although mesotrophic lakes were all below 10,000 ft (3,048 m) and ultra-oligotrophic lakes were located above 11,800 ft (3,597 m). TP concentrations explained 33 percent of the variability in chlorophyll *a* concentrations across the lakes, but total N explained much less. Ratios of DIN:TP* also suggest that P regulated algal growth more than N did. We classified sixteen lakes as having phytoplankton limited by P, three by N, and nine by some interaction of N and P.

Why don't nutrient levels correlate better with chlorophyll *a* or trophic status? Until now, the discussion has only included bottom-up control of algal biomass by nutrients and light. Top-down control, the "eating" of primary producers by herbivorous animals, can diminish the correlation between nutrient concentrations and algal biomass. Zooplankton are small invertebrates that feed mainly on algae, bacteria, and detritus (figures 7.2 and 7.3). A large biomass of zooplankton can consume enough phytoplankton such that even in a relatively nutrient-rich lake with high rates of photosynthesis and algal growth, there is little algal biomass. On the other hand, if there are few zooplankton to graze on phytoplankton, algal biomass can be higher than expected, even with relatively low nutrient input.

The presence of zooplankton-eating fish (*planktivores*) can reduce the biomass of zooplankton and shift the community to favor small-bodied species that can more easily escape capture. Similarly, predatory fish feed on planktivorous fish and can control their biomass. Therefore, the number of links (or *trophic levels*) in a food

---

* DIN:TP is the ratio of dissolved inorganic N ($NO_3^- + NH_4^+$) to total P, a common index used to estimate the nutrient most limiting to phytoplankton growth in lakes (Morris and Lewis 1988); N limitation was assigned to ratios < 1 and P limitation to ratios > 4.

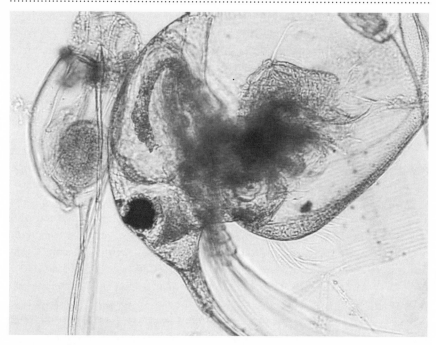

FIGURE 7.2 *Bosmina* cladoceran zooplankton from Highland Mary Lake, with a rotifer zooplankton shown on the left. Diatoms (a type of algae that makes intricate testes out of silica) are visible in the lower right. Magnification = 100x.

web is important in determining the strength of top-down control of algal biomass. In reality, however, food webs are much more complex. Many organisms are omnivorous and not necessarily characterized by a particular trophic level. Furthermore, fish in particular can shift levels throughout their life cycle—starting on algae and detritus, then moving on to zooplankton, and eventually feeding on other fish as adults.

The food webs of lakes also include species that inhabit the lake bottom, or *benthic zone*. Shallow areas receive enough light to allow growth of algae on sediment, downed logs, rocks, and rooted plants called *macrophytes*. Invertebrates, including insects, amphipods, mollusks, and zooplankton, feed on this algae. Benthic feeders also eat the dead organisms and fecal wastes that drop to the bottom of the lake.

Boreal toads were known to breed in shallow habitat of lakes and ponds (also marshes, fens, and wet meadows), at elevations of 7,500 to 12,000 ft (2,286 to 3,658 m), and had a historical range from southern Wyoming to northern New Mexico, including the San Juan Mountains. Their populations plummeted in the 1970s–1980s, and, by 1992, boreal toads were absent from 83 percent of their historical range in Colorado, including most areas in the San Juan Mountains. The boreal toad is an endangered species listed by both Colorado and New Mexico and is a

FIGURE 7.3 Copepod zooplankton from White Dome Lake. The red pigment is likely for protection from UV radiation.

protected species in Wyoming. Infection by the chytrid fungus *Batrachochytrium dendrobatidis* is thought to be the main culprit, although habitat loss, UV exposure, and non-native species introductions have also been investigated (Loeffler 2001). The fungus can be spread by humans and their vehicles, and care should be taken to avoid transporting sediments and vegetation among possible breeding habitats.

FISH IN MOUNTAIN LAKES AND RESERVOIRS

Many high mountain lakes were naturally fishless because of waterfalls and other obstacles that blocked their migration. Low productivity and poor spawning habitat made growth and reproduction difficult for fish that did make the trip. Settlers in the San Juan Mountains began stocking some of these high lakes in the late 1800s. For example, a fishery was developed above the current location of Vallecito Reservoir (Warlick 2003). Official government stocking efforts began shortly thereafter throughout the western mountains. As a result, fish have been stocked in about 60 percent of mountain lakes in the western United States (Bahls 1992). In the San Juans, stocking of rainbow trout, brook trout, and various subspecies of cutthroat trout (i.e., Colorado River, Pikes Peak, Rio Grande, Snake River, and Yellowstone) occurs today in many high mountain lakes (CDOW 2007). The ethics of stocking fish in naturally fishless lakes has been questioned, however, and there have been some efforts to remove fish from formerly fishless lakes (Eby et al. 2006; Pister 2001). The stocking of native Colorado River cutthroat trout has continued in an effort to reestablish populations that have drastically declined as a result of competition from non-native trout (CRCT Coordination Team 2006).

The effects of fish stocking have not been studied in San Juan Mountain lakes. Research in other mountain ranges has shown that trout introductions to fishless lakes typically cause major species shifts and can result in the loss of native species, such as amphibians, bottom-dwelling invertebrates (including larval insects), and large zooplankton (Knapp et al. 2005), but not likely boreal toads, as discussed earlier. In a fishless lake, large-bodied zooplankton species dominate and can graze phytoplankton to low levels. Stocked fish feed selectively on the larger, more visible zooplankton, however, allowing smaller-bodied zooplankton to increase in abundance and to coexist with otherwise competitively dominant large species. Suppression of large zooplankton can stimulate phytoplankton growth by reducing overall grazing pressure. Trout also transfer benthic nutrients to the water column by feeding on bottom-dwelling insects and excreting waste, particularly P, into the water column. Overall, the stocking of naturally fishless lakes tends to increase algae biomass (Schindler, Knapp, and Leavitt 2001).

Most, if not all, reservoirs in the San Juan Mountains have been stocked with fish. Water levels fluctuate seasonally in many reservoirs, which makes it difficult for plant and invertebrate communities to establish in the shallow shoreline areas. As a result, the food webs that support game fish are based primarily on zooplankton, particularly species of the crustacean cladoceran genus *Daphnia*, which inhabit open water. *Daphnia* abundance depends on a combination of nutrient status, phytoplankton production, water temperatures, and the abundance of fish and invertebrates that feed on them. Fishery managers must carefully balance the stocking of predator and prey fishes to maintain the zooplankton food base.

Trout stocked in San Juan reservoirs predominantly include rainbow and occasionally brook or brown trout, plus several subspecies of cutthroat trout. Additional species stocked in the larger reservoirs include kokanee salmon (a cold-water species), walleye and northern pike (cool-water species), and several warm-water species, such as smallmouth and largemouth bass, bluegill, black crappie, channel catfish, and green sunfish (CDOW Database 2007). Kokanee salmon are an economically important game species in Colorado cold-water reservoirs. The CDOW stocks kokanee fingerlings in Vallecito, McPhee, and Blue Mesa Reservoirs. In the fall, three- to four-year-old adult kokanee travel upstream, out of the reservoirs to spawn. Kokanee have not been able to reproduce in the wild in Colorado, however, so sport fishing depends on obtaining eggs for hatcheries. Since the early 1990s, Blue Mesa Reservoir just northeast of the San Juan Mountains has been Colorado's primary source of kokanee eggs (Martinez 2008), but some eggs are also harvested from Vallecito and reared in the Durango Fish Hatchery (Jim White, CDOW aquatic biologist, personal communication, October 2008). Vallecito's kokanee population declined following wildfires in 2002 and 2003 but began to rebound by 2005 (Jim White, CDOW aquatic biologist, personal communication, October 2008).

In contrast to the abundant examples of adding fish to water bodies in the San Juans, the CDOW has proposed to eradicate non-native fish—mainly rainbow trout—from Woods Lake, a small reservoir on the Uncompahgre National Forest. These fish were stocked as recently as 2000. The project aims to establish a Colorado River cutthroat trout population once the rainbow trout are removed (Schutzah 2008). Currently, invasive zebra and quagga mussels are not found in San Juan lakes and reservoirs, but populations have been discovered in a few reservoirs in northern Colorado (CDOW Website 2008).

MERCURY

Recently, the issue of mercury pollution has come to the forefront of concern. All reservoirs in the southwest-most corner of Colorado and the northwest portion of New Mexico that have been tested have been put on fish-consumption advisories for mercury (CDPHE 2008; NMED 2008). Mercury is a naturally occurring element, but it is also a well-know neurotoxin. Fetuses and the young are the most sensitive to its toxic effects. Fish advisories warn people to limit their consumption of specific game fish species, which vary by water body. The advisories are based on fish tissue analyses at concentrations above the limit of 0.5 μg/g. Mercury bio-accumulates in food webs, meaning concentrations increase with increasing trophic levels, so the fish species of concern to human health are predatory game fish. While limiting the amount of mercury entering a lake or reservoir is certainly important in controlling mercury levels in fish, other factors—such as high organic matter, low productivity, low pH, and low oxygen levels—play a critical role as well. Methyl mercury is the form of mercury that "sticks" best in tissues. It is produced by methylating bacteria in anoxic environments, including wetland soils and the bottoms of lakes.

Measuring mercury in game fish is directly tied to human health criteria for consumption of these fish, but mercury in zooplankton can serve as an index of mercury incorporation into the food web. While game fish may provide a more direct measure of mercury impact, collecting and analyzing fish are costly and time-consuming endeavors. In addition, many lakes do not have game fish, or even any fish. Thus, zooplankton can serve as a more cost-effective method that can be used in a variety of lakes and resurveyed every year as a long-term monitoring tool. This kind of survey can also point to factors that make certain lakes more sensitive to bio-accumulation of methyl mercury. Initial data collected by MSI show high variability in the medium to large zooplankton (> 300 μm) sampled. Total mercury concentrations ranged from 36 to 806 μm (average of 154), and methyl mercury concentrations ranged from 7 to 185 μm (average of 56). Despite low levels of organic matter in the San Juan lakes, these mercury levels were similar to those found in zooplankton of lakes in eastern North America and Europe (Driscoll et al. 2007; Garcia, Carigan, and Lean 2007; Monson and Brezonik 1998).

Studies of McPhee and Narraguinnep Reservoirs, whose headwaters are in the Rico area, indicate a variety of sources for the mercury, including past mining activity, atmospheric deposition, and naturally occurring background in local rocks and soils (CDPHE 2003). The sources of mercury for Vallecito Reservoir, which has its headwaters in the Weminuche Wilderness, are under question and are less likely to include geology and mining as major sources. Vallecito was listed with a fish consumption advisory based on sampling in summer 2004, following wildfires in its watershed in 2002 and 2003 (CDPHE 2008; Wright 2008). Sampling of streambed sediments in Vallecito's watershed showed concentrations less than 0.09 μg/g, the average global crustal abundance for mercury. Mercury concentrations in precipitation and meteorological modeling suggest that atmospheric deposition originating in the Four Corners area may contribute half of the mercury (Wright 2008).

Sediment cores from several lakes and reservoirs in the western San Juans suggest that mercury inputs have increased notably since about 1960–1970 (Gray et al. 2005; Nydick 2010; Wright 2008). The timing of increased mercury fluxes for these sediment cores suggests a source of mercury emissions from Four Corners coal-fired power plants, the largest of which began operations in the late 1960s and early 1970s. Concentrations of mercury in precipitation for the Four Corners region are indeed high (MDN 2007; Nydick 2010; Wright 2008). The effects of the 2002 and 2003 wildfires in the Upper Pine River Watershed on mercury concentrations in Vallecito Reservoir are currently uncertain, but studies are ongoing.

REFERENCES

Bahls, P. 1992. The Status of Fish Populations and Management of High Mountain Lakes in the Western United States. *Northwest Science* 66:183–193.

Baron, J. S., H. M. Rueth, A. P. Wolfe, K. R. Nydick, E. J. Allstott, J. T. Minear, and B. Moraska. 2000. Ecosystem Responses to Nitrogen Deposition in the Colorado Front Range. *Ecosystems* 3:352–368.

Bergström, A.-K., and M. Jansson. 2006. Atmospheric Nitrogen Deposition Has Caused Nitrogen Enrichment and Eutrophication of Lakes in the Northern Hemisphere. *Global Change Biology* 12:635–643.

Bove, D. J., M. A. Mast, J. B. Dalton, W. G. Wright, and D. B. Yager. 2007. Major Styles of Mineralization and Hydrothermal Alteration and Related Soil- and Aqueous-Phase Geochemical Signatures. In S. E. Church, P. von Guerard, and S. E. Finger, eds., *Integrated Investigations of Environmental Effects of Historical Mining in the Animas River Watershed, San Juan County, Colorado*. Professional Paper 1651. Reston, VA: US Geological Survey, 151–230.

Brown, P. D., W. A. Wurtsbaugh, and K. R. Nydick. 2008. Lakes and Forests as Determinants of Downstream Nutrient Concentrations in Small Mountain Watersheds. *Arctic, Antarctic, and Alpine Research* 40(3):462–469.

Campbell, D. H., J. S. Baron, K. A. Tonnessen, P. D. Brooks, and P. F. Schuster. 2000. Controls on Nitrogen Flux in Alpine/Subalpine Watersheds in Colorado. *Water Resources Research* 36:37–47.

Carlson, R. E., and J. Simpson. 1996. *A Coordinator's Guide to Volunteer Lake Monitoring Methods.* North American Lake Management Society (no city given).

Clow, D. W., and J. K. Sueker. 2000. Relations between Basin Characteristics and Stream Water Chemistry in Alpine/Subalpine Basins in Rocky Mountain National Park, Colorado. *Water Resources Research* 36:49–61.

Colorado Department of Public Health and Environment (CDPHE). 2003. *Total Maximum Daily Load for Mercury in McPhee and Narraguinnep Reservoirs, Colorado: Phase I.* Denver: Water Quality Control Division.

———. 2008. Fish Consumption Advisories. Available at http://www.cdphe.state.co.us/wq/FishCon/analyses/. Accessed September 22, 2008.

———. 2007. Database (described in H. Vermillion 2008. *Aquatic Data Analysis—Federal Aid Project F-239R-15, Progress Report.* Fort Collins: Colorado Division of Wildlife).

———. 2008. http://wildlife.state.co.us/NewsMedia/PressReleases/Press.asp?PressId=5699. Accessed September 22, 2008.

CRCT Coordination Team. 2006. Conservation Strategy for Colorado River Cutthroat Trout (*Oncorhynchus clarkii pleuriticus*) in the States of Colorado, Utah, and Wyoming. Fort Collins: Colorado Division of Wildlife.

Driscoll, C. T., Y. Han, C. Y. Chen, D. C. Evers, K. F. Lambert, T. M. Holsen, N. C. Kamman, and R. K. Munson. 2007. Mercury Contamination in Forest and Freshwater Ecosystems in the Northeastern United States. *BioScience* 1:17–28.

Eby, L. A., W. J. Roach, L. B. Crowder, and J. A. Stanford. 2006. Effects of Stocking-Up Freshwater Food Webs. *TRENDS in Ecology and Evolution* 21(10):576–584; doi: 10.1016/j.tree.2006.06.016.

Elser, J. J., M. M. Kyle, L. Steger, K. R. Nydick, and J. S. Baron. 2008. Nutrient Availability and Phytoplankton Nutrient Limitation across a Gradient of Atmospheric Nitrogen Deposition, Colorado, USA. Unpublished manuscript.

Garcia, E., R. Carigan, and D.R.S. Lean. 2007. Seasonal and Interannual Variations in Methyl Mercury Concentrations in Zooplankton from Boreal Lakes Impacted by Deforestation or Natural Forest Fires. *Environmental Monitoring and Assessment* 131:1–11.

Gray, J. E., D. L. Fey, C. W. Holmes, and B. K. Lasorsa. 2005. Historical Deposition and Fluxes of Mercury in Narraquinnep Reservoir, Southwestern Colorado, USA. *Applied Geochemistry* 20:207–220; doi: 10.1016/j.apgeochem.2004.05.011.

Hood, E. W., M. W. Williams, and N. Caine. 2003. Landscape Controls on Organic and Inorganic Nitrogen Leaching across an Alpine/Subalpine Ecotone, Green Lakes Valley, Colorado Front Range. *Ecosystems* 6:31–45.

Kling, G. W., G. W. Kipput, M. M. Miller, and W. O'Brien. 2000. Integration of Lakes and Streams in a Landscape Perspective: The Importance of Material Processing on Spatial Patterns and Temporal Coherence. *Freshwater Biology* 43:477–497.

Knapp, R. A., C. P. Hawkins, J. Ladau, and J. G. McClory. 2005. Fauna of Yosemite National Park Lakes Has Low Resistance but High Resilience to Fish Introductions. *Ecological Applications* 15:835–847.

Loeffler, C., ed. 2001. *Conservation Plan and Agreement for the Management and Recovery of the Southern Rocky Mountain Population of the Boreal Toad (Bufo boreas boreas).* Boreal Toad Recovery Team and Technical Advisory Group. Denver: US Fish and Wildlife Service and ten partners.

Martinez, P. J., ed. 2008. *Coldwater Reservoir Ecology: Federal Aid Project F-242-R15, Progress Report.* Fort Collins: Colorado Division of Wildlife.

Mercury Deposition Network (MDN). 2007. National Atmospheric Deposition Program. Available at http://nadp.sws.uiuc.edu/mdn/maps/2007/07MDNconc.pdf. Accessed October 1, 2008.

Monson, B. A., and P. L. Brezonik. 1998. Seasonal Patterns of Mercury Species in Water and Plankton from Softwater Lakes in Northeastern Minnesota. *Biogeochemistry* 40: 147–162.

Morris, D. P., and W. M. Lewis. 1988. Phytoplankton Nutrient Limitation in Colorado Mountain Lakes. *Freshwater Biology* 20:315–327.

Musselman, R. C., and W. L. Slauson. 2004. Water Chemistry of High Elevation Colorado Wilderness Lakes. *Biogeochemistry* 71:387–414.

New Mexico Environment Department (NMED). 2008. Fish Consumption Advisories. Available at http://www.nmenv.state.nm.us/swqb/Mercury2.html. Accessed September 22, 2008.

Pister, E. P. 2001. Wilderness Fish Stocking: History and Perspective. *Ecosystems* 4:279–286.

Nydick, K. R. 2010. Mercury in Precipitation and Lakes of Southwestern Colorado. Unpublished data. Mountain Studies Institute, P.O. Box 426, Silverton, CO 81433.

Schindler, D. E., R. A. Knapp, and P. R. Leavitt. 2001. Alteration of Nutrient Cycles and Algal Production Resulting from Fish Introductions into Mountain Lakes. *Ecosystems* 4:308–321.

Schutzah, J. 2008. Norwood, CO: US Forest Service, Norwood Ranger District, Fall Creek and Woods Lake Colorado River Cutthroat Trout Reclamation Project, Public Comment Period Notice File Code 9500/2600. March 24.

Taylor, J. G., K. J. Czarnowski, N. R. Sexton, and S. Flick. 1996. The Importance of Water to Rocky Mountain National Park Visitors: An Adaptation of Visitor-Employed Photography to Natural Resources Management. *Journal of Applied Recreation Research* 20:61–85.

Warlick, D. 2003. *Vallecito Country.* Lake City, CO: Western Reflections.

Williams, M. W., P. D. Brooks, A. Mosier, and K. A. Tonnessen. 1996. Mineral Nitrogen Transformations in and under Seasonal Snow in a High-Elevation Catchment in the Rocky Mountains, United States. *Water Resources Research* 32:3161–3171.

Wright, W. 2008. *Upper Pine River Watershed: State of the Watershed Report 2008.* Durango, CO: Pine River Watershed Group.

Wurtsbaugh, W. A., M. A. Baker, H. P. Gross, and P. D. Brown. 2005. Lakes as Nutrient "Sources" for Watersheds: A Landscape Analysis of the Temporal Flux of Nitrogen through Sub-Alpine Lakes and Streams. *Verhandlungen—Internationale Vereinigung fur theoretische und angewandte Limnologie* 29:645–649.

CHAPTER EIGHT

Fens of the San Juan Mountains

Rodney A. Chimner and David Cooper

TYPES OF WETLANDS

Types of mountain wetlands include marshes, riparian areas, wet meadows, fens, bogs, and salt marshes (Carsey et al. 2002). However, fens are the dominant wetland type above 9,000 feet (2,700 m) in the San Juan Mountains (Chimner, Cooper, and Lemly 2010). The fen is a distinctive wetland type that accumulates organic soil, or "peat," and can be termed a peatland. Peatlands are the most widespread wetland type in the world, occurring from the tropics to the arctic and covering an estimated area of 1.5 million mi² (4 million km²), or 3 percent of the Earth's land surface (Maltby and Proctor 1996). Although peatlands are widespread, over 90 percent of peatland area lies in the boreal regions (Maltby and Proctor 1996). In the United States, the majority of peatlands occur in Alaska, the Upper Great Lakes, the Northeast, and along the low-lying Gulf and Atlantic Coasts (Malterer 1996). Peatlands were once thought to be rare in the western United States because of regional aridity. However, it has recently become clear that peatlands are numerous at high elevations in western mountain ranges (Chimner, Cooper, and Lemly 2010; Cooper and Wolf 2006; Lemly 2007). We have calculated that at least 4,000–5,000 fens occur in the San Juan Mountains (Chimner, Cooper, and Lemly 2010).

There are two main types of peatlands—fens and bogs—which are differentiated by their hydrologic regimes or their water sources. Fens are supported by groundwater, while bogs are predominantly supported by precipitation. All peat-lands in the San Juan Mountains are fens because they are hydrologically connected to groundwater (Cooper and Andrus 1994; Woods, MacDonald, and Westbrook 2006). Fens form because soils therein are permanently saturated; these oxygen-free soils slow decomposition rates and allow organic matter produced by wetland plants to accumulate (Chimner and Cooper 2003). For example, water levels in fens in Prospect Basin, near Telluride, Colorado, rarely drop to 8 inches (20 cm) below the soil surface (Cooper, unpublished data). The saturated conditions in San Juan fens are caused by the cold high-elevation temperatures, deep winter snowpack, and summer monsoonal precipitation. These three factors help form groundwater flow systems with perennial discharge into the fens.

Fens are ancient ecosystems that started to form in the San Juans after the mountain glaciers began to melt around 12,000 years ago. Many of the fens in the San Juans are 6,000–10,000 years old, as determined by carbon dating (Carrara, Trimble, and Rubin 1991; Cooper, unpublished data). The typical thickness of fens in this region is 3–8 feet (0.9–2.4 m) (Chimner, Cooper, and Lemly 2010). Fens are slow-growing ecosystems that accumulate peat at a rate of about 8 inches (20 cm) per 1,000 years in Colorado (Chimner 2000; Chimner and Cooper 2002).

The slow but constant accumulation of peat over thousands of years makes fens globally important as carbon sinks. Peatlands play a critical role in the global carbon cycle because they either sequester or emit the greenhouse gases carbon dioxide (CO_2) and methane (CH_4) (Gorham 1991; Roulet 2000). Since the last deglaciation, peatlands have slowly sequestered atmospheric CO_2 as peat. Peatlands are estimated to currently store 224 to 455 Pg $(1\ Pg = 10^{15}\ g)$ of carbon, equal to 12–30 percent of the global soil-carbon pool (Clymo, Turunen, and Tolonen 1998; Gorham 1991; Lappalainen 1996; Moore, Roulet, and Waddington 1998; Zoltai and Martikainen 1996). Peat accumulates when carbon fixed through plant produc-tion exceeds losses from decomposition, leaching, disturbance, or some combina-tion of these. Currently, peatlands are thought to function globally as net sinks for atmospheric CO_2, sequestering approximately 76 Tg $(1\ Tg = 10^{12}\ g)$ C/yr (Vasander and Kettunen 2006; Zoltai and Martikainen 1996). However, peatlands simultane-ously serve as a major source of methane, which is twenty-three times stronger as a greenhouse gas than is CO_2 per molecule, on a 100-year time scale (Wahlen 1993). There is concern that global warming may cause peatlands to start losing their large pools of carbon, creating a significant positive feedback to global warming.

Peat in Colorado is generally about 50 percent carbon. In addition to stor-ing and cycling carbon, Colorado fens are also important locally as areas of high regional biodiversity and refugia for rare species (Cooper 1996). For example, an assessment of San Juan fens identified 188 species of vascular plants and 63 species of mosses (Chimner, Cooper, and Lemly 2010). Several species and one genus new to Colorado were discovered during recent fen assessments. Fens also support populations of plant species that are relicts of colder environments that occurred many thousands of years ago, at the beginning of the Holocene. Plant species, such as Altai cottongrass (*Eriophorum altaicum*), buckbean (*Menyanthes trifoliata*), purple cinquefoil (*Comarum palustre*), arctic peat moss (*Sphagnum balticum*), and many others, are disjunct by many hundred or thousands of miles from the main ranges of their species in the boreal forest regions of North America. Colorado peatlands are also important because of their habitat value, especially to elk (*Cervus elaphus*), moose (*Alces alces*), and birds, such as snipe. Fens provide other critical ecological functions as well, including filtering the large volumes of water that flow through them and providing base flow to streams (Bedford and Godwin 2002).

FEN TYPES IN THE SAN JUAN MOUNTAINS

Fens in the San Juans form in two major landform positions: basins and slopes. Basin fens occur in depressions and form through infilling of ponds or small lakes by sediment and peat (plate 26). Many basin fens still have central open-water areas, with tall sedges and floating mats along the edges. Basin fens can become totally filled and grown over, with no visible water present. Filled basin fens can contain water beneath the peat (as in quaking fens), or they can be completely filled with peat. Water levels tend to be fairly stable throughout the year, but they peak in early summer. The surface slopes of basin fens are usually less than 1 percent because of their origins as ponds or lakes.

Basin fens that are still open support animal populations not found in the other fens, including tiger salamanders, nesting mallards, green-winged teal, and muskrats. Vegetation of floating mats is usually dominated by sedges *Carex limosa* and *Carex magellanica*, buckbean, and occasional mosses (*Sphagnum spp.*). Outside of the floating mat is typically a zone of beaked sedge (*Carex utriculata*) in deep ponded water. A number of species are of secondary importance in this community and include *Galium trifidum*, *Carex vesicaria*, *Geum macrophyllum*, and *Carex aquatilis*.

Sloping fens are more numerous than basin fens and occur where groundwater discharges along a geologic discontinuity or at a break in slope (Chimner, Cooper, and Lemly 2010; Smith et al. 1995). Sloping fens can occur on slopes ranging from 1 to 20 percent (plate 27). Many plant communities occur primarily in sloping fens (Chimner, Cooper, and Lemly 2010). Some fens are dominated by several species of sedges, with few woody plants. Others are dominated by willows, with an understory of sedges and herbaceous plants. The most common shrubs are the plane leaf willow (*Salix planifolia*) and mountain willow (*Salix monticola*). The most common sedges are water sedge (*Carex aquatilis*), beaked sedge, and sheep sedge (*Carex illota*). Common flowering dicots include marsh marigold (*Psychrophila leptosepala*), elephant's head (*Pedicularis groenlandica*), and queen's crown (*Clementsia rhodantha*).

Spring fens are sloping fens where groundwater discharging from the aquifer forms small springs and streams. These areas have high water tables and flowing water that transports high concentrations of dissolved oxygen and nutrients. Most species occurring in this community type are also found along fast-moving streams and include alder (*Alnus incana*), beaked sedge, willowherb (*Epilobium lactiflorum*), arrowleaf ragwort (*Senecio triangularis*), Richardson's geranium (*Geranium richardsonii*), mountain willow, Tracy's rush (*Juncus tracyi*), Rocky Mountain hemlock (*Conioselinum scopulorum*), woodrush (*Luzula parviflora*), and Fendler's cowbane (*Oxypolis fendleri*) (plate 27).

Sloping fens can also form unique patterns of strings and flarks, also called patterned fens. Excellent examples of patterned fens with high ridges or strings and low swales, or flarks, can be seen in the Sand Creek/Mosca area, near Williams Creek Reservoir, northwest of Pagosa Springs, Colorado. Although the exact mechanisms of formation are unclear, the patterning is created by flowing water, which arranges the patterns perpendicular to water flow. These sloping fens are normally dominated by fewflower spikerush (*Eleocharis quinqueflora*), sheep sedge, and *Carex saxatilis*.

Natural water acidity varies greatly in San Juan fens because of the diverse geologic conditions found in the mountains (Blair 1996). The majority of fens range from slightly acidic (pH ~5.5) to slightly basic (pH > 7.5). Acidity is influenced by the bedrock with which the groundwater has been in contact (Cooper 1996; Cooper and Andrus 1994). [Acidity is also influenced by widespread acid mine drainage, a potent relic from the region's mining era.—Ed.] However, several rare acid fens with pH < 4.5 also occur in the San Juans. These are termed "iron fens" because they form in areas of weathering iron pyrite, which oxidizes to form sulfuric acid (Cooper, Andrus, and Arp 2002). Iron fens are common around Red Mountain, between Silverton and Ironton. The most common plants in iron fens are bog birch (*Betula glandulosa*), water sedge, and a dense cover of mosses (*Sphagnum angustifolium* and other *Sphagnum* species). Some rare plants also occur in iron fens. For example, in Chattanooga Fen near Silverton, a population of an arctic peat moss (*Sphagnum balticum*) occurs, disjunct by more than 1,200 miles (2,000 km) from its main range in Canada (Cooper, Andrus, and Arp 2002).

IMPACTS ON AND THREATS TO FENS

Fens have formed under relatively stable conditions over thousands of years. However, long-term growth can be interrupted by natural and human disturbances. Because fens are sensitive to disturbance, we studied impacts on and threats to San Juan fens (Chimner, Cooper, and Lemly 2010). The most common disturbance encountered was roads, which are numerous throughout the San Juans. Roads impact fens by intercepting the water flow to fens and bisecting fens, and they are commonly a source of mineral sediment that can bury fen soils. Most roads have limited impacts on fens, but a few roads have caused severe disturbance, especially where poor culvert placement has created channels and erosion in fens. In addition to the physical factors of altering water flow and sediment deposition, roads also allow vehicle access that can directly impact fens. Heavy four-wheel-drive use is common in many areas of the San Juans, where historic mining roads are heavily traveled. Several fens showed signs of off-road vehicles having driven through or adjacent to them. Off-roading adjacent to fens has usually had a minor influence unless it cut into the soil and altered groundwater flow paths. However, off-roading through fens has left deep tire tracks that cut ditches and drains into fens. In some fens, tire tracks had enough water moving through them to cause erosion and gully formation.

Development of golf courses, parking structures, condos, ski areas, and houses—especially around the adjacent towns of Mountain Village and Telluride—has created severe impacts, such as the filling of fens and the building of structures that altered groundwater flow to or from fens. Recreation has also impacted many fens, especially where heavily used by residents and visitors for skiing, hiking, mountain biking, camping, and golf. Many trails, ski runs, and campsites have been located adjacent to fens. Heavy recreational use can cause trampling of fen vegetation or alteration of fen hydrology.

Disturbances from mining have been numerous in the San Juans because of the long history of mining in the region. Some mining impacts consist of tailing piles located on or adjacent to fens. Tailing piles cover peat and disrupt surface and groundwater flow, altering chemical and mineral-sediment influx to fens. At the other end of the spectrum, we found mining impacts so severe that they destroyed many fens. Peat had been removed from these sites, and little or no vegetation remains.

Several fens were also found to have drainage ditches or water diversions that dried fens. Dewatering a fen is a severe disturbance because it lowers the fen water table, allowing peat to oxidize and the peat surface to subside as a result of increased decomposition (Chimner and Cooper 2003).

Grazing by cattle occurs in many fens. Cattle grazing can damage fens by trampling the peat and peat-forming plants (Cooper and Wolf 2006); in addition, deep cattle trails function as ditches, as do vehicle tracks. We ranked most cattle grazing in the San Juans as of low severity, although we found fens that were ditched to increase forage production for cattle. Grazing impacts to fens in the San Juan Mountains were minor when compared with grazing impacts in other mountain regions, such as California's Sierra Nevada and the Andes of South America (Chimner 2007; Cooper, Andrus, and Arp 2002; Cooper and Wolf 2006). Besides cattle, we also found disturbances from native animals, primarily elk and deer. Disturbances from animals were generally limited to small bare patches where elk wallowed or to trampled areas where elk grazed fen vegetation.

Editors' note: Peat is mined from fens in some Colorado highlands, such as South Park. The product ("peat moss") is packaged and sold for garden fertilizer.

REFERENCES

Bedford, B., and K. Godwin. 2002. Fens of the United States: Distribution, Characteristics, and Scientific Connection vs. Legal Isolation. *Wetlands* 23:608–629.

Blair, R. 1996. Origin of Landscapes. In R. Blair, ed., *The Western San Juan Mountains: Their Geology, Ecology, and Human History*, 3–17. Niwot: University Press of Colorado.

Carrara, P. E., D. A. Trimble, and M. Rubin. 1991. Holocene Treeline Fluctuations in the Northern San Juan Mountains, Colorado, U.S.A., as Indicated by Radiocarbon-Dated Conifer Wood. *Arctic and Alpine Research* 23:233–246.

Carsey, K., G. Kittel, K. Decker, D. Cooper, and D. Culver. 2002. *Field Guide to the Wetland and Riparian Plant Associations of Colorado*. Fort Collins: Colorado Natural Heritage Program.

Chimner, R. A. 2000. Carbon Dynamics of Southern Rocky Mountain Fens. PhD dissertation, Department of Ecology, Colorado State University, Fort Collins.

———. 2007. Exploring the Hydro-Ecological Conditions of High-Elevation Wetlands in the Andes Mountains. Final report to the National Geographic Society.

Chimner, R. A., and D. J. Cooper. 2002. Modeling Carbon Accumulation in Rocky Mountain Fens. *Wetlands* 22:100–110.

———. 2003. Carbon Balances of Pristine and Hydrologically Modified Southern Rocky Mountain Fens. *Canadian Journal of Botany* 81:477–491.

Chimner, R. A., D. J. Cooper, and J. M. Lemly. 2010. Mountain Fen Distribution Types and Restoration Priorities, San Juan Mountains, Colorado, USA. *Wetlands* 30:763–771.

Clymo, R. S., J. Turunen, and K. Tolonen. 1998. Carbon Accumulation in Peatland. *OIKOS* 81:368–388.

Cooper, D. J. 1996. Soil and Water Chemistry, Floristics and Phytosociology of the Extreme Rich High Creek Fen, South Park, Colorado. *Canadian Journal of Botany* 74: 1801–1811.

Cooper, D. J., and R. Andrus. 1994. Peatlands of the West-Central Wind River Range, Wyoming: Vegetation, Flora and Water Chemistry. *Canadian Journal of Botany* 72: 1586–1597.

Cooper, D. J., R. A. Andrus, and C. D. Arp. 2002. Sphagnum Balticum in a Southern Rocky Mountains Iron Fen. *Madrono* 49:186–188.

Cooper, D. J., and E. C. Wolf. 2006. Fens of the Sierra Nevada, California. Final report to the USDA Forest Service.

Cooper, D. J., E. C. Wolf, C. Colson, W. Vering, A. Granda, and M. Meyer. 2010. Wetlands of the Minas Congas Region, Cajamarca, Peru. *Arctic, Antarctic, and Alpine Research* 42:19–33.

Gorham, E. 1991. Northern Peatlands: Role in the Carbon Cycle and Probable Responses to Climatic Warming. *Ecological Applications* 1:182–195.

Lappalainen, E. 1996. General Review on World Peatland and Peat Resources. In E. Lappalainen, ed., *Global Peat Resources*. Jyväskylä: International Peat Society and Geological Survey of Finland, 53–56.

Lemly, J. M. 2007. Fens of Yellowstone National Park, USA: Regional and Local Controls over Plant Species Distribution. Masters thesis, Department of Ecology, Colorado State University, Fort Collins.

Maltby, E., and M.C.F. Proctor. 1996. Peatlands: Their Nature and Role in the Biosphere. In E. Lappalainen, ed., *Global Peat Resources*. Jyväskylä: International Peat Society and Geological Survey of Finland, 11–19.

Malterer, T. J. 1996. Peat Resources of the United States. In E. Lappalainen, ed., *Global Peat Resources*. Jyväskylä: International Peat Society and Geological Survey of Finland, 253–260.

Moore, T. R., N. T. Roulet, and J. M. Waddington. 1998. Uncertainty in Predicting the Effect of Climatic Change on the Carbon Cycling of Canadian Peatlands. *Climate Change* 40:229–245.

Roulet, N. T. 2000. Peatlands, Carbon Storage, Greenhouse Gases, and the Kyoto Protocol: Prospects and Significance for Canada. *Wetlands* 20:605–615.

Smith, D. R., A. Ammann, C. Bartoldus, and M. M. Brinson. 1995. *An Approach for Assessing Wetland Functions using Hydrogeomorphic Classification, Reference Wetlands, and Functional Indices*. Technical report WRP-DE-9, NTIS AD A307 121. Vicksburg, MS: US Army Engineer Waterways Experiment Station.

Vasander, H., and A. Kettunen. 2006. Carbon in Boreal Peatlands. In R. K. Wieder and D. H. Vitt, eds., *Boreal Peatland Ecosystems, Ecological Studies*. Heidelberg, Germany: Springer-Verlag, 165–188.

Wahlen, M. 1993. The Global Methane Cycle. *Annual Reviews in Earth and Planetary Science* 21:407–426.

Woods, S. W., L. H. MacDonald, and C. Westbrook. 2006. Hydrologic Interactions between an Alluvial Fan and a Slope Wetland in the Central Rocky Mountains, U.S.A. *Wetlands* 26:230–243.

Zoltai, S. C., and P. J. Martikainen. 1996. Estimated Extent of Forested Peatlands and Their Role in the Global Carbon Cycle. In M. J. Apps and D. T. Price, eds., *Forest Ecosystems, Forest Management, and the Global Carbon Cycle*. NATO ASI Series. Berlin: Springer I 40, 47–58.

CHAPTER NINE

Fungi and Lichens of the San Juan Mountains

J. Page Lindsey

T HE KINGDOM EUMYCOTA IS COMPOSED OF FUNGI—heterotrophic, terrestrial organisms that are evolutionarily more closely related to the animal kingdom than to any of the other life forms on Earth. Fungi often appear similar to plants, but they are not photosynthetic; instead, they must obtain nourishment from organic substrates. Fungi enter into a wide variety of interactions with other organisms and with their waste products. Fungi may decompose animal or plant material by being saprophytes (living on dead materials), or they may be parasites (subsisting on living hosts and causing the hosts to suffer while the fungi benefit). Plants, with few exceptions, use light energy from the sun to make their own food; this process is known as photosynthesis.

Some fungi are mutualists, a category that includes many of the mushrooms found on the forest floor during the rainy season. Spreading through the soil or its food source, the main fungal organism (of which mushrooms are its fruiting bodies) is a wooly perennial network of vascular filaments called a mycelium. Attached to the roots of trees and shrubs, the mycelium acts as an extension of plants' root systems, granting the plants greater access to soil nutrients and water. The fungus is paid for this service with carbohydrates the plants make through photosynthesis. This mutually beneficial (symbiotic) association is called a mycorrhiza.

The ultimate expression of mutualism and symbiosis is the lichen. A symbiont is an organism that lives together with another organism. Within a species of lichen live two symbiotic species, one of a fungus and the other of an alga (or a Cyanobacterium). Fungi are not photosynthetic, although algae and Cyanobacteria are. In this symbiosis the two partners, the mycobiont (the fungus) and the phycobiont (the alga, or the Cyanobacterium), are mutualists (i.e., each contributes fairly equally to the partnership). The alga photosynthesizes and makes food, and the fungus protects the alga from desiccation and from damage by small grazing animals.

OCCURRENCE OF FUNGI AND LICHENS

Both fungi and lichens occur almost anywhere in the world, from hot deserts to polar deserts, but they are especially prevalent in forest ecosystems.

Fungi occur in all of the Earth's biomes. Some fungi are tiny and cannot be identified without the aid of a microscope and detailed observation. Other fungi produce macroscopic reproductive structures. The gilled mushrooms, boletes, puffballs, coral fungi, and bracket fungi—all of which the laity refers to as "mushrooms"—represent the most conspicuous fungi and contain these reproductive structures.

In the San Juan Mountains, mushrooms emerge in great abundance in the late summer and fall, when nights are not freezing and the rainy season is in effect. Because they are about 95 percent water, a regular and continual supply of rain is needed to produce a good mushroom season. They can be found on soil above treeline on dead wood, living trees, forest litter, herbivore dung, and various other nutrient-rich substrates. Although the hard polypores that subsist on decaying wood may live for decades, mushrooms may last only a day or a week before they collapse and return to the soil. Some mushrooms live in fairly dry environments and can be found at any time of the year in specific microclimates.

Mushroom collectors know that a fungal species is often associated with a particular host plant, especially if it is parasitic or mycorrhizal. A mycorrhizal fungus has an association with the roots of a specific plant (or related groups of plants) and offers a benefit to the host by assimilating phosphates and other soil nutrients the host plant cannot gain for itself. For example, *Amanita muscaria*, the fly agaric, is usually found under Engelmann spruce in the subalpine zone (spruce-fir forests) (plate 28). *Cystoderma amianthinum*, a small orange mushroom associated with the roots of ponderosa pine, is found only in the montane zone. Seeds of trees not associated with the appropriate fungal partner may germinate, but the seedlings will not thrive and will eventually die.

In the San Juan Mountains, lichens can be seen as tightly appressed (crustose) patches on rock surfaces above and below treeline; as leaf-like (foliose) flakes on soil, rock, and tree trunks below treeline; and as small shrub-like (fruticose) clumps attached to, or draping from, tree limbs. Lichens are usually long-lived organisms; some live for millennia.

Lichens are rarely harmful to the host organisms they grow upon. If they are situated on plants, they are considered epiphytes, organisms that use another organism as a site of attachment. They do not take nutrients from the host. Lichens receive much of their mineral nutrition and all of their water from rain and snow and their carbon from the air. They are thus extremely sensitive to changes in air quality. For example, toxic compounds such as sulfates may become dissolved in rainwater and fall as acidic rain, affecting different lichen species in different ways. Lichens are referred to as indicator organisms because they often show responses to air pollution long before other organisms do. They are most sensitive to compounds in gaseous form, particularly sulfur dioxide (Huckaby 1993). The San Juan National Forest (SJNF) monitors air pollution using lichen species as indicators. Lichen species are categorized as most sensitive, sensitive, intermediate, tolerant, and most tolerant. Species in all categories are found in the SJNF. Different communities of lichens show specific proportions of these species. When air pollution affects the more sensitive species, the species composition will change over time. This change can be correlated with specific pollutants from defined sources (Geiser and Reynolds 2002).

FUNGI DISTRIBUTION IN MOUNTAINS

Mountains are islands in the sky and harbor many common species of mushrooms and related fungi. These fungi are found at different elevations, under different conditions, and sometimes in association with specific tree or shrub hosts. Undoubtedly, the mushrooms that occur at subalpine elevations are undergoing differential selective pressures in their sky-island environments because these same fungi will not grow in the warmer, more arid valleys between mountain ranges. However, because mushroom spores are microscopic and can be blown long distances (and into the stratosphere) before they land on appropriate substrates and germinate, the distribution of fungal species is usually less restricted than that of plant species, which have more difficulty transporting their reproductive structures from one subalpine zone to another.

For example, a polypore, *Phellinus torulosis*, was discovered in the sky island of the Santa Catalina Mountains, north of Tucson, Arizona; it was growing on the base of a *Pinus strobiformis* (southwestern white pine). (This tree also occurs in the montane zone of the San Juan Mountains, its only distribution in Colorado.) Other host species include *Pinus ponderosa* (ponderosa pine) and *Pseudotsuga menziesii* (Douglas fir). Since its original discovery in the Catalinas, the fungus has also been found in the Pinaleno Mountains in southeastern Arizona. Prior to its report in southern Arizona (Gilbertson and Burdsall 1972), it had only been reported from Europe. While little research has been done on mycogeography (the distribution of fungi), it is possible that species having worldwide distribution crossed large distances as microscopic spores. These distributions may also reflect the distribution of their hosts or historical factors.

Conversely, Cripps (2004) reported that the Southern Rocky Mountains are part of the southernmost distribution of a circumpolar arctic-alpine fungal flora distributed on high mountaintops in North America, Asia, and Europe. She determined that, in the true alpine areas above treeline, 80 percent of the macrofungi are mycorrhizal with willows, bog birch, and *Dryas*. This is true for the San Juans and other alpine areas of New Mexico and Arizona; however, the more southerly portions of the ranges overlap with subtropical mycoflora that extend northward from Mexico.

Elevation only indirectly affects many fungi, since they are associated with particular host plants. Therefore, fungi that are mycorrhizal with spruce will by definition be found in the spruce-fir zone, while those associated with oak will be found at much lower elevations in the montane forests. As mentioned earlier, precipitation is required for all fungi, especially mushrooms; because of a combination of day length, favorable temperatures, and available moisture, the fruiting season in the San Juans is commonly from August into early September. Slopes facing north and east will have more moisture retention and will therefore be more favorable habitats for most fungi. Some fungi, however, such as the polypores that fruit from live wood, may be less affected by ambient moisture, since they obtain most of their moisture from their host plant.

Some mushrooms are routinely associated with disturbed areas, such as compacted roadsides or campgrounds. *Cantherellus cibarius*, the delectable chanterelle, is found in spruce-fir and aspen stands in areas where the ground has been trampled. *Coprinus comatus*, the shaggy mane, comes up each year along roadsides in mixed-conifer zones (see chapter 8), sometimes through hard-packed gravel. Sediment and bedrock underlying the forest soil do not seem to affect most fungi, since they mainly derive nutrients from the organic layers of the soil.

On the other hand, lichens are often associated with only one type of rock or only specifically with conifer bark or hardwood bark. For example, the bright orange-yellow lichen *Xanthoria montana* is associated with aspen bark. A related species, *X. elegans*, is found on any kind of rock surface where birds or mammals have perched and nitrogenous material from their droppings has leached into the rock. Some lichens prefer granitic substrates, while others prefer sandstone (Corbridge and Weber 1998).

ECOLOGICAL RELATIONSHIPS

Lichens are often the first organisms to colonize bare ground and rock (pioneer organisms). In the canyon country west of the San Juans, much of the soil is covered with dark lichen communities—a condition known as cryptobiotic soil. These pioneer lichens stabilize soil, and some actually fix atmospheric nitrogen (N_2) into more complex compounds (NO_3) that plants can use as fertilizers.

Cryptobiotic communities exist in the San Juan Mountains, although they are less conspicuous than those in the desert areas to the west. Lichens and fungi, along with mosses and other small plants, stabilize the soil, prevent erosion, and recycle nutrients.

Some fungi are pathogens, causing disease in the host plant. One of the most notable of these in the San Juan Mountains is *Armillaria ostoyae*, the honey mushroom. This common edible fungus infects several hundred woody host plant species, including most of the commercially important tree species. This fungus attacks ponderosa pine, white fir, aspen, Engelmann spruce, Gambel oak, and many others. It causes a serious white root rot that will eventually kill the tree or cause it to windthrow. The US Forest Service monitors campgrounds in heavily visited areas in the SJNF for trees that are infected by this fungus and removes them.

Much of the Gambel oak in the ponderosa pine and warm-dry mixed-conifer zones of the San Juans is infected with a polypore fungus called *Inonotus andersonii*. This fungus fruits underneath the bark of living oaks, causing the bark to separate from the wood, eventually killing the tree. On the positive side, the fungus also rots the wood underneath the bark, creating a softer substrate in which flickers and other fauna can excavate holes for nests in the dead trees.

Many fungi harbor insect communities; the insects derive food and habitat from the fungi. Some mushrooms are notably free of insects; chanterelles are a fortunate example. Others are predictably infested with insects. *Sparassis crispa*, an eminently delectable fungus that looks like a cauliflower and causes root rot in ponderosa pine, white fir, and Douglas fir, is invariably full of fly maggots. This fungus is common on these hosts in the Piedra area of the eastern San Juans. Most *Russula* and *Lactarius* species are also good suspects for maggot infestation.

An interesting relationship occurs between a little polypore fungus, *Cryptoporus volvatus* (which fruits on recently killed conifers), and a number of beetle species. The pore surface of most polypores is exposed to the atmosphere so the microscopic spores can fall out easily and be caught up in the wind. The pore surface of *C. volvatus* is covered with a membranous layer. At maturity, beetles that feed on the fungus tissue will break open this membrane, allowing the spores to be freed. The beetles probably also carry spores on their bodies. *C. volvatus* can be found on all members of the genus *Pinus*, as well as on Douglas fir, spruce, and true firs in the eastern San Juan area.

Many animal species, in addition to insects, use lichens for food and nesting. Some birds, including hummingbirds, line their nests with lichens. Sharnoff and Rosenstreter (1998) list various large mammals that use lichens for food: elk, mule deer, mountain goat, pronghorn antelope, bighorn sheep, and moose. Small mammals, such as squirrels, voles, moles, and shrews, also supplement their diets with lichens.

POISONOUS FUNGI

Most people who are interested in collecting mushrooms and their fungal relatives want to know if their finds are edible or poisonous. A few important points should be considered.

Many mushrooms are inedible but not poisonous. They may be tough, have an unpleasant flavor, or cause allergic reactions. None of these characteristics makes a mushroom poisonous, only unfit for consumption. Poisonous mushrooms are classified into several general categories, described in the next paragraph. To gain a deeper insight into the nature of mushroom poisons and poisonings, consult one of the reputable mushroom guides available in most bookstores (Arora 1979; Benjamin 1995; Evenson 1997; Miller and Miller 2006; States 1990).

WARNING:

- When you collect wild mushrooms for eating, be sure they are clean, fresh, and worm-free.

- Collect more than one specimen of each species, and reserve a good specimen (do not cook or eat it) in case someone gets sick and the specimen needs to be identified.

- Do not mix collections of different species of mushrooms.

- Stick to the known and easily identified species of mushrooms. Do not experiment with those you cannot clearly identify.

- Never eat a large amount of wild mushrooms at one time because some are not easily digested. To test a species for the first time, clean, cook, and eat a teaspoon-size amount, then wait at least twenty-four hours to see if you have any reactions to the ingestion. If no adverse reactions have occurred, eating a larger portion may be tolerable. Note: Even a small amount of some species can kill in an agonizing way, and symptoms may not occur until after twenty-four hours, when successful intervention is impossible. The effects of some toxins, such as those of Type 5, may be cumulative over time.

- Never eat wild mushrooms raw. Some raw mushrooms will cause digestive upsets, even if they are not poisonous. Always cook them.

- There is no easy way to tell if a mushroom is poisonous. It must be identified to species.

Mushroom toxins can be classified in eight or more categories (Benjamin 1995; Miller and Miller 2006). Toxins in the Types 1, 5, and 7 groups can cause fatalities. Type 1 toxins (amatoxins) are usually considered the most serious because there is a delay in symptom development (as long as twenty-four hours), and the toxins have usually affected the liver by that time. The toxins deactivate important enzymes in the liver, causing a shutdown of protein synthesis. The most common mushrooms that possess this toxin group are in the genus *Amanita*, although many *Amanita* species in the United States do not contain these toxins. In Colorado, there have been no reports of the *Amanita* species that contain Type 1 toxins. Dangerous species include *A. phalloides*, *A. verna*, and *A. virosa*, which have been found in other areas of the United States, especially on the West and East Coasts.

Galerina autumnalis, a little brown mushroom that occurs on rotted and buried wood in the subalpine (spruce-fir) zone, can be found throughout Colorado and in the San Juans. This mushroom should never be eaten because a small amount contains enough Type 1 toxin to be deadly. Most mushroom guides suggest not eating any small brown mushrooms that grow on wood or moss.

Type 2 toxins (muscarine) are found in small quantities in other *Amanita* species, including one of the most common high-elevation mushrooms in the San Juans, *Amanita muscaria*, the fly agaric (plate 28). This beautiful mushroom is conspicuous and can become fairly large. It is mycorrhizal with spruce and other conifers and is common in areas such as Wolf Creek Pass and Molas Pass. The caps may enlarge to be as big as a dinner plate; in younger stages they are blood-red, with abundant white warts on top of the cap. The stipe (stalk) is white, as are the gills. There is an annulus (ring) around the stipe. While *A. muscaria* contains Type 2 toxins, it also contains Type 3 toxins, which are more of a concern (discussed later).

Muscarine poisoning is more commonly caused by certain species of *Clitocybe* and *Inocybe*, two genera that are very difficult for the amateur collector to recognize. These genera contain an abundance of little white, big white, and little brown mushrooms with features that are difficult to discern. It is best to avoid eating all mushrooms in these descriptive categories. Type 2 toxins cause SLUDGE symptoms: salivation, lacrimation, urination, defecation, gastrointestinal distress, and emesis.

Type 3 toxins include the two compounds called muscimol and ibotenic acid. The latter is the precursor of muscimol and is chemically slightly different. Muscimol is similar in chemical structure to GABA (gamma-aminobutyric acid), which inhibits certain functions of the nervous system. Symptom onset occurs within 30–120 minutes of ingestion. This toxin group causes poor coordination, dizziness, cramps, nausea, and sometimes hallucinations. The red- or yellow-capped *A. muscaria* and the brown-capped *Amanita pantherina* are common in the San Juan Mountains and are the likely culprits in Type 3 mushroom poisonings in the area.

Type 4 toxins (psilocybin) are not abundant in any mushrooms in the San Juans.

Type 5 toxins may occur in false morels, which grow predominately in the spring and early summer in the subalpine, near the edges of melting snow banks. *Gyromitra montana* is a large species that is common in the San Juans and that produces brown, brain-like, stalked fruiting bodies. Some authorities suggest that this species is edible and tasty if it is parboiled before being eaten, but the safe route would be to avoid all *Gyromitra* species, because many contain gyomitrin, which can be lethal.

Type 6 (coprine) toxins are found in a number of "inky cap" mushrooms, the best-known of which is *Coprinus atramentarius*. This mushroom is fairly common at lower elevations in the montane and can be found in lawns and other domestic situations, usually at the base of a living deciduous tree, such as aspen, box elder, or cottonwood. The symptoms (nausea and vomiting) are unpleasant although not injurious, simulating those alcoholics experience when taking Antibuse (disulfuram). Symptoms are brought on only when the person has also ingested alcohol in the twenty-four hours before or after eating the fungus.

Type 7 toxins (orellanine) are found in a few species of *Cortinarius* and produce an extremely delayed onset of symptoms (some appear after almost ten days). The toxins cause kidney failure and hemorrhaging and often result in kidney transplants as a last resort. Because there are over 900 species of *Cortinarius* in the United States and most are very difficult to identify, this is a genus to avoid, as far as edibility is concerned.

Type 8 toxins (gastrointestinal) are the cause of a majority of mushroom poisonings. They cause nausea, retching, vomiting, abdominal cramping, and diarrhea. They are rarely fatal, except in cases of extreme dehydration. Many unrelated groups of mushrooms contain these toxins, and the toxins are probably not related either. Common San Juan mushrooms that contain GI toxins include *Russula emetica* (the sickener), some species of *Agaricus, Pholiota squarrosa,* and *Hebeloma crustuliniforme.*

EDIBLE FUNGI

The San Juans provide many edible fungi. Use a field guide to help in identification, and be absolutely certain before eating any wild-collected mushroom. One of the easiest to identify is *Coprinus comatus*. This common mushroom is commonly called the shaggy mane, the lawyer's wig, or the inky cap. It is fairly large (sometimes approaching a foot tall) and often fruits along the sides of roads where the soil is compacted or gravely and there is little vegetation. It often grows in groups of tight clusters. Both *C. comatus* and *C. atramentarius* are unique in that they release their microscopic spores from the gills (underneath the caps) by autodigestion of the gills. Many of the spores drip onto the ground in a black, inky fluid and will germinate in the same location during the next season. Thus the shaggy mane is often found in the same area year after year. If these mushrooms are collected early in the day and carried for several hours, they will create a black, inky mess in the bottom of the collecting basket. These mushrooms should be cooked soon after collection, and alcohol should not be consumed within twenty-four hours of ingestion.

Boletus edulis (porcini, the cép, or king bolete) is technically not a mushroom. The difference is that boletes have a spongy pore layer on the underside of the cap instead of gills (plate 29). These large fungi are commonly collected throughout the San Juans under spruce and fir and also where aspen stands are abundant; they appear to be mycorrhizal with all these trees in this area. The caps of king boletes are often wormy and must be very carefully cleaned. Remove the soft tube layers (located underneath the cap), which often contain insects and will become very slimy upon cooking. Small "button" specimens are usually preferable for eating. Mature specimens can be up to 14.5 in (37 cm) in diameter (Miller and Miller 2006).

The king bolete should not be confused with the orange-capped bolete (*Leccinum insigne*) more commonly found in aspen forests. The Denver Poison Center has had several reports of toxicity in these mushrooms, which are more slender than the king bolete and turn black upon bruising or cooking.

Chanterelles (*Cantherellus cibarius*) are common at higher elevations and seem to fruit most abundantly under aspen, as well as spruce and other conifers. These mushrooms have a pale-yellow to orange cap and tangerine-colored ridges in place of gills; they are small, rarely exceeding 2 inches (5 cm) high in the San Juan area. They have a wonderfully fruity smell, especially when they have been enclosed in a paper bag for several hours. An added bonus is that they are usually worm-free.

Morels (*Morchella* species) occur in the spring rather than in late summer and fall, when most other mushrooms appear. Morels are most common after fires (such as the Missionary Ridge Fire in 2002) or in bottomlands such as the floodplain of the Animas River or the Mancos River, under willows and cottonwoods. They are delicious, but care should be taken to distinguish them from the false morels (such as *Verpa* and *Gyromitra* species). Some people are allergic to morels and experience various unpleasant reactions, often similar to those caused by mushrooms that contain GI toxins. Individuals should take the precautions mentioned earlier when preparing to sample a mushroom they have not previously eaten.

OVERCOLLECTING OF MUSHROOMS

Mushroom collecting is gaining in popularity in the San Juan Mountains as the area's population increases. The San Juan mushroom flora is, for the most part, available only during the late summer and early fall, with the exception of the late-spring morels. Rainfall and snowmelt early in the year do nothing to enhance the production of mushrooms in the summer months, and the season is dependent on substantial, regular monsoon moisture. In other parts of the western United States, such as the Pacific Northwest, mushrooms may be present during most of the year as long as freezing does not occur.

Mushrooms are reproductive structures that emerge from a hidden, usually underground mycelium. A mycelium is composed of hyphae, which are threadlike, microscopic filaments of the fungus that ramify throughout the nutrient-containing substratum. Picking a mushroom generally does not harm this mycelium, which is actually the "real" organism. The mycelium simply goes on growing as long as moisture and nutrients are available, assimilating these materials from the soil, log, or whatever substrate it is growing in. However, sustained and heavy picking of the mushrooms over a period of many years may ultimately affect the mycelium's ability to reproduce or disperse spores; perhaps more important, trampling of the area by mushroom pickers may affect the vitality of the mycelium.

At present, there is no indication that San Juan forests are suffering from over-collecting of mushrooms. Picking mushrooms is akin to collecting fruit from a tree; it does not kill the mycelium any more than removing an apple would kill an apple tree. However, in some areas of the Pacific Northwest, there are additional serious concerns (Molina et al. 1993). While there is no indication that mushroom populations have been depleted in the Northwest, the increasing pressure from commercial collectors and the conflicting pressures of logging and recreation in the forests have raised questions about how these factors might affect mushroom availability in the future. Scientists from the Pacific Northwest Research Station of the US Forest Service (Pilz and Molina 2001) list five concerns that need to be considered when examining whether mushroom flora are being harmed: (1) mushroom productivity; (2) mushroom-harvesting efforts; (3) forest management practices; (4) the fungi's biology, ecology, and ecosystem functions; and (5) visitor management. It would be wise for managers of land use in the San Juans to address these concerns. Global climate change could also affect mushroom distribution, either directly or by affecting host and partner organisms with which fungi associate.

LICHENS AS TOOLS OF MEASUREMENT

Weber and Wittmann (1992) list over 600 species of lichens in Colorado. Perhaps the most commonly noticed lichen in any Colorado forest is *Usnea*, or old man's beard. This is a fruticose lichen that hangs from the branches of trees in subalpine forests and resembles the "Spanish moss" of the Deep South (figure 9.3). Many lichens of all forms are common from the montane to alpine zones.

Perhaps the most spectacular lichen display is above treeline, where saxicolous (rock-inhabiting) lichens abound. These are usually crustose, slow-growing, brilliant in color, and detailed in texture. These pioneer organisms are among the first to establish in harsh habitats and on nutrient-poor substrates. Acids produced by the lichens as products of their metabolism and pressure of growth from the tiny rhizoids that anchor them eventually degrade the rocks, producing rudimentary soils in which alpine plants can take root.

Lichenometric dating technique employs characteristics of these saxicolous lichens. The technique requires that the lichen grows slowly, is roughly circular in outline, and has well-defined borders. Over fifty taxa of lichen have been used in lichenometric dating (Müller 2006). *Rhizocarpon geographicum* is the most commonly used species (plate 30).

Lichenometry allows scientists to use lichen growth to apply approximate dates to surfaces up to 500 years old. This has an advantage over C_{14} dating, which is

FIGURE 9.3 *Usnea* sp. on Engelmann spruce branches.

usually ineffective over such short time frames. In combination with dendrochronology, lichenometry has proved very useful in dating gravestones, moraines, floods, and landslides.

A growth curve is established for each lichen species by measuring incremental radial growth of the lichen colony and comparing it to measurements of colonies of known age. Growth curves for lichens are generally non-linear. Crustose lichens are best for this purpose because they tend to be fairly regular and circular in outline (Armstrong 2004). Information from worldwide lichenometry (and other techniques) supports the concept of global warming. These data confirm that, around the world, glaciers have retreated based on an average warming trend of 0.66 degrees Kelvin per 100 years (Oerlemans 1994).

Lichens play an important role in monitoring air pollution. Several coal-fired power plants in the Four Corners region (with another, Desert Rock, probably on the way) spew large amounts of sulfur-containing pollutants into the atmosphere, affecting the watersheds and forests of the San Juans, as well as other Western Slope ranges. Jackson and others (1996) have shown higher concentrations of sulfur in lichen tissues from the Mount Zirkel Wilderness in an area downwind from two coal-fired power plants and also that sulfur concentrations are higher in Western Slope lichens. It is known that "weedy" species of lichens grow well in the presence of certain pollutants (e.g., N-containing compounds), while others are less tolerant (U.S. Forest Service 2001). *Xanthoria*, a bright orange saxicolous lichen common in the San Juans and throughout the alpine West, is a notoriously weedy species that shows pollution tolerance in Colorado (McCune et al. 1998). Other lichens are completely intolerant of toxins in the atmosphere, rainwater, and snowmelt and will simply not grow if minute amounts of those chemicals are present. Specific indications of air pollution can be determined by observing the position of various species of lichens along a gradient of air pollution (McCune et al. 1998).

Fungi and lichens are often unnoticed, yet they are important parts of the environment of the San Juan Mountains. They carry out roles of nutrient recycling, decomposition, parasitism, primary productivity (lichens only), and more, as well as serving as food sources for humans and wildlife. These organisms' often misunderstood and poorly documented role belies their contribution to the entire San Juan ecosystem and demands more intensive study by future scientists.

REFERENCES

Armstrong, R. 2004. Lichens, Lichenometry, and Global Warming. *Microbiologist*. Available at http://www.blackwellpublishing.com/Microbiology/pdfs/lichens.pdf. Accessed July 2, 2007.

Arora, D. 1979. *Mushrooms Demystified*. Berkeley: Ten Speed Press.

Benjamin, D. R. 1995. *Mushrooms: Poisons and Panaceas*. New York: W. H. Freeman.

Corbridge, J. N., and W. A. Weber. 1998. *A Rocky Mountain Lichen Primer*. Niwot: University Press of Colorado.

Cripps, C. L. 2004. Alpine Mushrooms in the San Juans, Rocky Mountains, U.S.A.: A Southern Extent of the Arctic-Alpine Mycoflora. Paper presented at the State of the San Juan Conference, Mountain Studies Institute, Silverton, CO, September 21–24.

Evenson, V. 1997. *Mushrooms of Colorado and the Southern Rocky Mountains*. Boulder: Westcliffe.

Geiser, L., and R. Reynolds. 2002. *Using Lichens as Indicators of Air Quality on Federal Lands*. Corvallis, OR: USDA Forest Service, Pacific Northwest Region. Report on a workshop held at Arizona State University, Tempe, October 2–3, 2001. Available at http://ocid.nacse.org/research/airlichen/workgroup.

Gilbertson, R. L., and H. H. Burdsall Jr. 1972. *Phellinus torulosus* in North America. *Mycologia* 64:1300–1311.

Huckaby, L. S., ed. 1993. *Lichens as Bioindicators of Air Quality*. General Technical Report RM-224. Fort Collins, CO: USDA Forest Service.

Jackson, L. L., L. Geiser, T. Blett, C. Gries, and D. Haddow. 1996. *Biogeochemistry of Lichens and Mosses in and near Mt. Zirkel Wilderness, Routt National Forest, Colorado: Influences of Coal-Fired Power-Plant Emissions*. Open File Report 96–295. Denver: US Department of the Interior.

McCune, B., P. Rogers, A. Ruchty, and B. Ryan. 1998. *Lichen Communities for Forest Health Monitoring in Colorado, USA: A Report to the USDA Forest Service*. Internal report to Interior West Region, USDA Forest Service. Available at www.fia.fs.fed.us/lichen/pdf-files/co98_report.pdf.

Miller, O. K., and H. H. Miller. 2006. *North American Mushrooms*. Helena, MT: Globe Pequot.

Molina, R., T. O'Dell, D. Luoma, M. Amaranthus, M. Castellano, and K. Russell. 1993. *Biology, Ecology, and Social Aspects of Wild Edible Mushrooms in the Forests of the Pacific Northwest: A Preface to Managing Commercial Harvest*. General Technical Report PNW-GTR-309. Portland, OR: US Department of Agriculture, Forest Service, Pacific Northwest Research Station, 1–42.

Müller, G. 2006. Gregg Müller on Lichenometry and Environmental History. *Environmental History* 11(3), 18 parts. Available at http://www.historycooperative.org/journals/eh/11.3/muller.html. Accessed July 2, 2007.

Oerlemans, J. 1994. Quantifying Global Warming from the Retreat of Glaciers. *Science* 264:243–245.

Pilz, D., and R. Molina. 2001. Commercial Harvests of Edible Mushrooms from the Forests of the Pacific Northwest United States: Issues, Management, and Monitoring for Sustainability. *Forest Ecology and Management* 5593:1–14.

Sharnoff, S., and R. Rosenstreter. 1998. *Lichen Use by Wildlife in North America.* Available at http://www.lichen.com/fauna.html. Accessed June 13, 2009.

States, J. S. 1990. *Mushrooms and Truffles of the Southwest.* Tucson: University of Arizona Press.

U.S. Forest Service. 2001. Air Quality Monitoring in the Columbia River Gorge National Scenic Area by the U.S. Forest Service, 1993–2001. Available at http://gis.nacse.org/lichenair/doc/AQ_CRGNSA.pdf. Accessed October 7, 2007.

Weber, W. A., and R. C. Wittmann. 1992. *Catalog of the Colorado Flora.* Niwot: University Press of Colorado (revised electronic version, January 2000).

CHAPTER TEN

Fire, Climate, and Forest Health

Julie E. Korb and Rosalind Y. Wu

NATURAL DISTURBANCES, HISTORICAL RANGE OF VARIABILITY, AND FIRE REGIMES

WHILE DRIVING FROM PAGOSA SPRINGS TO SOUTH FORK over Wolf Creek Pass, we cannot help but notice changes in the vegetation that blankets the spectacular mountain scenery. The vegetation in the San Juan Mountains is dominated by different forest types, assemblages of species whose distributions are controlled by biological and physical factors (Spencer and Romme 1996). One factor rarely limits the distribution of a species; more often, an aggregate effect of many interacting factors—such as elevation, soil type, temperature, and precipitation—sets distribution limits. Natural disturbances—such as fire, insect outbreaks, wind, and avalanches—are the main influences that affect the reasons different patches of vegetation within a forest have different characteristics, such as species composition, density, age and size of trees, and amounts of downed wood. The combined effects of these influences, some of which are periodic, can be complex and can last for centuries. Some examples in the San Juan Mountains include the way drought magnified the Missionary Ridge Fire in 2002 and the killing of epidemic numbers of piñon pine by piñon engraver beetles (*Ips confusus*) during a drought that started in 2001. Logging, livestock grazing, and fire suppression also cause lasting changes to vegetation.

Knowing how a forest worked in the past and how different it is today helps us understand current management options and needs. The historical range of variability (HRV) is unique for each forest type and varies with different geographic settings. Land managers ask these questions pertaining to HRV: Are current forest densities and species compositions within the range of what was found in the past for a specific forest type? Is a given area one in which extinguishing a fire is the best way to maintain its ecological health, or is the forest healthier if we leave it alone? Is this area at risk for irreversible changes, such as invasion by non-native species?

Forest history—including how forest conditions have varied—is a benchmark for evaluating current conditions, identifying restoration potential, and evaluating our stewardship of the land (White and Walker 1997). Sources of information about historical forest conditions include field notes, photographs, information from sites that are not degraded or that have not changed significantly, and current observation of pollen, packrat middens, charcoal, fire scars, seeds stored in the soil, soil characteristics, and tree ages and locations. The way indigenous peoples burned, foraged, and planted crops also provides clues about climate change and forests' fire histories (figure 10.1) and helps us understand forest processes.

Fire regimes are characterized by typical fire behavior, frequency, severity, size, and season of burning. There are four main fire regimes in the San Juan Mountains: (1) frequent low-severity surface fires, (2) periodic mixed-severity fires, (3) periodic high-severity fires, and (4) infrequent high-severity crown fires. In the San Juan Mountains, this fire-regime classification is based on fire frequency and severity prior to Euro-American settlement (~1880) (table 10.1). Within these four fire regimes, fire behavior varies greatly based on fuels, topography, and climate. The type, size, quantity, arrangement, and moisture content of fuels are critical to the way fires burn. For example, the fuel supply in many ponderosa pine forests in the San Juan Mountains was historically dominated by fine fuels composed of needles, dry grasses, and shrubs. Each ponderosa pine produces, on average, 1 ton/acre of dry pine needles every fall, providing appropriate conditions for frequent low-severity surface fires. In higher-elevation forests, including the cool-moist mixed-conifer and spruce-fir, infrequent high-severity crown fires depend on tree-canopy fuels being dry enough and on weather conditions, especially wind, capable of moving fires into the canopies. These forest types generally lack fine fuels and are dominated by both rotten and sound coarse woody debris. These larger fuels dry out more slowly than fine fuels. In prolonged drought conditions, the large fuels become fairly dry, affecting fire behavior.

In addition, topography and its interaction with climate have a strong influence on the fire environment. Elevation, slope position, aspect, slope steepness, and natural and artificial barriers influence fire behavior and spread. Elevation affects the length of the fire season, type of vegetation, and weather patterns. Slope

Evidence of past fires can still be found in the forest today. Frequent surface fires left scars called catfaces. Catfaces are typically triangular in shape, usually found on the uphill side of trees. Flames from fires burning uphill will swirl and eddy on the uphill side of a tree, much like water eddies around the downstream side of a rock in the middle of a river.

A classically shaped catface. This catface is on a snag or dead standing tree. Notice that individual fire scars (marked by arrows) show up as ridges in the catface.

This catface is on an old log. Note the multiple ridges of fire scars. The gray color of weathered wood and lack of any bark is an indication of very old wood.

Old logging stumps can give a cross-section view of fire scars in a catface. Arrows mark the scars as they appear in the cross-section on top of the stump.

FIGURE 10.1 Fire history in the forest.

INFORMATION BOX: FIRE SCARS

The evidence for fire history can still be found in the forest today. Frequent surface fires left fire scars found in catfaces on the base of trees (Figure 10.1). Fire scars are typically triangular, being narrow at the top and having a broader base at the forest floor. They are usually located on the uphill sides of trees. As a fire burns uphill, flames and heat from fuels burning around the base of a tree swirl and eddy on the uphill side much like water eddies around the downstream side of a rock in the middle of a river. Not all trees are scarred by a fire. It takes considerable heat to penetrate a ponderosa's thick bark. Once the growing tissue of a tree is killed, the bark will slough off and the scar becomes an exposed wound, with thinner bark on the rapidly growing sides of the catface. Once a tree has been scarred initially, it often becomes an excellent recorder of subsequent fires.

Catfaces are often seen while hiking in ponderosa pine or mixed-conifer forests and can give the observant hiker an estimate of the number of fires that burned the base of that tree after the initial scarring. Key diagnostics for a fire scar are that the catface goes all the way to the ground, and, if scarred more than once, there usually is charcoal residue on the face of the scar. There are exceptions to these rules, however if both indicators exist you can have high confidence that you have found a fire scar.

Look for fire scars on old gray stumps, which were trees logged in the early 1900s. Cross sections of old stumps are especially interesting because they sometimes reveal otherwise hidden scars (Figure 10.1). Weathered old logs and dead standing trees may also display fire scars. The older the sample, the more valuable the information, because it can provide a record further back in time. Estimating fire history from scars only works for surface fires. Forests whose natural fire regime is predominantly crown fire rarely exhibit fire scars, because most trees are killed rather than scarred.

position affects temperature and relative humidity. Aspect influences solar radiation, soil moisture content, and wind patterns. Slope steepness has a direct effect on flame length and rate of spread of a surface fire, and barriers can influence the extent of fire spread. Variability within a specific biotic community's fire regime must be recognized when determining the historical role of fire on landscape-patch dynamics. Small spatial-scale differences in fuels, topography, and climate impact fire behavior.

Table 10.1. Mean fire intervals (MFI) by forest type and historical fire regime for major forest types in the San Juan Mountains.

Forest Type	MFI, Years	Historical Fire Regime
Piñon-juniper	> 200	Periodic or infrequent high-severity crown fire
Ponderosa pine Warm-dry mixed conifer	3–11	Frequent, low-severity surface fire
Cool-moist mixed conifer	14–63	Periodic mixed-severity fire
Aspen	70–100+	Periodic high-severity crown fire
Cold-wet mixed conifer Spruce-fir	> 200	Infrequent high-severity crown fire

Sources: Data compiled from Brown and Wu 2005; Grissino-Mayer et al. 2004; Romme, Floyd, and Hannah 2006; local research on the SJNF (Wu, unpublished).

SURVEY OF BIOTIC COMMUNITIES AND FIRE

Nine generalized vegetation zones (biotic communities) were identified in Spencer and Romme (1996). We focus on five of the major forested biotic communities and their relationships to fire, climate, and forest health. For a detailed discussion of the ecologies of these biotic communities, refer to the present book's companion volume, *The Western San Juan Mountains* (Blair 1996).

Piñon-Juniper Woodlands

Piñon-juniper woodlands grow between 5,000 and 7,750 feet (1,515–2,350 m) elevation in the San Juan National Forest (SJNF). At its lower boundary, piñon-juniper grades into sagebrush, desert shrub, and desert grasslands; at its upper boundary, it grades into ponderosa pine. Piñon pine (*Pinus edulis*) and Utah juniper (*Juniperus osteosperma*) dominate the piñon-juniper community. *Juniperus* spp. dominates lower-elevation xeric sites, and piñon pine dominates higher-elevation mesic sites. Both piñon pine and Utah juniper have low resistance to fire because of the relative flammability of their foliage, low and dense branches, and thin to moderate bark thickness; juniper bark is also "shaggy." Representative stands in the eastern San Juan Mountains are found south of Highway 160, near Bayfield, in the HD Mountains and surrounding Chimney Rock Archeological Area.

The piñon-juniper fire regime in southwestern Colorado was once thought to be a frequent fire regime, as evidenced by dense piñon-juniper forests. This evaluation was made by generalizing all piñon-juniper as one biotic community and not recognizing the inherent diversity of structures and fire regimes in piñon-juniper forests, which dominate millions of hectares in the western United States. In addition, determining HRV for piñon-juniper is methodologically difficult because of a lack of direct evidence, such as fire scars, and slow reestablishment after fire. A synthesis of current knowledge (Baker and Shinneman 2004) and regional study

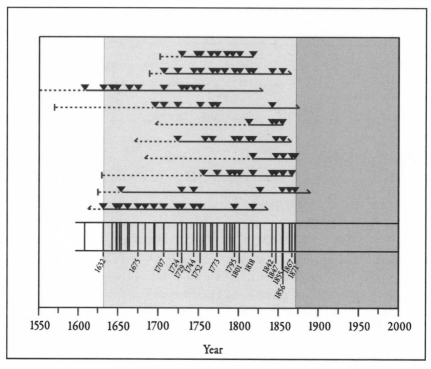

FIGURE 10.2 A fire history plot graphically displays a tree-ring fire history reconstructed from fire scars found in the forest. Each horizontal line represents a sampled log, snag, or stump with fire scars. Inverted triangles mark the fire scars. A scar year not found on any other samples is interpreted as a small isolated fire. A scar year recorded on many samples, visible where inverted triangles line up vertically, is interpreted as an extensive fire year, assuming samples are widely distributed across a large area. The composite fire chronology, located at the bottom of the plot, above the timeline, is a compilation of all fires that scarred all the samples in the collection. Dates are listed for fires found on a minimum of 25 percent of samples to filter out small isolated fires. This fire history is for the Saul's Creek area just west of the town of Bayfield.

(Romme, Floyd, and Hannah 2009) presents a different viewpoint regarding the HRV for piñon-juniper. Romme and others (2009) identified three major types of piñon-juniper stand structures and fire regimes for the western United States: piñon-juniper-grass savanna, piñon-juniper-shrub woodlands, and piñon-juniper forest.

In the eastern San Juan Mountains, piñon-juniper-shrub woodlands dominate at lower elevations. The piñon-juniper-shrub structure is characterized as having sparse to moderately dense herbs, shrubs, and trees and a periodic high-severity fire regime. Anthropogenic disturbances, such as fire suppression and grazing, along with recent climatic fluctuations in piñon-juniper, must be put into a longer climatic timeframe—from hundreds to thousands of years—to understand whether piñon-juniper types are outside their HRV. For example, a large increase in piñon recruitment in the western United States has occurred since 1976. We easily conclude that piñon recruitment is a result of human influences, such as fire suppression and grazing. However, when this recruitment pulse is put into a longer climatic time scale, it appears to be the result of high seedling survivorship following the 1950s drought in the western United States—one of the worst in the past 1,000 years—and warm, wet springs during most years since 1976 (Swetnam and Betancourt 1998).

A regional study in the Southwest by Breshears and colleagues (2005) illustrated that the 2000–2003 drought was warmer than (although not as dry as) the 1950s drought and therefore represented a global-change-type drought. Warmer temperatures coupled with drought initiated a rapid mortality of piñon trees as a result of trees having increased energy and water-stress demands in association with increased infestations of *I. confusus* (Breshears et al. 2005). Piñon mortality in the early twenty-first century is greater in magnitude and extent than mortality during the 1950s drought.

The cessation of drought conditions will not immediately restore piñon-juniper woodlands in the Southwest. Piñon trees take a minimum of several decades to reestablish into the piñon-juniper overstory structure. Therefore, the drought in the early twenty-first century, coupled with warmer temperatures and epidemic *I. confusus* populations, will have management implications that will last many decades. Another management concern is the invasion of cheatgrass (*Bromus tectorum*), an annual non-native. Cheatgrass is fundamentally changing the fire regime in piñon-juniper by introducing surface fire into a crown-fire regime through the establishment of continuous surface fuels. Repeated surface fires can prevent the establishment of piñon and juniper trees, which will alter the species composition and structure of piñon-juniper woodlands and move some of these woodlands into new plant community trajectories.

Ponderosa Pine

Ponderosa pine forests generally grow between 7,000 and 8,500 feet (2,134–2,591 m) elevation in the SJNF. Ponderosa pine (*Pinus ponderosa*) is the dominant tree species in this forest type, Gambel oak (*Quercus gambelii*) is the dominant shrub, and a variety of bunchgrasses and other herbaceous plants (forbs) grow in the understory. Ponderosa pine and Gambel oak are wonderfully adapted to frequent surface fire. A ponderosa pine's thick bark protects its trunks from fire. Its long needles protect sensitive buds. Fire also prunes the ponderosa's lower branches, increasing the difficulty for fire to jump from the ground to tree crowns. While Gambel oak lacks thick bark and is easily top-killed by very light surface fires, new stems sprout readily from the bases of burned stems, and its unharmed, intact roots give Gambel oak a competitive advantage over other plants that must generate from seeds.

Ponderosa forests in the San Juan Mountains have a frequent low-severity surface-fire regime. Average fire intervals range from three to eleven years; however, longer intervals have occasionally occurred. Typical fires burned through needles, duff, and woody debris on the forest floor, killing grasses, forbs, and Gambel oak. Fires also killed most tree seedlings and helped thin the forest while seldom killing large, mature trees. Longer fire intervals gave seedlings enough time to grow big enough to survive subsequent surface fires (Keane, Arno, and Brown 1990) and allowed a sufficient number of seedlings to survive and maintain the stand. These frequent surface fires had a major ecological role in maintaining open forest structure by keeping the forest floor relatively clear of coarse, woody debris and excessive litter buildup while regulating the densities of Gambel oak and ponderosa pine.

Frequent fires were caused mainly by lightning and were likely augmented by indigenous people at certain times of the year. The majority of lightning fires were small. In years of average precipitation, live grasses and forbs were too moist to burn during the summer fire season. Because fuels were too moist, fires were small and patchy and limited to flammable pine-needle litter directly under trees.

Extensive surface fires that burned hundreds to thousands of acres occurred on a decadal time scale (Brown and Wu 2005; Grissino-Mayer et al. 2004). On average, they occurred every dozen years or so. Such large landscape-scale fires were set up by one or more years of above-average precipitation followed by drought (Brown and Wu 2005; Swetnam and Baisan 1996; Swetnam and Betancourt 1998). Wet years created a flush of grasses and forbs. In the subsequent drought years, grasses and forbs were dead or had very low moisture content and provided large continuous fuel beds that burned easily. This climate pattern is linked to the El Niño–La Niña Oscillation, which occurs roughly every four–seven years (Swetnam and Baisan 1996). In these major fire years, many lightning ignitions across the ponderosa landscape burned thousands of acres throughout the summer, shrouding the region in smoke for weeks at a time. Surface fire was the dominant mode of fire behavior; however, crown fires could occur if a fire encountered a pocket of dense pine in its path. Crown fires were unusual because of ponderosa's open stand structure.

In ponderosa forests in the San Juan Mountains, large-scale surface fires stopped abruptly in the mid- to late 1800s (figure 10.2). Although the actual date of the last fire varies from site to site, tree-ring studies across the San Juan Mountains have shown that 1880 is a reasonable date to mark the general beginning of fire exclusion for the area. The term *fire exclusion* is often used to refer to the cessation of fire as a result of European-American settlement activities.

During European-American settlement, the sudden end of fires elsewhere in the Southwest has been attributed to widespread sheep and cattle grazing (Savage and Swetnam 1990; Touchan, Swetnam, and Grissino-Mayer 1995). In the San Juan Mountains, thousands of sheep and cattle chronically overgrazed, killing or stunting grasses and forbs that had formerly provided connected fuel beds for large surface fires. In the early 1900s, land management agencies adopted a policy of total fire suppression. At the time, fires were seen as a destructive force against which forest resources must be protected. Thus fires and the maintenance role they played in forests were effectively removed from all forest types. Without fire and competing herbaceous cover, trees and shrubs could fill in the once open ponderosa stands.

Historical ponderosa forests (plates 31 and 32) were predominantly open, with a patchy distribution of trees of various sizes and age classes (Cooper 1960; Fulé, Covington, and Moore 1997). Pockets of high-density pine could develop on moist sites, on protected sites, or in sites that escaped fire for unusually long periods through sheer chance. However, large dense stands of ponderosa pine were very uncommon prior to European-American settlement, given these forests' natural historical fire regime.

As a result of fire suppression, grazing, and logging, ponderosa forests are far outside their HRV. Today, in the early twenty-first century, the majority of ponderosa forests can be characterized as even-age, second-growth stands. These forests are visually very homogeneous and lack old-growth trees. Modern ponderosa stands are composed of trees of generally uniform size and age, as seen along Highway 160 between Durango and Pagosa Springs, along Saul's Creek east of Bayfield, and in the Piedra River drainage. These same patterns in structure can be found south of Pagosa Springs, along Highway 84 to Chama, New Mexico.

Instead of a predominantly open ponderosa landscape with infrequent dense stands, the landscape has developed into a mostly closed-canopy forest punctuated by some open stands. Herbaceous understories are greatly reduced or absent in the densest stands. Gambel oak dominates the understory and commonly grows directly underneath ponderosa pine. These shrubs act as ladder fuel, enabling surface fire to spread into the crowns of overstory trees. As a result, the ponderosa's low-severity surface-fire regime is being replaced by one of high-severity crown fires.

Recent fires demonstrate that the ponderosa pine's fire regime has changed fundamentally. The 2002 Missionary Ridge Fire burned during one of the most severe drought years on record. It was human-caused and burned over 70,000 acres, through all forest types. Of the roughly 12,000 acres of ponderosa pine that burned, 65 percent burned in a moderate- to high-severity crown fire, in which 50–100 percent of all trees and shrubs were killed (USFS 2002). If frequent fires had continued throughout the twentieth century, forest structure would have been more open and less susceptible to crown fires of moderate to high severity (table 10.2). The dead standing trees that remain from the Devil Creek Fire, north of the junction of Highway 160 and State Highway 151, and the Ute Fire, at the campground east of the highway junction, can easily be seen from the highway. The size and frequency

Table 10.2. Major fires in ponderosa pine and warm-dry mixed-conifer forests in the San Juan Mountains, 2000–2005. The sizes of these fires are not outside the HRV for ponderosa-pine or warm-dry mixed-conifer forests, but fire behavior is. All the fires, even the Ute Campground fire, had significant crown-fire acreage.

Year	Fire	Total Acres
2000	Cabezon	330
2001	Martinez Canyon	15
2002	Missionary Ridge	70,121
2002	Valley	939
2003	Devil Creek	235
2003	Trail Ridge	89
2003	Bolt	2,160
2004	Devil Mountain	60
2005	Ute Campground	5

Source: From USFS 2007.

of recent fires is within HRV, despite artificial constraint by suppression efforts, while these fires' crown-fire behavior is outside HRV.

Pockets of historical conditions can still be found in ponderosa forests today. In the Turkey Springs area north of Pagosa Springs, we can still see old-growth, orange- or yellow-bark ponderosa amid the dense infill of younger ponderosa. We can visualize a historically typical forest by mentally erasing the dense infill from the scene.

Ponderosa forests are the most in need of restoration because they have been the most greatly altered by fire exclusion. Continued fire suppression is ecologically detrimental. Even if ponderosa forests could be fire-proofed, forest health would continue to decline. Dense stands of ponderosa pine are at high risk for bark beetle (*Dendroctonus brevicomis* and *D. ponderosae*) attack. Today, beetle activity is still relatively light and focused on large old-growth trees and small pockets of trees. However, the potential for a major outbreak exists because of dense forest conditions, an abundance of trees in the ideal size range (> 6 in [15.2 cm] diameter) and age range (> 80 years), warmer-than-average temperatures, and recurring drought (Sartwell and Stevens 1975).

Thinning and burning would improve ecological health and reduce fire hazard (crown fire). Fire-hazard mitigation, however, is the social driver for ponderosa forest restoration, especially with heavy urbanization in these forests. In ponderosa forests where ecological and social values converge, many challenges remain to reintroducing fire and restoring these forests. First, the sheer scale of the problem is daunting. Ponderosa forests cover over 400,000 acres (1,618 km²) of US Forest Service and Bureau of Land Management land in the San Juan Mountains. Time, money, and workforce capacity are inadequate to handle such a workload. Second, social constraints (mainly smoke intolerance) persist. Most present-day residents of the West would not tolerate the smoky conditions that occurred in historically active fire years. While some residents have health issues, smoke is a nuisance for most. This public attitude can limit the amount of prescribed burning land managers conduct, thereby slowing forest-restoration efforts. Ultimately, smoke will not be a choice because wildfires cannot be prevented in ponderosa forests.

Aspen

In a sea of conifer, aspen is crucial for landscape and plant-species diversity, wildlife habitat, and ecosystem processes such as biogeochemical cycling (Turner et al. 2003). Aspen grow between 7,000 and 11,000 feet (2,100–3,300 m) elevation. It is most prevalent on north- and east-facing slopes and in drainages where deep, loamy soils offer ample nutrients (plate 33).

Fire managers often use aspen as a fuel break to stop fire, and some consider aspen an "asbestos" forest type. It is prevalent in wet, cool environments, and its deciduous leaves have low flammability (Jones and DeByle 1985). Aspen is frequently one of the first species to appear following disturbance and is dependent on periodic disturbances for regeneration. A stand of aspen that shares the same root system is called a clone and is considered to be one organism, because it has the same genetic makeup.

Recent genetic research on aspen suggests that an aspen clone identified solely on similar leaf size and color and branching habits may, in fact, be an aspen grove. Aspen groves are composed of two or more genetic varieties that are most likely the result of individual seedling establishment, previously considered rare for aspen (Mock 2008). Although aspen can generate from seeds, the primary natural disturbance that initiates aspen regeneration is fire. Geomorphic events and wind are natural disturbances that can also promote regeneration. When such a disturbance kills aboveground aspen stems, the source of auxin, a hormone that prevents sucker resprouting, is removed. Most aspen resprout rapidly in dense thickets from root suckers within 6 inches (15 cm) of the soil surface and can reach up to 3 feet (1 m) in height during the first year following fire (DeByle and Winokur 1985).

Aspen can be viewed as having both low and very high fire resistance. Individual aspen stems are considered to have low fire resistance because of their thin bark, which offers little protection from surface fires. In contrast, the aspen clone has very high fire resistance, thanks to its evolutionary adaptation to respond favorably to fire through resprouting. In contrast, individual stems of aspen are considered short-lived, with lifespans between 100 and 150 years. In the eastern San Juan Mountains, representative stands are found along Highway 160 on both sides of Wolf Creek Pass. Young reinitiating stands of aspen can be found north of Durango and around Vallecito Reservoir, in the Missionary Ridge area that burned in 2002.

Two major aspen community types are present in the San Juan Mountains: stable, or persistent; and seral, or transitional. Multilayered aspen stems of different ages, with no conifer invasion, characterize stable aspen stands. These stands tend to be located at lower elevations in areas adjacent to ponderosa pine, although stable stands also exist at higher elevations. Historically, fires within stable aspen stands were most likely frequent, low-severity surface fires that did not burn into tree canopies (Romme et al. 2001). Frequent fires prevent conifer seedlings from reaching reproductive age, thereby eliminating a conifer seed source. Even in the absence of fire, these stands have remained stable because of a paucity of conifer seed sources.

Fires of moderate intensity produce the highest amount of sprouting, which allows stable aspen stands to persist (Parker and Parker 1983). Stable aspen is more common in the western San Juan Mountains near Mancos and Dolores, where the climate is warmer and drier than in the eastern San Juan Mountains. In the eastern San Juan Mountains, the majority of aspen stands are seral and are characterized by a single-layer canopy of aspen, all of the same age, with conifer regeneration (Mueggler 1985). Seral stands are prevalent at higher elevations, where spruce and fir become established—typically decades after a significant disturbance event, such as a stand-replacing fire or an avalanche. Conifer trees provide the fuel for this periodic, high-severity crown-fire regime. For aspen to successfully regenerate, fire needs to be of moderate to high intensity (Bartos, Brown, and Booth 1994). Seral aspen stands also exist at lower elevations in ponderosa and mixed-conifer stands, where white fir and Douglas fir replace aspen.

Determining HRV is more difficult in aspen stands than in other forested biotic communities because aspen trees do not readily form fire scars, and fire is generally lethal to aspen trees. As a result, the ages of post-fire tree-regeneration stands are most often used to identify past fire timing and severity. Multiple lines of tree-ring evidence used to determine age cohorts, fire scars on other trees adjacent to a burned area, and other records are the best methods to determine HRV in aspen (Margolis, Swetnam, and Allen 2007).

Studies in the San Juan Mountains indicate that some seral aspen stands burned approximately every seventy years, while other stands did not burn for over a century or longer (Romme et al. 2001). When assessing whether seral aspen stands are outside their HRV, stand age should be taken into consideration. Numerous fires occurred in the 1880s because of a drought. These fires resulted in large areas of seral aspen stands that were established in one short period of time. Many of the stems in these aspen stands are about 120 years in age, are regarded as old and decadent, and are reaching maximum longevity before natural succession to conifer trees.

In a snapshot view of this current situation, aspen would appear to be declining as a result of fire suppression and other anthropogenic influences. Another interpretation that incorporates a longer temporal scale suggests that the lack of fire in the twentieth century resembles the lack of fire in the late 1700s to early 1800s in the

Southwest and that the high densities of old aspen currently seen in the San Juan Mountains with conifer invasion are most likely not outside their HRV (Romme et al. 2009). More detailed research is needed to confirm or reject whether fire suppression has impacted seral aspen stands.

Aspen can be managed using different techniques, depending on where it is geographically located. Management of aspen groves near existing roads may use harvesting and prescribed fire to initiate regeneration. Wildland fire should be allowed to promote regeneration in remote aspen groves. The extreme and sudden aspen decline (SAD) observed in the San Juan Mountains since 2004 is attributed primarily to the drought that started in 2001 along with the old age of aspen, coupled with secondary biological agents such as Cytospora canker (generally caused by the fungus *Valsa sordida*), aspen bark beetles (*Trypophloeus populi* and *Procryphalus mucronatus*), poplar borer (*Saperda calcarata*), and the bronze poplar borer (*Agrilus liragus*) (Worrall et al. 2008). Other biological and physical factors associated with SAD include low elevations, south and southwestern aspects, low-density stands, and open, flat topography (Worrall et al. 2008).

By definition, SAD is characterized by sudden synchronous branch dieback, crown thinning, and aspen mortality at a landscape scale not associated with aggressive pathogens and insects (Worrall et al. 2010). As of 2006, approximately 10 percent of the aspen biotic community on the Mancos-Dolores Ranger District, SJNF, had been affected by SAD, which may result in clones dying and long-term aspen-cover loss. Sudden aspen decline is differentiated from normal conifer succession in older, seral aspen stands because of how immediately and rapidly the decline has occurred. SAD is currently less prevalent in the eastern San Juan Mountains than in the western San Juans because the eastern side is cooler and wetter. More studies are needed to determine what biological and physical factors are responsible for SAD and what percentage of the landscape will be affected by SAD in the coming years. Research indicates, however, that warm-dry growing seasons will result in recurrence of SAD (Worrall et al. 2010).

Mixed-Conifer

Mixed-conifer is a generic term for any forest composed of several conifer species. The composition of such forests varies throughout the country, within a region such as the Rocky Mountains, and even within an area such as the San Juan Mountains. Mixed-conifer forests cover over 250,000 acres—approximately one-fifteenth of the San Juan National Forest—and grow between ~7,100 and 9,800 feet (2,164–2,987 m) elevation. Most lie within the eastern portion of the SJNF.

Mixed-conifer forests of the eastern San Juan Mountains are some of the region's most diverse and complex forests. Over half a dozen conifer species, plus aspen and a wealth of understory woody and herbaceous plants, are found in the mixed-conifer. It is a transitional forest between ponderosa pine and spruce-fir. As one moves up in elevation, ponderosa pine drops out until firs and spruces dominate the forest. Mixed-conifer forests in the eastern San Juans can be described along a continuum, from warm-dry to cool-moist to cold-wet (Romme, Floyd, and Hanna 2009). The differences in species composition and dominant fire regime between the phases are driven primarily by moisture and temperature.

Adjacent to ponderosa forests and higher in elevation are warm-dry mixed-conifer forests. In the mixed-conifer continuum, it grows at lower elevations and on southern aspects. Ponderosa pine is the dominant tree species, followed by white fir (*Abies concolor*) and Douglas fir (*Pseudotsuga menziesii*). Gambel oak is common in the understory. Warm-dry mixed-conifer share many characteristics with ponderosa pine forests, including a frequent low-severity fire regime.

Cool-moist mixed-conifer is the intermediate phase and makes up the largest component of the mixed-conifer forest type. Species from warm-dry mixed-conifer, plus blue spruce (*Picea pungens*), southwestern white pine (*Pinus strobiformis*), and Rocky Mountain juniper (*Juniperus scopulorum*), are found in the cool-moist mixed-conifer. The heart of the cool-moist mixed-conifer forest is dominated by white fir and Douglas fir. Ponderosa pine is no longer dominant, although individual trees can be found in small patches on warm sites, exposed ridges, and meadow edges. Gambel oak is present, although less common, and is joined by a host of other shrubs, such as serviceberry (*Amelanchier alnifolia*), snowberry (*Symphoricarpos rotundifolius*), chokecherry (*Prunus virginiana*), and Rocky Mountain maple (*Acer glabrum*).

The cold-wet mixed-conifer is the upper extreme of the mixed-conifer continuum, occurring at the highest elevations of any mixed-conifer in the area and on north aspects. Engelmann spruce (*Picea engelmannii),* subalpine fir (*Abies lasiocarpa*), corkbark fir (*Abies lasiocarpa var. arizonica*), Douglas fir, and white fir grow together in this phase. The presence and continued dominance of Douglas fir in many stands is what distinguishes this forest type from true spruce-fir forests. The fire regime of cold-wet mixed-conifer will be discussed with that of spruce-fir because these forests share the infrequent high-severity crown-fire regime.

An excellent place in which to see all three types of mixed-conifer forests is in the Piedra Area in the SJNF, north of Highway 160 and roughly midway between Bayfield and Pagosa Springs. It is a roadless area that has never been logged and is managed to retain its wilderness characteristics. Fire management, grazing, and recreation are the main management activities there. The southernmost tip and lowest elevations display pure ponderosa forests. As one moves north and upward in elevation, these stands quickly transition into warm-dry mixed-conifer forests. Cool-moist mixed-conifer cover the bulk of the Piedra Area. At the northernmost end of the Piedra Area, just below the Weminuche Wilderness boundary, one enters cold-wet mixed-conifer and spruce-fir forests. Aspen grows throughout all phases of the Piedra Area's mixed-conifer.

The boundaries between phases are not uniform across the landscape, because more than elevation controls moisture conditions. Topographic position and aspect also strongly influence the distribution of mixed-conifer forests. A hike down the Piedra River Trail provides an excellent view of the influence of aspect. The section of the Piedra River running through the northern portion of the Piedra Area runs roughly northeast to southwest, from Sally's Overlook through the Second Box Canyon. Cool-moist stands mixed with aspen occupy the north-facing aspects sloping out of the wide valley, while pure ponderosa and warm-dry mixed-conifer stands grow on the south-facing aspects.

Warm-Dry Mixed-Conifer

The historical fire regime in the warm-dry mixed-conifer was frequent and low severity, as was that of the lower-elevation ponderosa pine zone. Fire intervals were not significantly different from those in ponderosa forests, ranging from a few years to a few decades. Surface fires played the same maintenance role as in ponderosa forests. However, warm-dry mixed-conifer developed denser forest structure than ponderosa forests because of moister conditions. During intervals between fires, white fir and Douglas fir continuously recruit. Their prolific seeding would overwhelm ponderosa pine regeneration if it were not for frequent fires. White fir and Douglas fir are easily killed by fire because their thin bark and low-lying branches make them susceptible to even light surface burns. Some white fir and Douglas fir survive fire, however, and become part of the overstory tree canopy, to be the seed source for their continual regeneration in warm-dry mixed-conifer stands. Limited crown fire could occur in warm-dry mixed-conifer, killing dense patches of trees. Mature ponderosas could sometimes also be killed.

As in ponderosa forests, fire exclusion since Euro-American settlement has pushed warm-dry mixed-conifer forests outside their HRV. Their stand structure and fire regime have fundamentally changed without fire. White fir has increased and filled in stands throughout the warm-dry mixed-conifer. Tree-species dominance is shifting from ponderosa pine to white fir (plate 34). Dense forest conditions hinder ponderosa pine regeneration and favor shade-loving white fir. Significant to current conditions in warm-dry mixed-conifer without fire is the virtual absence of ponderosa pine regeneration. If this trend continues, ponderosa pine, the defining species of the warm-dry phase, will gradually disappear from these stands. The dense forest conditions also provide a fuel arrangement that supports large-scale crown fire, which is no longer limited to the most extreme weather conditions.

Management issues discussed for ponderosa pine forests also apply to warm-dry mixed-conifer forests. Fuel-reduction and forest-thinning goals are both ecologically and socially aligned. The conversion of warm-dry mixed-conifer forests to white fir gives its need for restoration greater urgency.

Cool-Moist Mixed-Conifer

Historically, fires in cool-moist mixed-conifer forests occurred decades to a century or more apart, with varied fire intensities. Its historical fire regime is often referred to as a mixed-severity regime. This is perhaps the most difficult and complex fire regime to characterize. Surface- and crown-fire behavior can occur during the same fire event. How large a fire can grow and which type of fire behavior dominates during a fire event depend on the moisture of live and dead fuels, as well as fuel continuity on the ground and in tree canopies. Because cool-moist mixed-conifer forests are very productive, fuel availability is not usually the limiting factor in fire occurrence and size. Long-term climate patterns, such as drought, and weather conditions, such as temperature, humidity, and wind, which influence fuel moisture and directly drive fire behavior, are more important to fire behavior and size.

Under average moisture and weather conditions, surface fire with passive crown fire (torching) would likely dominate fire behavior. Fuels too moist to burn would limit fire size. Burns are often described as "dirty," meaning the fire did not completely burn fuels and a great deal of dead and live fuel was left unburned. If the fire encounters a patch of heavy fuels, the surface fuels could ignite the tree crowns and cause a torching event, also referred to as passive crowning. If the live needles of the canopy are dry enough, fire can burn through the canopy, independent of surface fuels, for as long as the forest canopy is dense enough to carry fire. An independent crown fire is also referred to as active crowning. The drier the forest conditions, the greater the likelihood that active crown fire will dominate fire behavior in large fires. Prior to European-American settlement (~pre-1880), cool-moist mixed-conifer burned, on average, every 20 to 40 years. Extensive stand-replacing crown fires appear to have occurred about once every 100 years or more.

The effects of fire exclusion and suppression in cool-moist mixed-conifer forests are more difficult to identify and interpret than they are in warm-dry mixed-conifer and ponderosa forests. Although fires have been suppressed since the late 1800s, forest structure has not fundamentally changed. Cool-moist mixed-conifer forests are naturally dense. Large-scale crown fire is part of this forest's historical fire regime. However, surface fires have been removed from this mixed-severity fire regime, and large portions of cool-moist mixed-conifer forests now appear primed for crown fire.

Cool-moist mixed-conifer is outside its HRV, but not to the extent of ponderosa pine and warm-dry mixed-conifer. The primary concern for cool-moist mixed-conifer is a matter of landscape diversity. We do not see uncharacteristic conditions at the forest-stand level. The range of open to dense stand structures that presently exists probably also occurred before fire exclusion. What appears different in the present landscape is the lack of diversity, or narrowness in range, of stand age and structure compared to the past. At the scale of entire landscapes, the cool-moist mixed-conifer forests appear dominated by stands of similar age and structure.

Historically, variation in fire frequency, timing, and severity would have created a patchwork of diversely aged stands across the landscape, making some more prone to crown fire than others. This mosaic of stand ages also created diversity in plant and animal habitats across the landscape. Cool-moist stands have aged together for over a century, resulting in a more homogeneous landscape of mature forests than was probably found before 1880.

Management issues for the cool-moist mixed-conifer include the general loss of diversity in the landscape, the threat of major insect and disease outbreaks, and whether to allow fire to play its natural role. Lack of diversity makes cool-moist mixed-conifer forests less resilient to insect and disease outbreaks, fire, and other disturbances. Older trees and denser forests are more susceptible to insects and diseases. Insect and disease activity has been increasing across the San Juan Mountains since the late 1990s. Spruce budworm (*Choristoneura occidentalis*), root rot (*Armillaria mellea*), fir engravers (*Scolytus* spp.), and bark beetle (*Dendroctonus* spp.) are all active. Drought and warmer temperatures, on top of dense forest conditions, have triggered recent outbreaks. Current forest health issues are remarkable in that practically all species of trees are being attacked.

Spruce-Fir and Cold-Wet Mixed-Conifer

Spruce-fir and cold-wet mixed-conifer forests grow between 9,000 and 11,800 feet (2,743–3,597 m) elevation. Spruce-fir is the predominant forest type found within the Weminuche and South San Juan Wildernesses and lies at a higher elevation than bordering cold-wet mixed-conifer. Because cold-wet mixed-conifer has the same fire regime as spruce-fir, we discuss them interchangeably.

Engelmann spruce (*Picea engelmannii*) dominates most spruce-fir stands, although there are stands in which subalpine fir and corkbark fir are collectively more numerous. In general, the proportion of firs to Engelmann spruce decreases with elevation. Most spruce-fir stands at or near treeline are composed purely of spruce. Aspen grows throughout both spruce-fir and cold-wet mixed-conifer forests up to approximately 11,000 feet (3,300 m) (figure 10.3).

The historical fire regime in spruce-fir forests was infrequent high-severity crown fire. Lightning strikes were common and probably started numerous fires that never grew to any size. Major fires occurred about once every 200+ years (Alington 1998; Romme, Floyd, and Hanna 2009). Fire behavior ranged from smoldering duff fires to crown fires with flame heights over 100 feet. The 200+-year fire interval refers to large stand-replacing crown fires that killed practically all trees within a burn area. In contrast to the maintenance role surface fires performed in ponderosa stands, crown fires reset spruce-fir stands back to the beginning of their successional process.

Spruce-fir forests were usually too wet to burn often. Several conditions must coincide to facilitate a large-scale crown fire: stand structure must be dense enough

FIGURE 10.3 A typical spruce-fir landscape. Note the dense and continuous tree crowns. Aspen stands break up the spruce-fir in the background. Aspen stands are identifiable by their light-green color and smooth texture. Photo by Mark D. Roper.

to carry fire, drought conditions must allow dead and live fuels to become dry enough to burn, and a dry windy day with lightning is needed. Any one of these conditions is a common occurrence during the several centuries a stand lives; since all conditions rarely coincide, centuries pass between fires.

Of the necessary conditions needed for crown fire, stand structure is the most limiting because of the time required for spruce-fir forests to develop into stands dense enough to carry fire. Spruce-fir can take over a century to develop a dominant cohort (Romme, Floyd, and Hanna 2009). Once stand structure is dense enough, moisture becomes the most limiting factor. Unlike lower-elevation ponderosa pine, spruce-fir forests do not always dry out seasonally. In a typical season, the snow-pack can linger well into June, the hottest and driest month. The summer monsoon, which typically brings rain to the high country, usually follows. The high country can experience warm and dry periods in the fall, creating favorable fuel conditions for fire. By then, however, lightning is uncommon. In some instances, fires started during the summer survive, or hold over, into the fall to rekindle and burn large acreages if conditions allow.

Large crown fires require dry conditions associated with severe drought. Poor snowfall during a preceding winter resulting in a thin snowpack, followed by a hot summer and a warm, dry fall, is a climate recipe for large-scale fire. However, because high-elevation forests are generally wet environments, multiyear droughts are needed to break the moisture cycle in these forests to dry them sufficiently for large-scale crown fire. The drivers of long- and short-term drought have been the focus of recent research and appear to be linked to cycles in global and regional climate patterns (Sibold and Veblen 2006; Swetnam,and Betancourt 1998).

Most scientists and land managers feel that spruce-fir and cold-wet mixed-conifer forests are not outside their HRVs in spite of fire suppression. With average fire intervals of 200+ years, spruce-fir still appear to be within its average fire interval. In fires that burned in the San Juan Mountains in 2002 and 2003, fire behavior and forest structure in spruce-fir and cold-wet mixed-conifer did not appear to be uncharacteristic. However, past fire suppression might have had some impact in terms of coarse landscape diversity and interaction with insect epidemics. Although fire suppression in lower stands of spruce-fir and cold-wet mixed-conifer might have reduced fire occurrence there, fires coming from forests at even lower elevations could have the energy to burn into the higher elevations and burn out an area before losing energy in the cooler, wetter spruce-fir. The ecological significance of these fires for the spruce-fir landscape still needs to be determined. Some stands, especially lower ones, might have missed burning, but overall the dense spruce-fir forests on the landscape today are the same stands that covered the landscape in the pre-1880 era.

An HRV assessment of spruce-fir and cold-wet mixed-conifer forests is not complete unless other disturbances are also considered. Fires are so rare that windstorms and insect outbreaks could be considered the chronic disturbances that shape high-elevation forests. Windstorms blow down mature trees, and small-scale blow-down is a fairly common occurrence. Small-scale blow-down supports endemic insect populations, especially spruce beetle (*Dendroctonus rupifinis*), and can equal or surpass the force of a major crown fire. Current forest conditions, which are dominated by mature stands, and a warmer and drier climate are ideal for a spruce-beetle epidemic. Small to moderate spruce-beetle outbreaks and localized areas of increased beetle activity causing heavy mortality have been observed throughout the eastern San Juan Mountains in the early twenty-first century. Patches of beetle-killed trees are visible on Wolf Creek Pass, and extensive beetle kill is occurring in the Weminuche Wilderness. Beetle-killed stands can be removed by fire, but these stands are also just as likely to sit for decades without fire.

Fire management issues in spruce-fir and cold-wet mixed-conifer forests appear less complex than in lower-elevation forests. Unlike lower-elevation forests, wildland-urban interface is not a major issue in spruce-fir and cold-wet mixed-conifer, because most of these forests are in designated wilderness and buffered by national forest land. Because fire is uncommon and the forests are so remote, the ideal management strategy would be to not suppress any lightning ignitions and allow them burn. Most of these fires would not burn beyond their lightning-struck trees. Eventually, however, one of these lightning-induced fires could spread and burn hundreds of thousands of acres.

The major obstacles to allowing fires to burn in the spruce-fir are smoke and public perception. Spruce-fir fires need extremely dry conditions. It will require a great deal of outreach to explain to the general public why land management agencies will allow fires in high-elevation forests to burn while, at the same time, they aggressively suppress fires and institute fire bans in lower-elevation forests. High-elevation forests have the added disadvantage of being crown-fire regimes. Crown fires are unpredictable, scary events not widely welcomed by the public. Even if fires are remote, smoke can drain down valleys and create hazy skies. Therefore, smoke management and tolerance are crucial to allowing fires in spruce-fir and cold-wet mixed-conifer forests.

CONCLUSION

The shared history of all forest types in the San Juan Mountains includes the exclusion and active suppression of fire since the late 1880s. Fire suppression, in addition to the legacies of overgrazing and high-grade logging, has produced landscapes that are far outside the HRV of ponderosa and warm-dry mixed-conifer forests. Spruce-fir and other forests with longer average fire intervals appear only slightly affected and generally not outside their HRV.

Fire and other disturbances illustrate that forests change over space and time. Today's forests appear to be entering a period of potentially rapid change caused by a warming climate. Whether recent warming is the result of greenhouse-gas proliferation or natural long-term climatic fluctuation is irrelevant to the impacts a warmer climate will have on fire regimes. Since the 1980s, fire sizes have increased, and middle- and higher-elevation forests are burning with increased frequency across the western United States (Westerling et al. 2006). In 2002, the Missionary Ridge Fire burned 8,000 acres in high-elevation forests (and 70,000 acres overall) in the San Juan Mountains. Climate change underscores the need to restore forests using frequent fire regimes. Longer, hotter, and drier summers will put dense ponderosa pine and warm-dry mixed-conifer forests at risk of crown fire. Upper-elevation forests that have not burned will also be more likely to burn as crown fires under warmer climatic conditions. While crown fires are within the spruce-fir HRV, post-fire vegetation recovery in spruce-fir, along with all other forest types, may be altered. Forest communities may change in species composition, dominance, and density as trees and other plants seek refuge from temperature and water stress and redistribute geographically. Regardless of the new species distributions, however, fire will always remain a part of western forested landscapes. Forest community composition and structure and fire regimes may change in the future, but fire will always be a dominant natural disturbance in these landscapes (Westerling et al. 2006).

REFERENCES
Alington, C. 1998. Fire History and Landscape Pattern in the Sangre de Cristo Mountains, Colorado. PhD dissertation, College of Natural Resources, Colorado State University, Fort Collins.

Baker, W. L., and D. J. Shinneman. 2004. Fire and Restoration of Pinon-Juniper Woodlands in the Western United States: A Review. *Forest Ecology and Management* 189:1–21.

Bartos, D. L. J. K. Brown, and G. D. Booth. 1994. Twelve Years Biomass Response in Aspen Communities following Fire. *Journal of Range Management* 47:79–83.

Blair, R., ed. 1996. *The Western San Juan Mountains: Their Geology, Ecology, and Human History*. Niwot: University Press of Colorado.

Breshears, D. D., N. S. Cobb, P. M. Rich, K. P. Price, C. D. Allen, R. G. Balice, W. H. Romme, J. H. Kastens, M. L. Floyd, J. Belnap, J. J. Anderson, O. B. Myers, and C. W. Meyer. 2005. Regional Vegetation Die-off in Response to Global-Change–Type Drought. *Proceedings of the National Academy of Science* 102:15144–15148.

Brown, P. M., and R. Wu. 2005. Climate and Disturbance Forcing of Episodic Tree Recruitment in a Southwestern Ponderosa Pine Landscape. Ecological Society of America. *Ecology* 86(11):3030–3038.

Cooper, C. F. 1960. Changes in Vegetation, Structure, and Growth of Southwestern Pine Forests since White Settlement. *Ecological Monographs* 30:129–164.

DeByle, N. V., and R. P. Winokur, eds. 1985. *Aspen: Ecology and Management in the Western United States*. General Technical Report RM-119. Fort Collins, CO: USDA Forest Service.

Fulé, P. Z., W. W. Covington, and M. M. Moore. 1997. Determining Reference Conditions for Ecosystem Management of Southwestern Ponderosa Pine Forests. *Ecological Applications* 7:895–908.

Grissino-Mayer, H. D., W. H. Romme, M. L. Floyd, and D. Hanna. 2004. Climatic and Human Influences on Fire Regimes in the Southern San Juan Mountains, Colorado, USA. *Ecology* 85:1708–1724.

Jones, J. R., and N. V. DeByle. 1985. Fire. In N. V. DeByle and R. P. Winokur, eds., *Aspen: Ecology and Management in the Western United States*. General Technical Report RM-119. Fort Collins, CO: USDA Forest Service, 77–81.

Keane, R. E, S. F. Arno, and J. K. Brown. 1990. Simulating Cumulative Fire Effects in Ponderosa Pine/Douglas-Fir Forests. *Ecology* 71(1):189–203.

Margolis, E. Q., T. W. Swetnam, and C. D. Allen. 2007. A Stand-Replacing Fire History in Upper Montane Forests of the Southern Rocky Mountains. *Canadian Journal of Forest Research* 37:2227–2241.

Mock, K. 2008. Summary of Recent Aspen Genetic Research in Utah. In P. C. Rodgers, ed., Summary and Abstracts from Sudden Aspen Decline (SAD) Meeting. Fort Collins, CO, February 12–13, 2008. Unpublished document.

Mueggler, W. F. 1985. Vegetation Associations. In N. V. DeByle and R. P. Winokur, eds., *Aspen: Ecology and Management in the Western United States*. General Technical Report RM-119. Fort Collins, CO: USDA Forest Service, 45–55.

Parker, A. J., and K. C. Parker. 1983. Comparative Successional Roles of Trembling Aspen and Lodgepole in the Southern Rocky Mountains. *Great Basin Naturalist* 43:447–455.

Pyne, S. J., P. L. Andrews, and R. D. Laven. 1996. *Introduction to Wildland Fire*, 2nd ed. New York: John Wiley & Sons.

Romme, W. H., M. L. Floyd, and D. D. Hanna. 2009. *Historical Range of Variability and Current Landscape Condition Analysis: South Central Highlands Section, Southwestern Colorado and Northwestern New Mexico*. Fort Collins: Colorado Forest Restoration Institute.

Romme, W. H., M. L. Floyd-Hanna, D. D. Hanna, and E. J. Bartlett. 2001. Aspen's Ecological Role in the West. In *Sustaining Aspen in Western Landscapes: Symposium Proceedings RMRS-P-18*. Fort Collins, CO: USDA Forest Service.

Sartwell, C., and R. E. Stevens. 1975. Mountain Pine Beetle in Ponderosa Pine: Prospects for Silvicultural Control in Second-Growth Stands. *Journal of Forestry* 73:136–140.

Savage, M., and T. W. Swetnam. 1990. Early 19th-Century Fire Decline following Sheep Pasturing in a Navajo Ponderosa Pine Forest. *Ecology* 71(6):2374–2378.

Sibold, J. S., and T. T. Veblen. 2006. Relationships of Subalpine Forest Fires in the Colorado Front Range to Interannual and Multidecadal Scale Climate Variation. *Journal of Biogeography* 33(5):833–842.

Spencer, A. W., and W. H. Romme. 1996. Ecological Patterns. In R. Blair, ed., *The Western San Juan Mountains: Their Geology, Ecology, and Human History*. Niwot: University Press of Colorado, 128–142.

Swetnam, T. W., and C. H. Baisan. 1996. Tree-Ring Reconstructions of Fire and Climate History in the Sierra Nevada and Southwestern United States. In T. T. Veblen, W. Baker, G. Montenegro, and T. W. Swetnam, eds., *Fire and Climatic Change in Temperate Ecosystems of the Western Americas*, 158–195. New York: Springer.

Swetnam, T. W., and J. L. Betancourt. 1998. Mesoscale Disturbance and Ecological Response to Decadal Climatic Variability in the American Southwest. *Journal of Climate* 11:3128–3147.

Touchan, R., T. W. Swetnam, and H. D. Grissino-Mayer. 1995. Effects of Livestock Grazing on Pre-Settlement Fire Regimes in New Mexico. In J. K. Brown, tech. coord., *Proceedings: Symposium on Fire in Wilderness and Park Management; 1993 March 30–April 1, Missoula, MT*. General Technical Report INT-GTR-320. Ogden, UT: USDA Forest Service, Intermountain Research Station.

Turner, M. G., W. H. Romme, R. A. Reed, and G. A. Tuskan. 2003. Post-Fire Aspen Seedling Recruitment across the Yellowstone (USA) Landscape. *Landscape Ecology* 18:127–140.

US Forest Service (USFS). 2002. *Missionary Ridge and Valley Fire Burned Area Stabilization and Rehabilitation Plan* (BAER Report). San Juan Public Lands, 15 Burnett Court, Durango, CO 81301.

———. 2007. San Juan History Layer. GIS Library. San Juan National Forest, 15 Burnett Court, Durango, CO 81301.

Westerling, A. L., H. G. Hidalgo, D. R. Cayan, and T. W. Swetnam. 2006. Warming and Earlier Spring Increases Western U.S. Forest Wildfire Activity. *Science* 313:940–943.

White, P. S., and J. L. Walker. 1997. Approximating Nature's Variation: Selecting and Using Reference Information in Restoration Ecology. *Restoration Ecology* 5:338–349.

Worrall, J. J., L. Egeland, T. Eager, R. A. Mask, E. W. Johnson, P. A. Kemp, and W. D. Shepperd. 2008. Rapid Mortality of Populus tremuloides in Southwestern Colorado, USA. *Forest Ecology and Management* 225:686–696.

Worrall, J. J., S. B. Marchetti, L. Egeland, R. A. Mask, T. Eager, and B. Howell. 2010. Effects and Etiology of Sudden Aspen Decline in Southwestern Colorado, USA. *Forest Ecology and Management* 260:638–648.

CHAPTER ELEVEN

Insects of the San Juans and Effects of Fire on Insect Ecology

Deborah Kendall

HIDDEN BENEATH THE AUSTERE BEAUTY of the rock, water, soil, trees, and plants of the San Juan Mountains and environs lies a complex community of insects whose lives have shaped the habitats of this architecturally diverse ecosystem. Insects and their life cycles, feeding habits, and life-history strategies provide the underlying frameworks of all terrestrial and freshwater communities (Johnson and Triplehorn 2004). Insects are adapted for survival in the harsh terrestrial environments of the San Juans, where cold temperatures, rarified atmosphere, low water availability, and increased ultraviolet radiation deter most animals.

High elevations of the San Juans do not deter the persistent insect. Insects are one of the few animal groups well represented in alpine and arctic fauna because of their ability to withstand low temperatures and extreme desiccation (Mani 1968; Sinclair, Addo-Bediako, and Chown 2003). Freshwater ecosystems also teem with aquatic insects that live in still waters, such as lakes and ponds (lentic systems), and in fast-moving streams and rivers (lotic systems). The properties of water provide a different framework for organisms that must adapt to a very different medium. Water is approximately 800 times denser than air and has less oxygen. Aquatic insects adapt to these changes in both morphology (body shape) and physiology (body function and metabolism) (Merritt, Cummins, and Berg 2008).

Disturbances are important factors in the ecology of the San Juans and environs. Fire is one of the most dramatic disturbances, opening the vegetational canopy and stimulating plant growth and arthropod succession. Drought is another disturbance that reduces the abundance of mature vegetation, such as trees, and opens the environment for early successional plants.

ARTHROPODS AND INSECTS

An arthropod is an animal that has jointed legs and an exoskeleton. Arthropods are the most diverse organisms on the planet and account for 95 percent of all species. Arthropods include insects, centipedes, millipedes, spiders, mites, ticks, scorpions, and crustacea, such as lobster, crab, shrimp, and crayfish. Every ecosystem on earth, including oceans, freshwaters, and terrestrial habitats, provides ecological niches for arthropods. Crustacea dominate the oceans; insects dominate terrestrial and freshwater ecosystems. Insects compose the majority of organisms. One in every five living species of life on Earth (including bacteria, fungi, plants, and animals) is a beetle. Such enormous diversity comes with equally diverse life-history strategies, and insects form the basic network of food webs that support all terrestrial life.

Several factors contribute to insects' incredible diversity. Adults of most insects are winged and take advantage of the amazing adaptation of flight, which allows them to escape from predators and exploit new habitats. Insects also have exoskeletons (outer support units) that offer protection from abrasion, predators, and ultraviolet radiation. This unique armor also provides inner attachment sites for leg and wing muscles and prevents desiccation in terrestrial species. Because the outer body covering cannot grow with the insect, it must be discarded, or molted, as the insect's body increases in size. A new, larger body covering is formed below the old exoskeleton, and this renewal provides the insect with an even thicker, more protective armor. Once most insects reach the adult stage or sexual maturity, they no longer molt. Primitive insects, such as silverfish, continue to molt throughout their entire lives (Berenbaum 1995).

Insects also owe their incredible success to metamorphosis, the ability to change body forms during their lives. The most dramatic examples of this transformation are insects that change from larval forms to very different adult forms. The most familiar example of this type of metamorphosis is the butterfly, which changes from a leaf-eating caterpillar, to a dormant pupa, to an adult butterfly that feeds on nectar (Berenbaum 1995). This strategy reduces competition for resources because the larvae and adults feed on different foods. In addition, the adult form is winged and can fly to new habitats to oviposit (lay eggs). With so many adaptations for success, it is no surprise that insects form the foundation of most terrestrial and freshwater ecosystems.

TERRESTRIAL ECOSYSTEMS

Insects are found virtually everywhere in terrestrial ecosystems, which offer myriad feeding opportunities. However, insects must deal with desiccation, and the exoskeleton is exquisitely designed to reduce moisture loss. Insects have further exploited this rarified world with the ability to fly. This ability to soar to the upper reaches of the atmosphere has contributed to insects' incredible diversity. The remaining arthropod classes, which include centipedes, millipedes, spiders, and crustacea, are wingless.

Temperate terrestrial ecosystems are producer-based; plants produce organic material in the presence of light and carbon dioxide. Herbivores are primary consumers because they feed directly on the producers. Secondary consumers feed on the primary consumers; tertiary consumers feed on secondary consumers, and the sequence continues to the apex of the food web, which is occupied by birds of prey and large carnivorous mammals, such as pumas and bears. In addition, there are numerically greater numbers of primary consumers than secondary, or higher-order, consumers (Price 1997).

THE PLANT FEEDERS

Herbivorous, or phytophagous, insects dominate the invertebrate community of the San Juans and adjacent regions. All parts of the plant, including leaves, stems, roots, flowers, and seeds, are subject to consumption. Herbivores employ a variety of feeding strategies adapted for efficiently ingesting plant tissues. Grasshoppers, butterfly and moth larvae, and beetles are defoliators, which chew pieces of leaf or stem material with jaw-like mandibles that excise tissues from the plant. True bugs, such as box-elder bugs and stink bugs (Order Hemiptera), and aphids and cicadas (Order Homoptera) employ sharp, hypodermic-like mouthparts to suck plant fluids from leaves and stems (Johnson and Triplehorn 2004).

Because herbivores spend long periods feeding on plants, they must avoid detection by predators. Birds are insects' most common enemy, although many amphibians, reptiles, mammals, and carnivorous insects also feed on insects. Most herbivores employ camouflage; body coloration is often brown, black, and green. Disruptive patterns, such as spots and lines, contribute to this subterfuge. Mimesis (mimicry), in which the herbivore's body resembles the environmental background, is also used—sometimes so convincingly that even birds have difficulty detecting their prey (Berenbaum 1995).

If camouflage is not successful, insects have other defensive mechanisms. Flight allows insects to physically escape predation. Other escape tactics involve jumping, or saltatorial locomotion. The speckled-winged grasshopper (*Arphia conspersa*) is a common saltatorial insect of the San Juans and surrounding areas. "Speckled-winged" refers to the disruptive coloration of its forewings, which are covered with spots. The beautiful hindwings are red and click audibly to further deter predators and attract mates (Willey and Willey 1969).

If physical escape is impossible, insects may resort to chemical warfare. Distasteful chemicals are extruded from mouths, in the case of grasshoppers, or from specialized pores in some species, such as the stink bug. The green stink bug (*Acrosternum hilare*) is a common San Juan insect that has large repugnatorial glands and pores on the sides of the body. When startled, it emits a noxious cloud of chemicals that deters most predators (Berenbaum 1995).

Plants also employ chemical warfare to avoid herbivore attack. Milkweed (*Asclepias* spp.) is a native plant that produces alkaloid poisons and is shunned by most herbivores. Only the larvae of the monarch butterfly (*Danaus plexippus*) can consume its deadly leaves. Tissues of monarch larvae and adults contain milkweed alkaloids, which are distasteful to birds (Berenbaum 1995). Monarchs have been common in the San Juans in summer.

Some of the most interesting herbivores are those that induce galls on the leaves and stems of plants. Galls are enlarged growths that are most visible in the fall and winter when leaves are absent. Female insects insert eggs into the leaves or stems of plants. When the larvae hatch from these eggs, they feed on the surrounding plant tissues. The larval saliva contains auxin-like substances that promote tissue growth in plants. As the larvae continue to feed, more auxins are pumped into the plant, which responds by producing a protective case—a gall—around the larvae. Species of flies (Order Diptera) and wasps (Order Hymenoptera) are the most common gall-inducing insects (Berenbaum 1995). Many San Juan plant species, including big sagebrush (*Artemisia tridentata*), rubber rabbitbrush (*Chrysothamnus nauseous*), juniper (*Juniperus* spp.), and Gambel oak (*Quercus gambelii*), are attacked by gall-inducing insects.

THE MEAT EATERS

Carnivorous insects use some of the most inventive methods to capture prey, which consists primarily of other insects. The carnivorous-insect body plan is designed for a predatory way of life. Chewing mouthparts consist of razor-sharp mandibles. The praying mantis is a formidable predator equipped with raptorial forelegs lined with sharp spikes. The mantid's forelegs have lightning-fast reflexes capable of plucking flies from midair. The prey is impaled upon the spikes and then consumed. The praying mantis receives its name from its folded forelegs, which appear to be in a posture of prayer; praying is a word play on "preying" mantis. Another interesting feature of the mantid is its predilection for sexual suicide. Approximately 25 percent of matings terminate in the female beheading the male with her mandibles, allowing her to receive his entire sperm mass by removing sexual inhibitory centers located in the male's brain (Berenbaum 1995). The most common mantid species in the San Juans and environs are the green Halloween mantis (*Stagmomantis limbata*) and the brown ground mantis (*Litaneutria minor*). Both species are camouflaged well, and their elongate bodies resemble plant stems (Berenbaum 1995; Johnson and Triplehorn 2004).

Other carnivorous insects that feed on insects and vertebrates have hypodermic-like mouthparts that are used to pierce the body coverings of prey. Possibly the most common sanguinivore (blood feeder) is the mosquito. Several species of adult female mosquitoes take blood from every terrestrial vertebrate, including amphibians, reptiles, birds, and mammals. The adult males feed on floral nectar. The mosquito is the most dangerous insect to humans because of the numerous types of pathogens it transmits. They include *Plasmodium*, a protozoan that causes malaria; *Flavavirus* spp., which cause yellow fever, dengue fever, and West Nile virus; and *Brugia* spp. and *Wuchereria bancrofti* nematodes, which cause elephantiasis. These debilitating diseases are increasing in prevalence worldwide because of an increase in mosquito-larvae habitats, such as stagnant ponds (Spielman and D'Antonio 2001).

THE SNOW ARTHROPODS

During the winter in the San Juans and environs, a snow-based group of arthropods survives on the harsh landscape of snow and ice. The snow, or nival, community is often overlooked, because cold temperatures deter most humans from venturing into this inhospitable world. The most common arthropods on snowfields include springtails (Class Insecta, Order Collembola), mites (Class Arachnida, Order Acari), spiders (Order Arachnida, Order Araenae), and predatory beetles (Class Insecta, Order Coleoptera, Families Carabidae and Staphylinidae) (Mani 1968).

The ecological basis of the nival food web consists of dead insects and detritus windswept from nearby terrestrial habitats onto the snowfields. Springtails and mites feed on this detrital "rain," and predatory spiders and beetles prey on the unsuspecting detritivores. In the summer, these arthropods depend on nearby terrestrial habitats, which form refuges of food, temperature, and water (Mani 1968).

The nival arthropods are especially well adapted to cold temperatures. A common misconception is that insects die in the winter in temperate regions. Many insects, such as the adult stages, succumb to the vicissitudes of winter. However, eggs and larvae survive freezing temperatures through a process called *diapause*, which results in a lower metabolic rate (Sinclair, Addo-Bediako, and Chown 2003). Nival insects, however, become super-cooled and can be active even at extremely low temperatures. Antifreeze chemicals prevent destruction of body tissues at subfreezing temperatures. The winter-adapted spider, *Bolyphantes index,* is active down to −5°C, and the nival springtail, *Isotoma hiemalis,* continues its life activities down to −6°C (Jones and Pomeroy 1999).

HIGH-ELEVATION INSECTS

Insects are also well adapted to the low temperatures and reduced moisture of high elevations (Mani 1968). Alpine insects are warmed by the sun during the day; at night, cold temperatures slow most species. Some species, such as moths, are beautifully designed to fly, mate, feed on floral nectar, and otherwise exploit the nocturnal world. Moths are clothed in a dense coat of hairs that covers the thorax, which houses the wing muscles. Moths shiver when cold, warming flight muscles and producing enough body heat to fly (Kendall and Kevan 1981). They also have increased quantities of fat, which is used to stay warm and to fuel continued flying (Kevan and Kendall 1997).

Resident and migratory insect communities can be found at high elevations. In the summer, many insects exploit the food resources of alpine regions and migrate to other habitats before winter. In the winter, however, resident species may remain active on snowfields or overwinter as eggs, larvae, or pupae (Mani 1968). One of the most interesting species of the San Juans and environs is the miller moth (*Euxoa auxiliaris*), which hatches from eggs in lowland areas and migrates to high elevations in the summer. In this way, the moths escape high lowland temperatures and forage on alpine flowers. After mating at high elevations, miller moths return to lower-elevation habitats, where they oviposit. Shortly thereafter, the adults die (Kendall and Kevan 1981) (figure 11.1).

THE POLLINATORS

Pollination of flowers is one of the most important functions of insects in terrestrial ecosystems. Flowering plants and insects have co-evolved for millennia. Diverse floral shapes, colors, and scents attract an endless diversity of flower visitors. Pollinators transport pollen between flowers, ensuring cross-fertilization and genetic diversity. Bees, butterflies, moths, and some flies and beetles are adapted for feeding on floral nectar and transporting pollen on body hairs. Pollinators, such as bees, butterflies, and moths, have specialized mouthparts designed for imbibing floral nectar (Faegri and Van Der Pijl 1979). Many species of butterflies and moths of the San Juans are active pollinators throughout the spring, summer, and fall.

Although beetles and flies lack specialized nectar-feeding mouthparts, they are able to partake of rich floral nectar and transport pollen on body hairs. The tachina fly is one of these species. Its body is covered with hairs that easily transport pollen between flowers, effecting the cross-fertilization of plants. One of the most conspicuous fall flower visitors in the San Juans and environs is the spiny tachina fly (*Paradejeania rutiliodes*) (figure 11.2).

FRESHWATER ECOSYSTEMS

Beneath the calm surface of aquatic San Juan ecosystems lie complex and diverse communities that teem with life. Insects are the consumers in this ecosystem and are extremely vital to the health of all freshwaters. Unlike the producer-based terrestrial ecosystems of temperate North America, freshwater ecosystems are detritus-based.

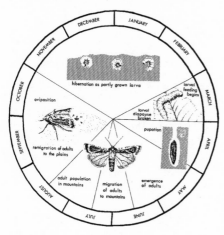

FIGURE 11.1 Life cycle of the miller moth (*Euxoa auxiliaris*) in the San Juan Mountains and surrounding regions.

Although plants and algae occur in these waters, their numbers are sparse and abundance is localized. Detritus is decayed organic matter (DOM) from dead plants and animals and can be classified as Coarse Particulate Organic Matter (CPOM) or Fine Particulate Organic Matter (FPOM). The majority of freshwater insects feed on detritus, and the remaining species are predatory. There are different strategies for gathering, collecting, and consuming CPOM and FPOM (Merritt, Cummins, and Berg 2008).

Freshwaters are classified as still or moving. Lentic systems are still waters and include ponds, reservoirs, and lakes. Lotic systems are moving waters and include creeks, streams, and rivers. Insects exhibit different morphological and physiological adaptations in these two ecosystems.

Lentic Insects

Lentic-adapted insects have several adaptations for survival in still waters. Most species live on the bottoms of, or on vegetation in, lakes or ponds and can be classified as burrowers, climbers, crawlers, clingers, swimmers, and divers (Merritt, Cummins, and Berg 2008). A common diving insect of San Juan lakes and ponds is the water boatman (Order Hemiptera, Family Corixidae). This insect comes to the surface to capture a bubble of air, then uses oar-like legs to dive and swim. The adults are winged and often fly nocturnally to new lentic habitats. In the San Juan region, one can observe the adults flying in large numbers around lights at night.

Lotic Insects

Lotic ecosystems of the San Juans are found in fast-moving streams and rivers that originate at high elevations and flow down into lower-lying habitats. Lotic-adapted insects have strategies that allow them to remain in one area without being

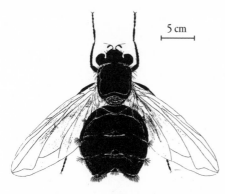

5 cm

FIGURE 11.2 The spiny tachina fly (*Paradejeania rutiliodes*) is a common species in the San Juan Mountains and surrounding regions.

swept away by the current. Many species, such as stoneflies (Order Plecoptera), are dorso-ventrally flattened (flattened from back to belly), allowing them to remain on the bottom more efficiently. Many species of caddisflies (Family Trichoptera) use silk from their mouthparts to glue small pebbles and twigs around their fragile bodies. These cases remain on the bottom or can be attached beneath stones (Merritt, Cummins, and Berg 2008).

EFFECTS OF DROUGHT ON INSECT ECOLOGY

Drought is one of the most important factors affecting insect diversity and abundance. Terrestrial insects depend on moisture in both the air and the soil. Drought has significantly reduced insect populations across North America (Hawkins and Holyoak 1998). Insects require moisture for physiological functioning, and the plants on which herbivores depend are also negatively affected by drought. Some species, including the pine beetle and several ant species, are drought-adapted.

The current drought in the San Juans and environs has severely affected plant and insect species. The piñon pine has decreased in population as a result of indirect effects of drought that reduce its ability to produce adequate quantities of pitch, a black material that acts as an anti-herbivory mechanism. The pine beetle can more easily attack the stressed tissues of the piñon pine, which has led to extremely large numbers of beetle-damaged trees in the San Juans and environs (Leatherman and Kondratieff 2003).

Several ant species in the San Juans and environs are adapted to drought conditions. They include the western harvester ant (*Pogonomyrmex occidentalis*), bigheaded ant (*Pheidole bicarinata*), cocktail ant (*Crematogaster punctulata*), pyramid ant (*Dorymyrmex pyramicus*), gravel ant (*Iridomyrmex pruinosus analis*), and field ant (*Formica perpilosa*) (Gregg 1963). Dry soil conditions and lack of environmental moisture do not deter these vital insect species. Ants are important decomposers in the environment, and they shape terrestrial ecosystems by their nest-making and feeding behaviors (Holldobler and Wilson 1990).

Table 11.1. Mean numbers of insects collected in 1991 and 1992 from burned and unburned habitats following the Long Mesa burn in 1989 in Mesa Verde National Park, Colorado.

Insect Order	Burned	Unburned
Hemiptera (true bugs)	39.64 a	6.31 b
Coleoptera (beetles)	12.13 a	1.07 b
Homoptera (aphids and kin)	3.53 a	4.32 a
Orthoptera (grasshoppers)	3.20 a	2.00 b
Hymenoptera (bees)	0.19 a	0.05 b

Means followed by the same letter are not significantly different. Means followed by different letters are significantly different. Kruskal-Wallis statistical tests; $p \leq 0.05$.
Source: Kendall 2003.

EFFECTS OF FIRE ON INSECT ECOLOGY

Terrestrial Ecosystems

Fire is an important disturbance factor in the San Juans and surrounding areas. The altered landscape of a fire-ravaged ecosystem appears bleak at first, but this is only temporary. Early successional plants and arthropods colonize theses disturbed areas, where vegetational canopies have been opened. Increased sunlight stimulates growth of fast-growing plant species that are called early successional pioneers because they are the first to colonize newly opened ecological "gaps." Insects appear soon after a fire. The majority of cursorial (crawling) insects can navigate through these gaps more quickly, and herbivorous species are attracted to the early successional plants (Kendall 2003). Plants appear to grow more rapidly following fire, perhaps because of increased nitrogen from burned vegetation or increased soil microbial activity (Woodmansee and Wallach 1981). Herbivore populations respond favorably to increased herbaceous plant abundance (Kendall 1992).

Several factors affect insect recolonization of burned habitats. They include the migratory abilities of insects living in adjacent unburned areas, severity of disturbance, and diversity of returning vegetation (Kendall 1992). Weakly flying insects are not able to invade burned habitats quickly, and their presence in recolonized communities may be a useful indicator of the ecosystem's returning health. Early plant communities that invade burned soils are characterized by herbaceous forbs (non-woody leafy plants) with increased productivity. Plant-fluid feeders, such as aphids and stink bugs, increase in diversity and abundance in these early pioneer plant communities. However, the numbers of hiding places are reduced following a fire, and some species are exposed to predators.

Following fire in Mesa Verde country in 1989, insects recolonized new vegetational communities in specific ecological patterns. Fire removes leaf litter, and cursorial (crawling) insects, such as beetles (Order Coleoptera), increased in number, because they can more easily reconnoiter early successional habitats (table 11.1).

Jumping insects, such as grasshoppers (Order Orthoptera), were also significantly more numerous in burned habitats, possibly because of ease of movement (table 11.1). True bugs (Order Hemiptera), which take fluids from plants, also increased in abundance following the fire, although homopterans, such as aphids and cicadas, did not exhibit significant population changes following the fire (table 11.1). Bees (Order Hymenoptera) also increased significantly in burned habitats as a result of increases in the populations of flowering herbaceous forbs (table 11.1).

Aquatic Ecosystems

Freshwater ecosystems are devastated by fire in nearby terrestrial areas. Ash from fire runs into the streams, rivers, reservoirs, and lakes. Soil is no longer held in place by roots, and the resulting increase in silt runoff clogs gills, suffocating insects and fish. Numbers of aquatic insects decrease dramatically following fire in adjacent terrestrial habitats.

Silt runoff is only one factor that contributes to fire's devastating effects on freshwater life. Fire-suppressant chemicals cause direct mortality of both invertebrates and fish (Boutton, Moss, and Smithyman 2003). Burned timber and other plant material also enter the water and cause eutrophication, or increased nutrient content of the freshwater ecosystem. This added bioload increases the nitrogen content of the water, increasing algal growth and subsequent decay; this decay results in oxygen depletion of the water as a result of increases in microbial activity in decomposing matter (Minshall, Brock, and Varley 1989).

REFERENCES

Berenbaum, M. R. 1995. *Bugs in the System: Insects and Their Impact on Human Affairs.* Reading, MA: Addison-Wesley.

Boutton, A. J., G. L. Moss, and D. Smithyman. 2003. Short-Term Effects of Aerially-Adapted Fire-Suppressant Foams on Water Chemistry and Macroinvertebrates in Streams after Natural Wild-fire on Kangaroo Island, South Australia. *Hydrobiologia* 498:177–189.

Faegri, K., and L. Van Der Pijl. 1979. *The Principles of Pollination Ecology.* Oxford: Pergamon.

Gregg, R. E. 1963. *The Ants of Colorado, with Reference to Their Ecology, Taxonomy, and Geographic Distribution.* Boulder: University of Colorado Press.

Hawkins, B. A., and M. Holyoak. 1998. Transcontinental Crashes of Insect Populations. *American Naturalist* 152(3):480–484.

Holldobler, B., and E. O. Wilson. 1990. *The Ants.* Boston: Belknap, Harvard University.

Johnson, N. F., and C. A. Triplehorn. 2004. *Borror and DeLong's Introduction to the Study of Insects.* St. Paul: Brooks/Cole.

Jones, H. G., and J. W. Pomeroy. 1999. The Ecology of Snow-Covered Systems: Summary and Relevance to Wolf Creek, Yukon. In J. Pomeroy and G. Granger, eds., *Wolf Creek Research Basin: Hydrology, Ecology, Environment.* Saskatoon, Canada: National Water Research Institute, Minister of Environment, 1–14.

Kendall, D. M. 1992. *Patterns of Invertebrate Recolonization in Long Mesa Burn, Mesa Verde National Park, Colorado.* Project MEVE-W91-0157. Washington, DC: US Department of the Interior, National Park Service.

———. 2003. Effects of Fire on Insect Communities in Piñon-Juniper Woodlands in Mesa Verde Country. In M. L. Floyd, ed., *Ancient Piñon-Juniper Woodlands.* Boulder: University Press of Colorado, 279–285.

Kendall, D. M., and P. G. Kevan. 1981. Nocturnal Flight Activity of Moths (Lepidoptera) in Alpine Tundra. *Canadian Entomologist* 13:607–614.

Kevan, P. G., and D. M. Kendall. 1997. Liquid Assets for Fat Bankers: Summer Nectarivory by Migratory and Non-Migratory Moths. *Arctic and Alpine Research* 29(4):478–482.

Leatherman, D. A., and B. C. Kondratieff. 2003. Insects Associated with the Piñon-Juniper Woodlands of Mesa Verde Country. In M. L. Floyd, ed., *Ancient Piñon-Juniper Woodlands.* Boulder: University Press of Colorado, 167–180.

Mani, M. S. 1968. *Ecology and Biogeography of High-Altitude Insects.* New York: Springer.

Merritt, R. W., K. W. Cummins, and M. B. Berg. 2008. *An Introduction to the Aquatic Insects of North America.* Dubuque, IA: Kendall/Hunt.

Minshall, G. W., J. T. Brock, and J. D. Varley. 1989. Wildfires and Yellowstone's Stream Ecosystems. *Bioscience* 39(10):707–716.

Price, P. W. 1997. *Insect Ecology.* New York: Wiley.

Sinclair, B. J., A. Addo-Bediako, and S. L. Chown. 2003. Climatic Variability and the Evolution of Insect Freeze Tolerance. *Biological Reviews* 78:181–195.

Spielman, A., and M. D'Antonio. 2001. *Mosquito: A Natural History of Our Most Persistent and Deadly Foe.* New York: Hyperion.

Willey, R. B., and R. L. Willey. 1969. Visual and Acoustical Social Displays by the Grasshopper *Arphia conspersa* (Orthoptera: Acrididae). *Psyche* 76:280–305.

Woodmansee, R. G., and L. S. Wallach. 1981. Effects of Fire Regimes on Biogeochemical Cycles. *Ecological Bulletins* 33:649–669.

Wildlife of the San Juans
A Story of Abundance and Exploitation

Scott Wait and Mike Japhet

A thing is right when it tends to preserve the integrity, stability, and beauty of the biotic community. It is wrong when it tends otherwise.

(LEOPOLD 1949:262)

THIS CHAPTER DESCRIBES A FEW SPECIES of wild animals that occur in southwestern Colorado. What makes these species' existence and natural histories in the San Juan Mountains noteworthy when any of hundreds of other species and their unique stories could have been chosen? These selected species are easier to portray than others because much has been documented about their life histories and interactions with humans. The puma has had a long antagonistic relationship with European settlers, elk and mountain sheep "fueled" the settlement of the region, and lynx and cutthroat trout have become the foci of recent reintroduction efforts.

Although any of the eighteen species of mice and rats found in southwestern Colorado could have been chosen, only a few have been studied enough locally to generate more than a paragraph. The literature on the heather vole (*Phenacomys intermedius*), for example, is scant, although it does occur throughout the region described in this book and likely has an interesting tale to tell.

This chapter is also written on the assumption that most readers no longer adhere to the fairy-tale beliefs that nature is at peace and that animals live in harmony with one another. The truth is that nature is harsh, and all species are eaten by other species. Even the puma ends up as food for something.

Anyone who believes environmental selection is not harsh should closely examine the way a puma kills a deer. Bighorn sheep, for their part, are exposed to sheer cliffs, howling winds, temperature extremes, and predators during their struggle to find food, maintain body condition, and procreate. During recent geologic times, countless species have lost this struggle.

Until the advance of the last Pleistocene ice sheets 22,000–20,000 years ago, camels, saber-toothed tigers, huge bears, and giant ground sloths roamed western North America. The horse became extinct in the New World and was reintroduced—albeit domesticated—by European settlers. The puma also became extinct in North America, only to recolonize the continent from South America following the Pleistocene.

The tale of a species through time oscillates between abundance and paucity. Some of those cycles have short periods (four years for some lemmings), and some have longer periods (ten years for northern hares), although nearly all species exhibit abundance and rarity when examined over thousands of years. Such natural cycles can be easily dismissed, assuming the habitat remains suitable to continue the cycles. When a cycle is broken by the actions of humans, we should evaluate what we can do differently.

Ultimately, humans must decide which resources are important to conserve and which are less important. However, it is dangerous to judge historical events only in today's context of ethics and knowledge. We may not like the actions of some of our forebears, but we must remember that they lived in a different time—a time when elk, bison, and mountain sheep were abundant and when the meat for dinner did not come from the local grocery store.

LYNX

The lynx (*Lynx canadensis*) is a long-legged cat with big feet and a short black-tipped tail. Its fur is typically grizzled gray to reddish-brown, displaying scattered black spots with no obvious pattern. A pronounced facial ruff and long dark ear tufts are typical. The lynx is a resident of the boreal forest, which is common throughout Canada and Alaska, with southern extensions through the Rocky Mountains into Colorado and Utah, as well as in New England and the Great Lakes. Its large paws act as snowshoes, allowing it to run easily across deep snow in pursuit of its primary prey, the snowshoe hare. Historically, the lynx occurred sparsely in Colorado's mountainous areas above 9,000 feet (2,745 m) in the Park, Gore, San Juan, and La Plata Mountains and in the White River Plateau (Armstrong 1972; Cary 1911; Fitzgerald, Meaney, and Armstrong 1994; Seton 1929). Confirmed lynx sightings were identified prior to reintroduction in La Plata Mountains, Silverton, the Bayfield-Vallecito region, areas north of Pagosa Springs, and around the Conejos

River in south-central Colorado. Cary (1911) referred to lynx as "tolerably common" in 1905 but "rapidly decreasing" by the time of his publication. The last confirmed lynx in Colorado were killed in 1969 and 1973, in the Loveland Pass and Vail areas, respectively. In 1991 credible experts observed lynx tracks near the site of the proposed East Fork ski area, northeast of Pagosa Springs.

In 1998, the Colorado Division of Wildlife (CDOW) initiated a lynx reintroduction program in the Upper Rio Grande Watershed west of Creede and along the west side of the Continental Divide, from Silverton to Wolf Creek Pass. At the time, lynx were feared to have been extirpated from Colorado. Even if some animals remained, the population was unsustainable because it was negligible and disconnected from any other population. Reintroduction was seen as the only option to reestablish the species in the state.

From 1999 through 2005, 218 lynx were released, all wearing radio collars for monitoring. Keeping close track of the animals is critical for the reintroduction effort. By monitoring lynx, researchers can check their health, movements, and habitat preferences. In the spring, females are tracked to determine the locations of birthing dens and numbers of kittens. True to their nature, the reintroduced lynx have established themselves largely in the higher boreal forests throughout the San Juan Mountains, as well as in adjacent mountain ranges in Colorado and surrounding states.

From the onset of the reintroduction, many obstacles had to be overcome as managers gained knowledge about lynx behavior. Capture and release protocols had to be developed, and the cat needed to survive long enough to first develop a home area and then to find a scarce mate. (A few years after the initial release, lynx density was two to four adults per 500 mi^2 [1,295 km^2].) Finally, they needed to successfully reproduce and raise kittens, which, in turn, would survive and reproduce.

That milestone was achieved in 2003, when researchers documented at least 16 lynx kittens from 6 known dens. Additional dens might have been present in remote, inaccessible areas, and some might have been occupied by females whose radio transmitters had failed. In 2004, at least 11 dens with 30 kittens were noted; in 2005, the figures were at least 16 dens and 46 kittens. During the next two years, however, a reversal occurred. In 2006, the number of dens found decreased to 4, with 11 kittens; no dens or kittens were found in 2007.

Although the birth rate declined, the reintroduction reached an important milestone in 2006, through native recruitment. A female cat born in the San Juan Mountains gave birth to two second-generation Colorado natives.

Field efforts also show that more reproductive activity may be occurring than biologists can confirm. During regular winter trapping operations in January 2007, two Colorado-born lynx that had never been handled were caught—evidence that more lynx might have been born and survived than the minimums reported. At present, survival exceeds mortality, yet reproduction must continue for long-term survival of the population.

From the onset of this project, one important question has remained unanswered: What is different now from the time when lynx disappeared? Many other questions have been posed as the reintroduction effort has continued: Are lynx populations cyclical? Are the populations of the snowshoe hare—the lynx's primary prey—cyclical? How are the natural histories of lynx and hare different in Colorado from those in well-studied regions of Canada?

The causative factors of their disappearance may never be known. Even though a simple explanation is desirable, relationships in nature are complex. Many of the possible reasons for lynx decline have changed since the early to mid-1900s. For example, widespread poisoning of predators no longer occurs, conservation of wildlife habitat has a higher priority, trapping and hunting are strictly regulated, and the passage of time has allowed the landscape to recover from past abuses.

However, other detrimental factors are now in operation, such as increased human density and development, the introduction of exotic diseases (plague), and habitat loss and fragmentation. Lynx populations are also vulnerable to over-harvest, especially along the fringes of the existing lynx range (Bailey et al. 1986).

One of the justifications for the Colorado reintroduction effort was to generate information on lynx behavior and natural history in the southern extremity of its range, to complement the abundant literature from Canada. Without this additional knowledge, management decisions regarding lynx and lynx habitat would be based on information from significantly different latitudes and habitats.

In addition, the primary prey of the lynx, the snowshoe hare, is being studied. Dolbeer and Clark (1975) hypothesized that hare populations in Colorado were stable over long periods because of habitat fragmentation. New evidence (Scott Wait, unpublished data) indicates that hare populations oscillate with smaller amplitude in Colorado than has been documented in western Canada. Malloy (2000) found the same trend in Montana.

Lynx are primarily solitary animals, except when a mother may be accompanied by kittens from June through mid- to late winter. Researchers have found, however, that lynx also form social networks based on prey abundance, relationships, and the age and sex of individuals. While lynx are seldom seen in the San Juans, they have occasionally been seen along roads and crossing cabin decks in mountain communities.

ELK

Rocky Mountain elk, or wapiti (*Cervus elaphus*), moved from Asia across the Bering Strait at least 120,000 years ago. The species arrived in Colorado about 8,000–10,000 years ago and in southern Colorado about 4,000–5,000 years ago (O'Gara and Dundas 2002). North American distribution and abundance ebbed and flowed in relation to glaciations. Ultimately, North America proved to be a suitable habitat. Elk roamed the entire continent, from coast to coast and from northern Canada into Mexico. Historical estimates suggest that the elk population might have exceeded 10 million prior to European arrival (Seton 1929).

Native Americans and elk shared the continent's vast landscape. Where elk appeared in greater abundance, they were consumed for food, clothing, and tools. Navajo, Utes, Jicarilla, Havasupai, Hopi, Zuni, and other Puebloan tribes harvested elk in the Four Corners region and the southern end of the Rocky Mountains (McCabe 1982). The various Ute bands had a longer relationship with elk than did other tribes and were reputed to be more skilled as hunters.

The pre-European period of abundance was followed by a time of exploitation. As settlers moved westward and the railroads were established to service the mines, elk, bison, deer, and mountain sheep were hunted to feed the growing population and to suppress the Native Americans. This exploitation reduced the elk population from an estimated 10 million to fewer than 100,000 in 1907 and approximately 90,000 in 1922, one-third of which were in the Yellowstone area and Canada (Bryant and Maser 1982).

The story in Colorado was similar. The US Forest Service estimated that Colorado's 1910 population was only 500–1,000 elk, with the largest herds in the White and Gunnison watersheds (Swift 1945). The March 17, 1953, edition of the *San Juan Lookout* [Durango, CO] quoted retired forest ranger Tom Price, who claimed, "There were a skimpy 28 elk in the Hermosa subdrainage in 1912, and two cows over in the Pine River country."

Swift described the situation:

> Mining expansion brought thousands of people to the mountain areas so, during the heyday of Leadville, Ouray, Creede, and many another mining town, there was hardly a part of the Colorado Rockies that was not subjected to scrutiny by prospector or promoter. Associated with all of this was the hunter who was unhindered by game laws and encouraged by a belief that there was plenty of game for all. By the time the rush of development had subsided, and the people of Colorado had more time to inventory their wildlife and other resources, some species had been nearly exterminated. Thus, even as late as 1910, there was good reason to suspect the eventual extermination of the Rocky Mountain elk. (1945:114)

The resource was running out, and, as a result, Colorado stopped elk hunting in most of the state from 1903 to 1933. From 1912 to 1928, CDOW's predecessor reintroduced 350 elk from Jackson Hole into fourteen areas in Colorado, including 25 elk into the Hermosa Creek drainage north of Durango in 1912. During the 1930s, after elk populations had rebounded, elk from abundant herds near Gunnison, Pagosa Springs, and Creede were trapped and transplanted to other states to begin new herds.

Elk have been successfully restored to Colorado. In fact, with an estimated elk population of 280,000, the state hosts the largest elk population in North America. About 250,000 hunters pursue elk every year in Colorado, harvesting nearly 50,000 per year. The elk population has cycled through abundance, exploitation, and back to abundance.

The 45,000 elk in the San Juans are migratory and make significant changes in elevation in response to seasonal snowfall, temperature, and disturbance. Although about six identifiable herds exist in this region, individual elk have been documented moving from Durango to Gunnison, from Gunnison to Saguache, and from the San Juan River to the Conejos River. All of those routes connect near the Continental Divide in the Upper Rio Grande drainage. Elk are most readily seen in the winter when they are at lower elevations, in more open habitats, and closer to highways and population corridors.

BIGHORN SHEEP

Bighorn sheep (*Ovis canadensis*) entered North America from Asia during the Pleistocene by way of the Bering land bridge. Wild sheep appear to have crossed the land bridge concurrently with humans, but not in association with them; these sheep were not related to those domesticated in early European or Asian cultures. No evidence of any wild sheep progenitors exists in North America prior to the Pleistocene, as there is for puma, horses, mammoths, and camels.

North American sheep might have evolved in Berengia (Cowan 1940). Following that, North American sheep split into two groups: the thinhorn sheep of northwestern Canada and Alaska (e.g., Dall's and Stone's sheep) and bighorn sheep farther south. In addition, the bighorn sheep split into several subspecies, with both desert and mountain subspecies native to, and still present in, Colorado. Fossil records exist in Utah and were found in Late-Pleistocene rock formations (Karpowitz and Stewart 2000). Numerous skeletal remains, pictographs, and petroglyphs from ancestral Puebloan (Anasazi) and more recent sites (Grant 1980) provide evidence of bighorns in Colorado.

Bighorn sheep were a source of both food and clothing for Native Americans. Hunting evolved from entrapping bighorns in corrals, to ambushing them, then to hunting them with spears and atlatl. Hunting with bows and arrows began about 2,000 years ago (Grant 1980). Native Americans never domesticated wild sheep.

Historical numbers of wild sheep are widely disputed, yet they are all far greater than present numbers. Seton (1929) estimated that 2 million wild sheep existed during what he termed "pristine" times (context unknown, possibly pre–European settlement) in the continental United States, plus another 2 million in Canada and Alaska. Although this estimate is widely cited as reliable, it falsely assumes uniform distribution throughout montane areas (Valdez 1988). Valdez (1988) estimated a historical population of a half-million wild sheep in North America, based on known habitat preferences.

In 1540, the Spanish explorer Francisco Coronado documented wild sheep in what is now New Mexico and Arizona. In 1776–1777, on the Colorado River in Utah, Father Silvestre Velez Escalante reported seeing abundant sheep tracks (Karpowitz and Stewart 2000). These were likely desert bighorns, which typically thrive in lower population densities than do mountain bighorns.

Fitzgerald and others (1994) reported that historical evidence indicates that wild sheep were much more widely distributed in Colorado than at present and were even found on the eastern plains, near the foothills. Buechner (1960) estimated that 7,300 wild sheep occupied Colorado in 1915, although that number likely represents a significant drop from pristine times. Valdez and Krausman (1999) suggested that mountain sheep's major population decline in Colorado occurred in the second half of the nineteenth century and continued to decrease, to 3,000 in the 1950s and 2,200 in 1970 (Fitzgerald, Meaney, and Armstrong 1994).

The era of abundance might have ended earlier for mountain sheep than for other species, and the era of decline might have occurred for different reasons. The most significant impact on mountain sheep probably occurred because of the huge numbers of domestic cattle and sheep introduced to western ranges, leading to overgrazing and the spread of disease (Valdez and Krausman 1999). Population was also reduced as a result of the impacts of hunting by miners and habitat loss from logging, dams, canals, and roads.

The wild sheep of North America are particularly vulnerable to a wide variety of diseases and parasites and are unable to recover from epidemics. Their reproductive potential is lower than that of deer or pronghorn and is roughly equivalent to that of elk (one lamb per year, with a slightly shorter reproductive life). Although many other wildlife species were exploited in the past 150 years and have been successfully restored, wild sheep have yet to recover and currently number about 7,300 in 88 herds in Colorado.

Shackleton and others (1999) characterize bighorn populations as numerically stable, although occasionally subject to catastrophic die-offs. They found no evidence that food shortage or inter-specific competition limit wild sheep populations. Food is not a range-wide limiting factor, even though winter foods may be deficient and significant diet overlap may occur with the diets of several other wild and domestic herbivores. Predation is identified as a limiting factor when bighorn populations lack adequate escape terrain (therefore marginal habitat), but they concluded that there is no evidence that predation regulates bighorn sheep populations in suitable habitat. Accidents are a minor mortality factor as a result of falls and spring avalanches.

Disease is identified as an infrequent but major limiting factor and is characterized by large (> 50%) and sudden (< 12 months) die-offs, as reported since the

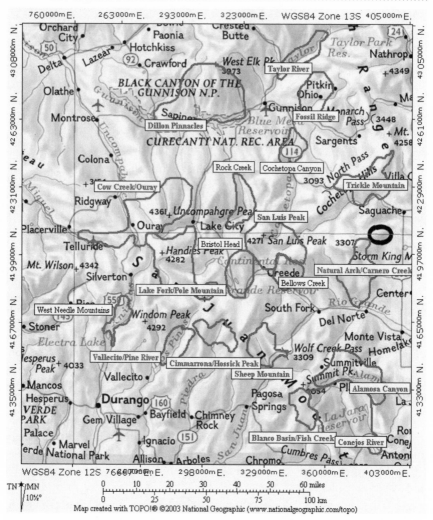

MAP 12.1 Approximate distribution of bighorn sheep in the San Juan Mountains, Colorado, 2009. Populations with red names have had documented or suspected disease-related die-offs. Populations with green names have not been augmented. Populations with black names are recently established or augmented.

1800s (Buechner 1960). Whether such die-offs occurred before this time is not known, because those are not reported in historical journals. Populations that have die-offs often display a predictable pattern of die-off, recovery to a lower level, then another die-off and recovery to an even lower level. Consequently, sheep have never attained a population level comparable to that in earlier times.

Population self-regulation has been observed in bighorn sheep, as the survival rate of juvenile females declines while the population increases (Jorgenson et al. 1997). The age of first reproduction in females may be delayed at higher densities, which also reduces productivity and recruitment (Festa-Bianchet et al. 1995).

Since 1945, nearly 2,700 bighorns have been introduced into suitable Colorado habitats. At least thirteen other state wildlife agencies have transplanted bighorns onto more than 200 historical ranges. Map 12.1 displays wild sheep populations in the eastern San Juans. Wild sheep in the past likely occupied most or all of the transplanted locations. Likewise, none of the native populations has suffered recorded die-offs, but they might have experienced unrecorded die-offs and stabilized at intermediate plateaus.

Historically, migration along drainage divides (e.g., the Continental Divide) connected many of these populations. Today, there is little consensus among wildlife managers about whether these populations should be connected. If connected, a larger population with higher genetic diversity could result, while a given population could be vulnerable to the spread of disease from another.

Valdez and Krausman (1999) summarized the status of the bighorn sheep as facing a precarious future. They concluded that the wild sheep are ecologically fragile because of limited, fragmented habitats. Bleich and others (1990) concluded that future conservation efforts will only be successful if land managers are able to minimize fragmentation.

PUMA

The puma (*Puma concolor*), also known as mountain lion, panther, cougar, painter, and catamount, displays spots as a kitten, along with blue eyes and a long tail. In the first year of life, its eyes turn to yellow and the spots fade until the entire body is tawny. Its underside turns whitish or pale yellow, and the long tail sports a black tip. The puma is the largest cat in the United States, except for the jaguar, which rarely makes an appearance from Mexico. In Colorado, the puma can weigh up to 225 pounds, although 150–175 pounds is the adult average.

Logan and Sweanor (2001) and Anderson (1983) extensively discuss the evolution and paleogeography of the puma. The species evolved from ancestral felids of the Western Hemisphere during the Miocene and Pliocene, deviating from the cheetah, jaguar, and lynx during the Pliocene, 2 to 5 million years before present. Puma-like felids are common in Pleistocene deposits in the United States.

Pumas inhabit nearly every ecosystem in Colorado and are most common in rough, broken foothills and canyon country. They are also associated with montane forests, shrublands, and piñon-juniper forests. The distribution of pumas is similar to the distributions of their primary prey—deer and elk (Anderson 1983). Pumas may make seasonal shifts in distribution to follow prey. Some researchers have found that females often behave differently from males. Anderson (1992) observed that females with young cubs in a den might have avoided areas of high deer densities. He speculated that such behavior might have been a tradeoff by mothers to avoid other pumas while raising their cubs in an effort to enhance cub survival.

Pumas will eat almost any animal available. Russell (1978) compiled a list of at least forty species, in addition to insects, grasses, and berries. Pumas are mostly crepuscular and may shift to more nocturnal patterns in the proximity of human disturbance (Pierce and Bleich 2003; Van Dyke et al. 1986).

Pumas are polygynous, and males do not contribute to the rearing of the young. A male tends to have a relatively large home range that overlaps the home ranges of several females. Puma cubs remain with their mothers for twelve–eighteen months. Cubs as young as eight weeks participate in stalks and kills. Dispersal of cubs is possibly triggered by abandonment or by aggression by the mother when she comes into estrous, because an adult male will kill any kittens that remain with her.

A puma can travel up to 300 miles (480 km); therefore, puma habitat can be thought of as contiguous across most of North America. In fact, its historical distribution in the Western Hemisphere is exceeded only by the distribution of humans and ranges from northern British Columbia to Patagonia, from the Pacific to the Atlantic (Logan and Sweanor 2001).

The general theory regarding feeding habits is that predators, especially solitary predators such as pumas, prey upon the young, old, and sick of a given species of prey. In reality, selection of prey is complex, and the age, sex, and reproductive status of the individual predator must be considered, as must the availability and vulnerability of prey (Ackerman, Lindzey, and Hemker 1984; Anderson 1992; Hornocker 1970; Logan and Sweanor 2001; Ross, Jalkotzy, and Festa-Bianchet 1997). The life of a transient puma (as opposed to a resident) is risky because survivorship declines with greater exposure to accidents, depredation, and conflicts with humans as young pumas attempt to locate prey in unfamiliar surroundings.

Puma males are physically bigger than females and have larger home ranges, which may be far removed from their natal areas (Pierce and Bleich 2003). Female home ranges may overlap significantly, and a male's home range may overlap the home ranges of several females. A juvenile female may carve a home range from her mother's territory, but one male's range rarely overlaps that of other males. This is important to managers because home-range size and suitable habitat area often translate into regional or statewide density estimates.

Demographic information is difficult to obtain for populations of large cryptic predators, such as pumas (Pierce and Bleich 2003). Colorado wildlife managers use a home-range size (derived from intensive published studies) of 5.3 to 12.1 pumas/100 mi^2 (2.0 to 4.6 pumas/100 km^2). Prey abundance and relative quality of puma habitat were evaluated in each of twenty-five puma management areas, and the density estimate was adjusted upward or downward within this range to develop local management plans. Colorado's San Juan River Basin supplies good to excellent habitat for pumas and may have nearly double the density of Colorado's Rio Grande drainage. The species is common in much of the western two-thirds of the state. Pumas were largely eliminated from the eastern plains during the pioneer days of the mid- to late 1800s (Fitzgerald, Meaney, and Armstrong 1994).

Settlement of the Americas has brought persecution and habitat loss since the 1500s. Pumas were killed to protect livestock and wild ungulates and because they were perceived as "bad" animals. They were eliminated from the eastern United States and severely reduced in the West by the early 1900s (Logan and Sweanor 2000). In 1903, the Colorado legislature listed species not protected by any law, including puma, eagles, hawks, falcons, and owls, among others. A similar list produced in 1921 included puma, wolf, lynx, coyote, and wolverine. In addition to having no protection, incentives were created to kill pumas. In 1881, a $10 bounty was offered per cat, compared with $1.50 for a wolf. In 1919, a cooperative effort was developed between the Colorado Game and Fish Department and the US Biological Survey to "destroy predatory animals." By 1929, the bounty had grown to $50 per animal, and *The Denver Post* added another $25 to bring the total to $75. Finally, in 1965, the puma was classified as a big-game species in Colorado in recognition of the recreational value of hunting the animal, as well as economic and social values (Barrows and Holmes 1990).

As rural and urban populations increase and expand farther into puma habitat, wildlife managers will be challenged to balance human values with the puma's conservation value. Conservationists see the puma as an integral part of the ecosystem. Stock growers are concerned that pumas can impact their businesses. The public is concerned that pumas may attack humans. At present, most of the public accepts the selective removal of individual pumas that prey on livestock or domestic animals and the elimination of those involved in threats to human safety (Pierce and Bleich 2003). This illustrates the dramatic change in values over the past 100 years, from indiscriminate killing in the early 1900s to limited killing in the early 2000s.

NATIVE CUTTHROAT TROUT

Two of Colorado's four subspecies of native cutthroat trout are found in the San Juan Mountains of southwestern Colorado. The Rio Grande cutthroat trout (*Salmo clarki virginalis*) is native to the Rio Grande Basin of New Mexico and Colorado, including the headwaters of the Rio Grande in the San Juan Mountains. The Colorado River cutthroat trout (*Salmo clarki pleuriticus*) is native to the Upper Colorado River Basin in the states of Utah, Wyoming, and Colorado, including the Upper Gunnison, Dolores, and San Juan River Basins of southwestern Colorado. Colorado is also home to the greenback cutthroat trout (*Salmo clarki stomias*), native to waters east of the Continental Divide in the South Platte River Basin. The yellowfin cutthroat, now extinct, is believed to have been native to the Upper Arkansas River Basin.

All of Colorado's native cutthroat trout are descendants from a common ancestral cutthroat species that inhabited the Columbia River Basin about 1 million years ago. The Colorado River cutthroat trout is believed to have arrived in Colorado from the Pacific Northwest during the last Ice Age. Rio Grande cutthroat may be a descendant of an archaic form of Colorado River cutthroat that entered the Rio Grande Basin through headwaters transfer as the Continental Divide shifted westward with time (Behnke 1992).

As the glaciers retreated, cutthroat trout in the Rio Grande and Colorado River Basins became geographically isolated from each other and developed along different evolutionary tracks. Both of the resulting subspecies possess a crimson slash along the lower jaw, which gives the cutthroat trout its name. The Rio Grande subspecies differs in appearance from the Colorado River subspecies by having fewer scales along the lateral line and irregularly shaped black spots concentrated toward the caudal, or tail, fin. While generally similar in appearance, the Rio Grande subspecies in spawning condition takes on a deep crimson color, while its Colorado River cousin is slightly less colorful but still fairly distinctive. Both have black spots concentrated on the posterior half of the body. At the genetic level, the subspecies can be readily differentiated from each other. Recent advances in molecular genetics show great potential for revealing more about the phylogenetic origins of Colorado's cutthroat subspecies.

At the time of the first European exploration of the San Juan Mountains, mostly during the 1800s, the historical range occupied by Rio Grande cutthroats in the San Luis Valley/Rio Grande Headwaters and by Colorado River cutthroats in the Upper San Juan, Dolores, and Gunnison Basins are believed to have included all rivers and most perennial streams above 6,000 feet (1,830 m) in elevation, or 5,500 feet (1,675 m) on north-facing slopes (Alves et al. 2007; Hirsch, Albeke, and Nesler 2005). However, many of the high-elevation lakes in the San Juan Mountains that presently contain cutthroats were historically barren of fish because of waterfalls and other barriers that blocked the upstream movement of fish in streams connected to these lakes.

The current status of native cutthroat trout in the San Juan Mountains has been greatly influenced by the opening of the Colorado frontier during the nineteenth century. Cutthroats are easily caught and require clean, cold, well-oxygenated water for spawning and survival. They evolved without competition from introduced brook and brown trout and without the risk of hybridization from introduced rainbow trout. While cutthroats are superbly adapted to their natural environment, they were ill-prepared for the habitat changes brought by white settlers. The discovery of rich mineral deposits in the San Juan Mountains in 1860 led to the construction of railroads and settlement of the mining towns of Creede, Lake City, Silverton, and Durango (O'Rourke 1980). Placer mining in streams eventually led to hard-rock mining and large-scale ore-crushing operations, all of which introduced toxic heavy metals into the water and damaged aquatic habitat. Those actions impaired water quality and the natural reproduction of cutthroat trout.

In addition, the early miners relied on cutthroat trout for food. Indiscriminate over-fishing was common and included the use of dynamite. Oliver North, an English sportsman traveling through the southwestern Colorado frontier, found few trout in the Upper Animas River (North 1878). In the May 4, 1878, edition of *The London Field: The Country Gentleman's Newspaper*, North reported, "A good angler may expect to catch his 40 pounds a day" on the Rio Grande near Del Norte, and he found trout weighing as much as four pounds in the San Juan and Gunnison Rivers. Although North did not say, it is presumed that they were all "black-spotted trout" (i.e., cutthroat trout).

By the 1880s, native cutthroat trout were nearly gone from many streams in the San Juan Mountains, and rainbow trout were widely stocked in their place. In 1885, brook trout were brought to the area by railroad from the eastern United States and stocked in the Animas River above Silverton (Wiltzius 1985). Following the same trend throughout the Rocky Mountain region, native cutthroat trout declined in the San Juan Mountains as a result of habitat alteration, over-fishing, and introduction of non-native trout. By the early twentieth century, rainbow trout, brown trout, and brook trout had been widely introduced throughout southwestern Colorado. Prominent hatcheries were located in Creede, Emerald Lake, and Durango. As a result, cutthroat trout were no longer found in many of the larger streams and rivers in the San Juan Mountains.

The modern concept of wildlife conservation dawned in the early 1900s, none too soon for the native Rio Grande and Colorado River cutthroat trout of southwestern Colorado. In 1897, the Colorado General Assembly created the Department of Forestry, Game, and Fish. One of the fledgling department's first actions was to implement daily limits on how many trout could be caught by fishing (Barrows and Holmes 1990). An era of widespread stocking with hatchery-reared trout followed. With advances in general aviation after World War II, many high-elevation alpine lakes were stocked with non-native rainbow and brook trout using fixed-wing aircraft and helicopters.

In the 1970s, CDOW biologists discovered isolated aboriginal populations of Rio Grande cutthroat trout on private property associated with the old Spanish land grants in the San Luis Valley. This discovery led to the creation of a brood stock of pure Rio Grande cutthroats. The progeny of those fish are now used to stock fifty-nine high-elevation lakes in the San Juan and Sangre de Cristo Mountain ranges and reclaimed Rio Grande cutthroat streams with the use of aircraft.

In the 1990s, CDOW biologists found a few remaining wild populations of genetically pure Colorado River cutthroat trout in isolated streams within the Weminuche and South San Juan Wilderness Areas. Low stream-flow conditions during a widespread drought in 2002 and 2003 enabled CDOW spawning crews to collect and fertilize trout eggs from these pure Colorado River cutthroat populations. The Weminuche and Navajo brood stocks of pure Colorado River cutthroats were founded in this manner. Progeny from these genetically pure brood stocks are used for conservation stocking of newly founded Colorado River cutthroat streams and to provide recreational fishing opportunities in over sixty high lakes in the San Juan Mountains, west of the Continental Divide. Today, fishing for native cutthroat trout is possible at most named high lakes in the San Juan Mountains as a result of active conservation efforts spanning the last thirty years. The streams where pure aboriginal cutthroats remain are typically protected by natural or manmade fish-migration barriers, catch-and-release fishing regulations, and a strict policy that prohibits fish stocking.

In 2007, Rio Grande cutthroat trout were found in 12 percent of historically occupied habitat, while Colorado River cutthroats were found in 13 percent of that habitat (Alves et al. 2007; Hirsch, Albeke, and Nesler 2005). While this is far different from Oliver North's findings 130 years ago, native cutthroats that remain in the San Juan Mountains are secure and in no danger of extinction. With ongoing conservation efforts, the current distribution of native cutthroats will continue to expand.

REFERENCES

Ackerman, B. B., F. G. Lindzey, and T. P. Hemker. 1984. Cougar Food Habits in Southern Utah. *Journal of Wildlife Management* 48:147–155.

Alves, J. E., K. M. Patten, D. E. Brauch, and P. M. Jones. 2007. Range-Wide Status of Rio Grande Cutthroat Trout (Oncorhynchus clarki virginalis): 2007. Unpublished report, Colorado Division of Wildlife and New Mexico Department of Game and Fish.

Anderson, A. E. 1983. *A Critical Review of Literature on the Puma (*Felis concolor*)*. Special Report 54. Denver: Colorado Division of Wildlife.

———. 1992. *The Puma on the Uncompahgre Plateau, Colorado*. Technical Publication 40. Denver: Colorado Division of Wildlife.

Armstrong, D. M. 1972. Distribution of Mammals in Colorado. *University of Kansas Museum of Natural History Monograph* 3:1–415.

Bailey, T. N., E. E. Bangs, M. F. Portner, J. C. Malloy, and R. J. McAvinchey. 1986. An Apparently Overexploited Lynx Population on the Kenai Peninsula, Alaska. *Journal of Wildlife Management* 50:279–290.

Barrows, P., and J. Holmes. 1990. *Colorado's Wildlife Story*. Denver: Colorado Division of Wildlife.

Behnke, R. J. 1992. *Native Trout of North America*. Monograph 6. New York: American Fisheries Society.

Bleich, V. C., J. D. Wehausen, and S. A. Holl. 1990. Desert-Dwelling Mountain Sheep: Conservation Implications of a Naturally Fragmented Distribution. *Conservation Biology* 4:383–390.

Bryant, L. D., and C. Maser. 1982. Classification and Distribution. In J. W. Thomas and D. E. Toweill, eds., *Elk of North America: Ecology and Management*. Harrisburg, PA: Stackpole Books, 1–60.

Buechner, H. K. 1960. The Bighorn Sheep in the United States, Its Past, Present, and Future. *Wildlife Monographs* 4:1–174.

Cary, M. 1911. *A Biological Survey of Colorado.* North American Fauna 33. Washington, DC: US Department of Agriculture, Bureau of Biological Survey.

Cowan, I. M. 1940. Distribution and Variation in the Native Sheep of North America. *American Midland Naturalist* 24:505–580.

Dolbeer, R. A., and W. R. Clark. 1975. Population Ecology of Snowshoe Hares in the Central Rocky Mountains. *Journal of Wildlife Management* 39(3):535–549.

Festa-Bianchet, M., J. T. Jorgenson, M. Lucherinin, and D. W. Wishart. 1995. Life History Consequences of Variation in Age of Primiparity in Bighorn Ewes. *Ecology* 76: 871–881.

Fitzgerald, J. P., C. A. Meaney, and D. M. Armstrong. 1994. *Mammals of Colorado.* Boulder: University Press of Colorado.

Grant, C. 1980. The Desert Bighorn and Aboriginal Man. In G. Monson and L. Sumner, eds., *The Desert Bighorn: Its Life History, Ecology, and Management.* Tucson: University of Arizona Press, 7–39.

Hirsch, C. L., S. E. Albeke, and T. P. Nesler. 2005. *Range-Wide Status of Colorado River Cutthroat Trout (*Oncorhynchus clarki pleuriticus*): 2005.* Denver: Colorado Division of Wildlife.

Hornocker, M. G. 1970. An Analysis of Mountain-Lion Predation upon Mule Deer and Elk in the Idaho Primitive Area. *Wildlife Monographs* 21:1–39.

Jorgenson, J. T., M. Festa-Bianchet, J. M. Gaillad, and W. D. Wishart. 1997. Effects of Age, Sex, Disease, and Density on Survival of Bighorn Sheep. *Ecology* 78:1019–1032.

Karpowitz, J., and R. Stewart. 2000. *Bighorn Sheep.* Wildlife Notebook Series 16. Salt Lake City: Utah Division of Wildlife Resources.

Leopold, A. 1949. *A Sand County Almanac.* New York: Oxford University Press.

Logan, K. L., and L. Sweanor. 2001. *Desert Puma: Evolutionary Ecology and Conservation of an Enduring Carnivore.* Washington, DC: Island Press.

Malloy, J. C. 2000. Snowshoe Hare, Lepus americanus, Fecal Pellet Fluctuations in Western Montana. *Canadian Field-Naturalist* 114:409–412.

McCabe, R. E. 1982. Elk and Indians: Historical Values and Perspectives. In J. W. Thomas and D. E. Toweill, eds., *Elk of North America: Ecology and Management.* Harrisburg, PA: Stackpole Books, 61–124.

North, O. 1878. Sport, Travel, and Scenery on the S.W. Colorado Frontier. *The London Field: The Country Gentleman's Newspaper* 51:540–541.

O'Gara, B. W., and R. G. Dundas. 2002. Distribution: Past and Present. In D. E. Toweill and J. W. Thomas, eds., *North American Elk: Ecology and Management.* Wildlife Management Institute. Washington, DC: Smithsonian Institution Press, 61–120.

O'Rourke, P. M. 1980. *Frontier in Transition: A History of Southwestern Colorado.* Denver: Bureau of Land Management.

Pierce, B. M., and V. E. Bleich. 2003. Mountain Lions. In G. A. Feldhammer, B. C. Thompson, and J. C. Chapman, eds., *Wild Mammals of North America: Biology, Management, and Conservation.* Baltimore: Johns Hopkins University Press, 744–757.

Ross, P. I., M. G. Jalkotzy, and M. Festa-Bianchet. 1997. Cougar Predation on Bighorn Sheep in Southwestern Alberta during Winter. *Canadian Journal of Zoology* 74:771–775.

Russell, K. R. 1978. Mountain Lion. In J. L. Schmidt and D. L. Gilbert, eds., *Big Game of North America*. A Wildlife Management Institute Book. Harrisburg, PA: Stackpole Books, 207–226.

Seton, E. T. 1929. *Lives of Game Animals*. Garden City, NY: Doubleday, Doran.

Shackleton, D. M., C. S. Shank, and B. M. Wikeem. 1999. Natural History of Rocky Mountain Bighorn and California Bighorn Sheep. In R. Valdez and P. R. Krausman, eds., *Mountain Sheep of North America*. Tucson: University of Arizona Press, 78–138.

Swift, L. W. 1945. A Partial History of the Elk Herds of Colorado. *Journal of Mammalogy* 26(2):114–119.

Valdez, R. 1988. *Wild Sheep and Wild Sheep Hunters of the World*. Mesilla, NM: Wild Sheep and Goat International.

Valdez, R., and P. R. Krausman. 1999. Description, Distribution, and Abundance of Mountain Sheep in North America. In R. Valdez and P. R. Krausman, eds., *Mountain Sheep of North America*. Tucson: University of Arizona Press, 3–22.

Van Dyke, F. G., R. H. Brocke, H. G. Shaw, B. B. Ackerman, T. P. Hemker, and F. G. Lindzey. 1986. Reactions of Mountain Lions to Logging and Human Activity. *Journal of Wildlife Management* 50:95–102.

Wiltzius, W. J. 1985. *Fish Culture and Stocking in Colorado, 1872–1978*. Special Report. Denver: Colorado Division of Wildlife.

Part 3

Human History of the San Juan Mountains

CHAPTER THIRTEEN

A Brief Human History of the Eastern San Juan Mountains

Andrew Gulliford

The eastern San Juan Mountains include some of the most rugged terrain in the United States. Within these "Shining Mountains" lie Ute vision-quest sites, old trails and mines, stock driveways, and so few roads that, in 1975, Congress established the Weminuche Wilderness (in the San Juan and Rio Grande National Forests)—the largest in Colorado, with over 500,000 acres (200,000 hectares [ha]). This wilderness is home to the last authenticated grizzly bear trapped and killed in Colorado, in 1952. The bear was trapped close to the headwaters of the Rio Grande near a mountain named the Rio Grande Pyramid, not far from a valley the Spanish appropriately named Rincon de la Osa, or "Hiding Place of the Bear."

Human, natural, and geologic histories are intertwined in the San Juans. Three different cultures—Native American, Hispanic, and Anglo—have called these mountains home. Every peak and drainage has been explored by Ute hunters, Hispanic shepherds, and Anglo prospectors; however, just as the last grizzly remained hidden for a time, many encampments, mines, and settlements have fallen into ruin and await rediscovery.

THE SHINING MOUNTAINS AND THE UTES

Any history of the eastern San Juan Mountains must begin with an understanding of the cultural landscape and the way the San Luis Valley sweeps north from inside

New Mexico, rimmed by the Sangre de Cristo Mountains on the east and the San Juan Mountains on the west. Despite its 100-mile (160-km) length, no rivers run through most of the valley. For millennia, creeks have disappeared into the northern two-thirds of the valley, depositing precious water into a vast aquifer. For centuries, the northern San Luis Valley and the eastern San Juan Mountains provided prime bison-, elk-, and deer-hunting grounds for the Utes, who traversed the landscape at first on foot and later, after the Pueblo Revolt (1680), on horseback.

Human history at the eastern edge of the San Juan Mountains is much older, however. Smithsonian archaeologists have identified 5,000-year-old paleo-Indian sites near what is now Great Sand Dunes National Park and at the northern edge of the San Luis Valley, near Poncha Pass. In the broad sweep of human history, the Moache, Capote, and Weminuche bands of Utes were relative latecomers, occupying the valley and the vast mountains on its western edge for perhaps 500 years. They hunted in seasonal rounds, moving into the high country during late spring and summer and wisely leaving before winter snows closed mountain passes. The Utes fought off other tribes and kept to themselves in their Shining Mountains. They remained solitary until the arrival of the Spanish.

The Spanish introduced significant cultural changes, including guns, horses, metal cooking pots, knives, and colored beads. As they settled in Santa Fe in 1610 and ventured slowly northward, the Utes held fast to the San Juan Mountains. By the mid-1700s Spanish explorers were aware of precious minerals in the San Juans. A catalyst for Don Juan Maria de Rivera's 1765 expedition appears to have been a piece of silver, possibly a smelted ingot, traded by a Ute at Abiquiú, in what is now New Mexico.

Original Spanish place names, such as La Ventana (the Window), resonate in what is now the Weminuche Wilderness. Rivers that retain Spanish names include the Piedra, San Juan, Los Pinos, and Rio Grande, the headwaters of which lie in the heart of the eastern San Juans. La Garita (the Cabin) Creek, which flows near Penitente Canyon, was christened by Spanish colonists who bravely moved north to farm and raise sheep on Spanish land grants, far from the safety of Santa Fe.

After centuries of isolation, the Utes were drawn into a Spanish mercantile orbit, which included fall trade fairs in Taos, where their women, children, and smoked and tanned deer hides were highly prized. As the Spanish connected Santa Fe with California, trade developed along the Old Spanish Trail, with horses and mules moving east from California and hides, Navajo blankets, and slaves moving west, along the southern edge of the San Juan Mountains.

One of the most famous Utes, Chief Ouray, was such a slave. Born around 1833, he grew up in a household in what is now northern New Mexico, where he learned English, Spanish, and Indian languages. His cultural understanding and linguistic fluency made him extremely valuable to Anglo prospectors and politicians who came to covet the San Juans.

Far from American population centers and north of the main Hispanic communities, the San Luis Valley and the eastern San Juans remained rugged, remote, and relatively unknown—except to the Taos trappers who moved northward in search of beaver pelts, riding up narrow drainages. Just as Spanish conquistadors and miners secretly entered the San Juans to mine without paying the king's royal fifth on whatever they discovered, mountain men also avoided trapping licenses and entered the San Juans on the sly.

In 1824, six major fur-trapping parties passed through the San Juans on their way north to the Green River. By the mid-1830s, the market for beaver had collapsed, eclipsed by the new fashion of black silk hats for men. Taos trappers—those who had survived the numbing cold, avalanches, and Indian attacks—drifted out of the San Juans and traded and farmed, often marrying into Hispanic families or bringing Indian wives into the small Spanish-speaking adobe villages. Trappers such as Kit Carson knew the country well.

Spanish hegemony in the Southwest eventually gave way to Mexican dominion. After the Mexican War of 1848, the United States officially ruled all of the San Juans. However, Spanish land grants complicated ownership; and troublesome Utes, Apaches, and Navajos routinely raided and stole from the settlers. Next came military explorers trying to map and understand this new American acquisition.

In 1848, Captain John C. Frémont searched in vain for a transcontinental railroad route; in the process, a third of his men died in heavy San Juan snows. Five years later, Captain John Gunnison crossed Cochetopa Pass and confirmed that Colorado would never have an east-west rail route through the mountainous San Juans. Dramatic changes occurred with the establishment of Indian agencies. Newly appointed agents included the short-in-stature Kit Carson and the alcoholic Albert H. Pfeiffer, with his red hair and beard.

In July 1859, Pfeiffer guided Captain John N. Macomb to the Pagosa Hot Springs, which measured above 110°F. Macomb described the springs as having "magnificent dimensions." He recalled: "We looked down through its tepid waters until the power of vision was lost in the cavernous depths whence the waters flow" (Oldham 2003:53–54). Later, Camp Lewis was built at that location, followed by Fort Lewis—home of the Eighth Cavalry—on what would become the Pagosa town site.

With the fur trade diminished and armed and mounted Utes in possession of their Shining Mountains, there was little motivation for others to enter the eastern San Juans until the "Pike's Peak or bust" gold rush began, in the same year Captain Macomb visited Pagosa. A year later, in 1860, a party of between 100 and 300 armed prospectors led by Charles Baker traveled north to Del Norte, west to South Fork, into the mountains, and down Cunningham Gulch into Baker's Park, now the town site of Silverton. The prospectors found traces of gold and silver ore and left the valley before the winter's deep snows. The pending Civil War postponed mining exploration.

In the Ute treaty of 1868, skillfully negotiated by Chief Ouray, the Utes retained title to all of western Colorado—one-third of the territory. The treaty was to be "final and forever." Their territory's eastern boundary was the eastern San Juans, which had been a prime hunting territory with access to hot springs along the West Fork of the San Juan River and at Pagosa Springs. Otto Mears built a wagon road from Saguache about 60 miles (96 km) west to the Los Pinos Indian Agency (located along Los Pinos Creek on the north side of the eastern San Juans, about 30 miles [48 km] south of the modern town of Gunnison) and, under government contract, hauled in flour, beans, hay, oats, and potatoes. For seven years Utes received goods and annuities at the Los Pinos Agency, where Ouray lived in a log cabin chinked with mud and buffalo hair. After listening to tall tales of lost gold and silver mines, eager prospectors ignored the Indian ownership and flooded into the Shining Mountains.

THE BRUNOT CESSION AND THE MINING BOOM

If treasure tales of lost mines seem more fable than fact, enough gold and silver ore had been found to pressure the US government to clear the Utes' title to the San Juans. In 1871, miners discovered gold in Arrastra Gulch near Silverton in what became known as the Little Giant Mine. A year later Congress passed the Mining Act, providing for legal 10-acre (4-ha) mining claims wherever any prospector attempted to develop a mine. By 1874, the Hayden Survey had begun to map Colorado's mineral resources, and the territorial legislature began to charter wagon roads across high passes. With Civil War veterans swarming into the San Juans, the federal government and Indian agents felt they had to negotiate a new treaty.

A year earlier, Felix Brunot, with the able assistance of Civil War veteran and road builder Otto Mears, negotiated the Brunot Treaty, or San Juan Cession, which vacated Ute ownership of the high country. He may have encouraged Utes to sign the treaty by giving them newly minted, shiny silver dollars. Originally, Utes had opposed the treaty. Ouray finally signed it at the Los Pinos Agency because the government promised to locate his lost son—a promise never kept. The Utes agreed to give up the mountaintops on which the minerals were located but not the excellent hunting grounds in the valleys. Despite the misunderstanding (or fraud) under which the Utes thought they were only selling the peaks, Anglos took ownership of the entire San Juan range, including peaks and valleys. The Utes were then forced onto reservations considerably south of their traditional summer hunting ranges.

As miners raced into the San Juans from all directions, rough wagon roads were built—including the road over Stony Pass, which linked the bustling boomtown of Del Norte with Baker's Park. Along the route, miners built cabins, way stations, and corrals. In the spring, the route was muddy and boggy and full of bone-jarring rocks, ruts, landslides, steep grades, and tree falls. In stretches, it narrowed between tight canyon walls, providing perfect cover for highway robbers. One traveler wrote that Stony Pass "is [a] humbug of the first water, and has been the author of more profanity than any other evil we endure. It is paved with the curses of the poor unfortunates who have climbed it, and it is darkened with the shadows of equine ghosts given up on Timber and Grassy Hills. Oh! Sir! It is a develish [*sic*] bad road" (Kindquist 1987:28).

Despite the horrible routes, miners flocked to the San Juans. Towns developed to serve their needs and became small commercial centers for the distribution of groceries, dry goods, guns, ammunition, and mining supplies—including Giant Powder and, later, TNT. Miners lived on whatever they could hunt and purchased canned goods, especially tomatoes, peaches, and condensed milk for coffee. Newly formed towns in the eastern San Juans included Pagosa Springs, Del Norte, South Fork, Saguache, Lake City, and Creede. Visiting in 1874, Ernest Ingersoll wrote, "The inhabitants of these towns were a curious lot—bearded miners, dirty laborers, strong-armed bullwhackers, thin-lipped gamblers, men of every character, and women of no character" (1994:31). Ingersoll claimed that for breakfast "they toss [flapjacks] up the chimney, and catch them right-side up outside the door" (1994:45).

In this remote part of Colorado, a host of mining camps sprang up, connected by stage lines and raucous, rambunctious stage drivers who forced their Concord coaches along narrow passes without regard for the weather or fallen trees. They sped into each camp, stopping with a flourish, horses lathered and breathing hard, hooves splattering mud onto wooden sidewalks. The Barlow and Sanderson Stage Company took intrepid travelers, mail, and freight from Alamosa to Lake City in six-horse stages along the rutted, rocky roads.

Mining camps founded in the late 1870s and early 1880s—some of which even had post offices—included Burrows Park, Whitecross, Tellurium City, Argentum, Sterling, Sherman, Mineral Point, and Ellwood. The July 17, 1875, issue of *The Silver World* newspaper, published in Lake City, reported that Tellurium had a "population [of] less than 500 men and enough valley vocalists [burros] to make the night hideous." Ores at Burrows Park were largely copper pyrite and argentiferous galena with gray copper, ruby silver, and sulphurets of silver. At Sherman, lead, gold, silver, and copper flowed from the New Hope, Smile of Fortune, Minnie Lee, and Black Wonder Mines. In 1881, the Sherman House provided "good accommodations for travelers, liquors, wines, St. Louis beer, and cigars" (Interpretive sign text on the Alpine Loop Interpretive Trail).

Miners toiled optimistically in the summer and often left before the onset of deep snows. If prospectors stayed on in the camps, one of the tales they told during the long winter nights came from the camps' early days—before the stage lines and the prim clapboard storefronts, even before the earliest log cabins. This was the 1874 story of Alferd G. Packer and deep San Juan snows, a story of desperate cold, starvation, and the psychological disintegration of a mining party hoping to strike it rich.

Having left the hospitality of Ouray's camp near Montrose, Packer's party became disoriented in deep snows and began to starve. Packer stated, "Our matches had all been used, and we were carrying our fire in an old coffee pot. Three or four days after our provisions were all consumed, we took our moccasins, which were made of raw hide, and cooked them ... Our trail was entirely drifted over. In places, the snow had blown away from patches of wild rose bushes, and we were gathering the buds from these bushes, stewing them and eating them" (quoted in Oldham 2005:13).

The weary party trudged for about 5 miles (8 km) beyond Lake City, into what today is known as Deadman's Gulch. One of the prospectors went mad, killed the others, and began to cook and gnaw at their bones. Packer returned from trying to find wild game and was confronted by Shannon Wilson Bell, who attacked him with a hatchet. Packer killed Bell with a pistol. He, too, began to eat his comrades and to steal their belongings. Eventually, Packer made his way to the Los Pinos Agency, then Saguache, where he was arrested and then escaped. Later, when he was found and returned for trial, the March 1883 *Saguache Chronicle* headline read, "Cannibal Packer—After Nine Years a Fugitive from Justice, the Capture Is Effected of the Human Ghoul Who Murdered and Grew Corpulant [*sic*] on the Flesh of His Comrades."

In a confession, Packer explained that after killing Bell in self-defense, "I tried to get away every day, but could not, so I lived on the flesh of these men the greater part of the sixty days I was out. Then the snow began to have a crust and I started out up the creek" (quoted in Oldham 2005:31).

After his 1877 trial in the Hinsdale County Courthouse, a two-story wood building that still stands, Packer served time in the state penitentiary. The Cannibal Plateau northeast of Lake City commemorates the saga of Alferd Packer, a cautionary tale for other miners about the need for adequate provisions and the difficulties of winter travel in the San Juans.

From his headquarters in Saguache, Otto Mears built toll roads across the San Juans to connect the far-flung camps. One toll road linked Ouray, Animas Forks, and Lake City, climbing over Engineer Pass at an elevation of 12,800 feet (3,900 m), for a fare of $2.25. Halfway between Lake City and Ouray stood Corydon Rose's one-story inn, which served as stagecoach stop, hotel, saloon, restaurant, and post office. Rose kept sixty burros to carry supplies to miners and to bring their ore back for wagon transport to Lake City mills.

With the discovery of rich silver deposits, Creede was the site of Colorado's last big silver strike, with a boom that began in 1889. According to John Canfield in *Mines and Mining Men of Colorado*:

> Miners resolved that, when $70,000 is paid for a hole not much deeper than a cellar, in a district before unheard of, there was something going on ... Dozens of

men tramped up the long road from Del Norte. They came from all parts of the state, and bore upon their brawny backs their tools, bed, and kitchen ... When night came they rolled themselves in their blankets and slept in the cradle of the silent hills. (Canfield 1893:59–60)

In the narrow canyon that houses Creede's Main Street, one poet-newspaperman quipped, "It's day all day in the daytime, and there is no night in Creede" (Smith 2000:128). Adjacent mineral deposits swelled the town to almost 10,000 citizens. The town's fortunes seemed bright when the tracks of the Denver and Rio Grande, coming up through Wagon Wheel Gap, arrived in 1891, a year after the town was founded. During its boom days, local residents included Robert Ford, the saloon-keeper who murdered Jesse James, and bar manager and quick-draw expert William Barclay ("Bat") Masterson, whose gun never left its holster because of his fearsome reputation. Citing a March 1892 booklet titled "Creed Camp," published by *The Rocky Mountain News,* Richard Huston wrote: "Drunken men come out occasionally and empty their guns into the air or somebody's legs. The latter process indicates a cultivated softening of old brutal habits. There are a few 'bad' men in Creede, and many who are reckless, but thus far there has not been a murder. This is a source of constant amazement, but it is true" (2005:97).

Over 60 mining companies were incorporated in 1892; by 1897, there were 1,896 lode claims—325 of them patented—and 10 placer claims. Famous mines included the Holy Moses and the Last Chance. The town's founder, Nicholas C. Creede, went from a penniless, itinerant miner to a millionaire. He married a Del Norte dance-hall girl, with whom he fought bitterly, and died of a morphine overdose in Los Angeles in 1897.

Mining camps were frequently "boomed" by local and regional newspapers. On May 18, 1892, *The Rocky Mountain News* crowed, "The new town of Carson [near Lake City] is being rapidly improved, and lots are being staked off. The drug business, general merchandise, clothing, livery, and saloons are all represented. A large number of people arrive daily, two coaches coming in today, heavily loaded with capitalists seeking investments." Today, almost nothing remains of Carson except a few tumbledown cabins. Most mining camps met the same fate.

Prospectors, miners, and settlers sought livelihoods in the remote eastern San Juans. Just as gold and silver mining began to show promise, however, the 1893 national depression and the demonetization of silver devastated the fledgling towns. Without the federal government buying silver at a fixed price, the value plummeted, and overnight the small camps began to fail. The card sharks, the prostitutes, and dance-hall girls all moved on. A few prospectors stayed, hoping to find better-paying gold ore, but the first phase of San Juan mining had ended, leaving a legacy of can dumps, mine dumps of broken rock, and stumps four feet high where trees had been cut for winter fuel.

HERDERS AND HOMESTEADERS

New economic enterprises included lumber cut and milled near South Fork, once the railroad arrived, and vast herds of sheep grazing in high mountain meadows. Key access routes into the eastern San Juans included La Garita Stock Driveway, used seasonally by Hispanic sheepherders who, beginning in the spring, brought 10,000–20,000 sheep from the San Luis Valley to the high mountains near Silverton by summer. They traveled westward from the mouth of La Garita Canyon, following La Garita Creek into the high country. Herders gathered herbs and high-altitude plants and watched carefully over their sheep. Stock driveways from the lowlands to alpine pastures often utilized the same trails established by prospectors and miners.

Tall grass in the valleys attracted homesteaders and their cattle into the eastern San Juans. Pioneer families went west on the newly established railroads to farm or ranch, although many starved or froze to death. Just as mining laws encouraged development of hard-rock lode claims, the 1862 Homestead Act allowed each head of household to claim 160 acres or less of the public domain, provided he or she would fence the homestead, erect a 10-by-12-foot (3-by-3.7-m) cabin, live there for five years, and plant crops. Ranchers subsequently grazed cattle on thousands of acres of adjoining public lands.

Ernest and Edith Shaw homesteaded northwest of Pagosa Springs in Hinsdale County, along Weminuche Creek, in 1902. They struggled to put up a cabin, survived sickness and marauding coyotes that ate their chickens, planted rhubarb and cabbages, and raised horses and cattle. In letters mailed home to a Boston suburb, Edith Shaw wrote, "I went to Pagosa last trip, and the long hard ride in the bitter cold was too much for me. I shan't go again until spring. I rode twenty miles of the distance [on] horseback so as to keep warm, and much of the way down it was a wet, blinding snowstorm" (quoted in Gulliford 2003:39).

She began writing letters in April 1902, and they ended in December when there was a snowstorm with three feet of snow on the ground and temperatures 14° below zero outside their log cabin. Yet the Shaw family persevered on their Sunset Ranch.

The civilized response to the mountain landscape had been extraction—extracting gold and silver, timber, and grass—while in the East a conservation movement had begun that focused on preservation and multiple uses of the vast public lands in the American West. The San Juan Mountains were no exception. In 1905, under President Theodore Roosevelt, federal forest reserves became national forests within the Department of Agriculture. Afterward, land use was regulated, ranchers paid grazing fees, and a new breed of westerner—the forest ranger with his regulation green shirt, flat-brimmed hat, and gold badge—guided private use of public land.

Desperately in need of money, rancher Ernest Shaw read about a Civil Service exam he could take to become a US forest ranger. He took the exam and, in 1907, became forest supervisor for the Montezuma Forest. In 1908, he was appointed supervisor of the San Juan National Forest. A pioneer family had succeeded in the San Juans.

REFERENCES

Canfield, J. G. 1893. *Mines and Mining Men of Colorado: The Principal Producing Mines of Gold and Silver, the Bonanza Kings and Successful Prospectors, the Picturesque Camps and Thriving Cities.* Denver: John G. Canfield.

Gulliford, A., ed. 2003. *Edith Taylor Shaw's Letters from a Weminuche Homesteader, 1902.* Durango, CO: Durango Herald Small Press.

Huston, R. C. 2005. *A Silver Camp Called Creede: A Century of Mining.* Montrose, CO: Western Reflections.

Ingersoll, E. 1994. *Knocking Round the Rockies.* Norman: University of Oklahoma Press (originally published by Harper & Bros., New York, 1882).

Interpretive sign text on the Alpine Loop Interpretive Trail on a promontory point overlooking the former townsite of Sherman. Text from Crofutt's *Grip-Sack Guide to Colorado* (1885).

Kindquist, C. E. 1987. *Stony Pass: The Tumbling and Impetuous Trail.* Silverton, CO: San Juan Book Co.

Oldham, A. 2003. *Albert H. Pfeiffer: Indian Agent, Soldier and Mountain Man.* Pagosa Springs, CO: self-published.

————. 2005. *Alferd G. Packer: Soldier, Prospector, Cannibal.* Pagosa Springs, CO: self-published.

Smith, D. 2000. *Song of the Hammer and Drill: The Colorado San Juans, 1860–1914.* Boulder: University Press of Colorado.

CHAPTER FOURTEEN

Disaster in La Garita Mountains

Patricia Joy Richmond

A NINETEENTH-CENTURY HERO

JOHN CHARLES FRÉMONT'S FOURTH EXPEDITION into La Garita Mountains at the eastern fringe of Colorado's San Juan Mountain Range (1848–1849) bears all the intrigue found in classic Greek drama. The hero, Frémont, had become a popular champion among his countrymen. Dubbed the "Pathfinder" by contemporaries burning with the fever of Manifest Destiny, Frémont's adventures fired the imaginations of young and old. His writings brought the worlds of science and exploration into the homes of average Americans and inspired dreams of venturing west to see the "Shining Mountains." Among those not prone to read scientific jargon, dime novels—the literary fad of the day—extolled the exploits of Frémont and his mountaineering companion and friend, Kit Carson.

Frémont had stepped onto the road to fame at a young age. Although a brilliant student, a spate of adolescence in the spring of his seventeenth year led to his dismissal from Charleston College in 1830. Unable to continue his formal studies, Frémont, an avid reader, found inspiration in two books. One recounted character tales of men famous for bravery and honor. The second contained extraordinary maps of the heavens. Frémont studied these maps so thoroughly that he became adept at using astronomical observations to determine both the latitude and longitude of western landmarks. Frémont's longitudinal readings, used to refine maps essential to successful travel through the western wilderness, still bear accuracy within a few seconds (Graham 2000).

Frémont's career as a frontier explorer began with the survey of a wilderness road between Charleston, South Carolina, and Cincinnati, Ohio. During the winter of 1837, at age twenty-four, he conducted a fast-paced reconnaissance of the mountainous Cherokee Strip in Tennessee, Georgia, and North Carolina. By May 1838, he was second in command to the well-respected Joseph N. Nicollett, with the assignment of surveying uncharted wilderness between the Mississippi and Missouri Rivers. Nicolett carried high praise for Frémont's skills to his superiors in Washington, DC. By January 1839, Frémont held the rank of second lieutenant in the US Army Corps of Topographical Engineers.

Frémont's career was now on a fast track. Following a second expedition with Nicollett, Frémont headed to Washington, DC, to file the official reports. A quick courtship led to an elopement with Jesse Benton, daughter of the prominent and powerful Senator Thomas Hart Benton of Missouri. Benton controlled US Senate funding for western exploration. Although often at odds with his son-in law, Benton would facilitate assignments that brought Frémont's name to public attention.

Positive responses to Frémont's first western expedition to pinpoint South Pass in southern Wyoming (1842) garnered him a more important assignment—determining an alternate route for migrants bent on crossing the Rocky Mountains to settle the Pacific Northwest. Frémont's account of his second expedition (1843–1844) became a guidebook for emigrants headed not only to Oregon but also to California. After reaching the mouth of the Columbia, Frémont, Carson, and their party explored the interior desert region that Frémont correctly identified as a Great Basin. In searching for a route through the Sierra Nevada, the expedition became trapped by a severe snowstorm. Men and animals suffered terribly before Carson found a pass that now bears his name, leading across the mountains. After recuperating among settlers on the American River, Frémont chose to return to the states by way of the Old Spanish Trail across the Mohave Desert, so he could explore even more uncharted terrain in Utah and central Colorado. The amount of territory covered, as well as the saga of the ordeals the men endured, put the expedition on par with that of Lewis and Clark.

Frémont's popularity with the general public was at its zenith. But fame also brought detractors, some fueled by jealousy and others by politics. West Point graduates like Lieutenant J. H. Simpson chafed at Frémont's rapid rise in rank. While many praised Frémont, letters, editorials, and testimonials debunking his accomplishments began to appear in newspapers—especially those that served a constituency at odds with Frémont's anti-slavery sentiments.

As the United States moved toward war with Mexico, President James Polk and others eyed the fabled land of California. Frémont, who had been promoted to colonel within the Corps of Topographical Engineers, received a directive to explore the Klamath Lake region and the border country between Oregon and Northern California. Whether the purpose of this expedition (1845–1846) initially went beyond the realm of discovery is as enigmatic as Pike's trek into Spain's San Luis Valley (1807). Frémont carried out his duties as a surveyor and cartographer. Then, just as the expedition prepared to mount Oregon's Cascade Range, a courier intercepted Frémont. The content of the message remains unknown, but the circumstances plying Frémont's destiny changed. Frémont the scientist and explorer, the Frémont who had professed his love for the solitude of the wilderness, transformed into a military leader—a role for which he had neither formal nor practical training. The United States and Mexico had gone to war.

Frémont's small band marched south into California. Joined by settlers in the Sacramento Valley, they waged guerrilla warfare against Mexican loyalists. To some, Frémont's escapades in California made him a hero. The ocean-to-ocean borders of the Manifest Destiny dream became reality. But fate played upon folly. When other American forces arrived, Frémont cast his lot with Commodore Robert F. Stockton. General Stephan Watts Kearny charged Frémont with insubordination, had him arrested, and transported the Bear Republic's hero to Washington, DC, for court-martial.

The momentum that had carried Frémont forward so rapidly seemed to spiral into reverse. Chagrined by the court-martial, Frémont resigned his commission to become an ordinary citizen. Having experienced the pleasant weather in California and seen the potential for wealth in land and resources, he devised to move his family there (Bigelow 1856; Frémont 1887; Jackson and Spence 1970, 1973; Spence 1984).

THE JOURNEY UNFOLDS

As political and economic posturing by the northern and southern states intensified, rivalries began festering toward civil war. Missouri—in the nation's midriff, the center of commerce for the Santa Fe Trail, and the point of embarkment for other westward travel—would lose its critical political and economic status if either a northern or a southern railroad spanned the continent. Benton, eager to keep Missouri in the nation's forefront, harangued the US Congress to fund exploration of a "Central Route" to the Pacific. Instead, Lieutenant Edward F. Beale received a commission to explore a southern route through New Mexico and Arizona Territories. Benton turned to his son-in-law to promote his cause—one last expedition, a last hurrah to set the record straight and rekindle public empathy. Jesse and the children would travel by way of Panama and meet Frémont in California. Frémont succumbed to the proposal—possibly out of loyalty, possibly with a desire to restore his dignity and regain his integrity, possibly because he wanted one more adventure in the wilderness.

Benton solicited businesses throughout St. Louis to equip the expedition. Many claimed later that they were promised government compensation after the unofficial, private expedition had been completed. The tally of supplies and equipment suggests that Frémont's fourth expedition set forth well equipped: the most recent design in pack saddles, the latest scientific instruments, plenty of horses and mules, and a small cannon of the type Frémont traditionally carried to divert Indian threats (Richmond 1990).

The participants included many seasoned veterans of previous expeditions. Ralph Proue had accompanied Frémont on three expeditions; this would be his fourth and last trek. About a third of the party had been with Frémont in California during the rebellion. Others had weathered the travails of the second expedition or clambered through the Wind River Mountains during the first expedition. Most were Missouri muleteers. A few, such as the German cartographer Charles Preuss, held expedition positions that required formal or skilled training (Preuss 1958).

Approximately half of those who joined the expedition were greenhorns or had not traveled with Frémont previously. George Ruxton had hoped to join the expedition, but he died before the departure. His friend, Andrew Cathcart, left another expedition to return to St. Louis after hearing of Ruxton's death. Cathcart decided to join the fourth expedition in Ruxton's place. The Kern River commemorates Frémont veteran Edward Kern's place in California's Bear Republic history. His letters home inspired his brothers Richard and Ben to join him in following Frémont westward again. Micajah McGehee, an ailing Mississippi gentleman, wanted a great adventure before his life ended. What inspired others, like father and son Amos and Elijah Andrews, is unknown. Three Indians—Gregorio, Juan, and Manuel—as well as Saunders Jackson, a freed servant, accompanied the expedition (Colville 1996; McGeHee 1891; Richmond 1990).

As the expedition readied for departure, the Frémonts' young son died. Both parents blamed the infant's death on the stress that had overtaken the family since Frémont's arrest in California. Jesse would forever hold Kearny accountable for her child's death. The grieving period pushed the expedition's departure date from late summer into fall. The men waited. Time passed—critical time. Finally, on October 3, 1848, Frémont and his thirty-five men departed from St. Louis. Although he believed the expedition could safely pass the Rockies before the heaviest snows, Frémont admitted that his heart was no longer in the journey. Nevertheless, he felt confident that, with the guidance of his old friend Kit Carson, the expedition could make a winter crossing through the Sierra Nevada.

The passage across the Kansas prairie flowed with the sense of care-free adventure the expedition members had anticipated. A few snow flurries and encounters with friendly Indians gave the Kern brothers something to write about in their diaries (Hafen and Hafen 1960; Spencer 1929).

At Big Timbers on the Arkansas River, Frémont conferred with his old friend and onetime guide, Thomas "Brokenhand" Fitzpatrick. Having been appointed agent for 600 lodges of Apache, Arapaho, Kiowa, and Comanche camped at Big Timbers, Fitzpatrick could not accompany Frémont this time. Stories about deep snow in the mountains sparked uneasiness among some members of the expedition. Frémont confided in a letter to Benton that he intended to ascend to the head of the Rio del Norte, descend onto the "Colorado" (today's Gunnison River), and then proceed across the Wahsatch Mountains and the Great Basin near the 37th parallel (Gwyther 1870; Preuss 1848).

At Bent's Fort, Frémont's faith in the expedition's success suffered a setback when he learned at Bent's Fort that Carson had been conscripted by the army to track Apache marauders. Alexis Godey, Frémont's right-hand man, sometimes flattered as a "young Kit Carson," had no firsthand knowledge of the route through the San Juan Mountains. A watercolor Richard Kern sketched at Bent's Fort on November 17, 1848, proved the stories about snow already blanketing the mountains were true. White mantles shrouded the summits of the Spanish Peaks, longtime landmarks designating what had once been Spain's northern frontier (Hine 1982 [1962]; Weber 1985).

Upon reaching the mountain man fortress El Pueblo de San Carlos at the juncture of the Fountain and Arkansas Rivers, Frémont and his men moved into abandoned cabins Mormon soldiers had built after leaving Kearny's "bloodless" New Mexico campaign. An old-timer who had previously traveled with Frémont as far as the Great Salt Lake approached Frémont and offered to serve as a guide. Claiming that he knew the route as well as Frémont knew the back of his own hand, Bill Williams declared that the expedition would experience no problems crossing the Rockies if they pressed through before Christmas. According to Preuss, Frémont was hesitant to accept Williams's offer (Ruxton 1973).

The expedition's departure from Missouri so late in the season might have set the stage for the portending disaster in La Garita Mountains, but the unusual circumstances of the winter of 1848–1849 would become the primary factor that jeopardized the expedition's success. Perhaps, as some historians have suggested in hindsight, Frémont should have heeded the stories at Big Timbers; however, entries in trappers' and traders' journals, as well as Ruxton's campfire remarks, demonstrate that tales about weather, hostiles, and wild beasts were more often fabrication than reality. Whatever influence other factors might have had in the subsequent events in the eastern San Juan Mountains, using Williams as guide brought detours and caused delays that eventually led to the expedition's collapse.

Stopping at the summer settlement of Hardscrabble in the mountains west of El Pueblo, the men shelled 130 sacks of corn—a two-week supply for the animals—in case Williams's prediction of clear passage through the San Juan Mountains proved wrong. Fresh horses and mules brought the *remuda* to 120 animals. Late in the afternoon of November 25, the expedition entered North Hardscrabble Canyon. That night they camped beneath monolithic cliffs. From an elevation, the Kern brothers watched the moon rise over Lake San Isabel. The elderly Amos Andrews had given up the quest at Bent's Fort; now Longe, eying the snow in the mountains, turned back after prophesying doom and failure.

TROUBLES BEGIN

Had the expedition proceeded up North Hardscrabble Creek into the Wet Mountain Valley and then followed Grape Creek to its head, the route would have led directly to Médano Pass, an easily visible notch through the Sangre de Cristo Mountains. Instead, Williams turned south toward Ophir Creek to pass behind Greenhorn Mountain onto Williams Creek, a tributary of the Huerfano River. While crossing Promontory Ridge, the men and animals wallowed in deep snow at elevations reaching 10,000 feet (3,048 m). With no grass or browse, the muleteers resorted to feeding the corn procured as a precaution for their passage through the San Juan Mountains. The expedition finally descended into the Huerfano Valley and intercepted "Road Robidoux" near the bastion-like Gardner Butte. (Trader Antoine Robidoux regularly used the Huerfano River route to transport wagonloads of Taos flour for trade with the Plains tribes.) Following Robidoux's Road, the expedition reached the saddle-like summit of Mosca Pass and gazed into the "valley of the del Norte"—the San Luis Valley. Both Médano and Mosca Passes lie in proximity to the 37th parallel. However, whereas Médano Pass is relatively wide and open, fallen timber and jumbles of boulders choke the west side of Mosca Pass. Men and beasts suffered terribly navigating past these obstacles in deep snow.

On December 4 the expedition camped at the edge of Zebulon Pike's "sea of sand." The temperature that night was 17° below zero. Come morning, icicles hung from the men's beards and hair. Frost coated their eyelashes. With the corn reserves consumed, the hungry, cold animals tried to bolt back the way they had come. While the muleteers rounded up the renegade livestock, Frémont, accompanied by Preuss, Williams, Henry King, and Frederick Creutzfeldt, explored the western mouth of Médano Pass. Preuss commented in his diary that it was strange that Williams had missed the "real pass."

For two days Godey tried to find a route between the sand hills and the base of the Sangre de Cristo Mountains. Finally, the men and animals moved directly up and over the crest of North America's highest interior dunes. Frémont's Big Timbers letter to Benton tends to direct focus away from the significance of the expedition's endeavors to proceed to the north side of the Great Sand Dunes. Such determination made sense only if Frémont's objective was to reach the Saguache Canyon and Cochetopa Pass (Heap 1853 [reprinted in Hafen 1957]; Richmond 1990).

Trappers and traders had trailed through the Cochetopa country since the early 1820s. Military personnel traveling between California and Santa Fe during the war with Mexico had also utilized this route. Carrying dispatches from General Kearny, Lieutenant George Brewerton, with Carson as his guide, had followed the trail across the Cochetopa in the spring of 1848. Although known to men such as Carson and Antoine Leroux, who frequented the mountain regions north of Taos, the Cochetopa route could be elusive for travelers unfamiliar with the terrain. (In 1857, Captain R. B. Marcy's party became lost and almost perished on the Cochetopa after a storm deposited several feet of snow.)

Snow became the bane of Frémont's expedition, too. Just as the weather can produce hundred-year floods, so, too, can exceptionally severe winters set records in the San Luis Valley. Week after week, snow fell and accumulated into measurements that doubled the normal snowpack. Day and night, temperatures registered below zero, without the San Luis Valley's usual midday thaw. Winds howled at gale force. Generally, the valley floor, as well as the sand dunes, remains snow free throughout most winter months. Yet the fourth expedition trudged through a foot of snow at the dunes and camped in two to four feet of snow while crossing the valley floor. Hoar frost glistened in the air. A white fog persistently curtained the San Juan Mountains. The shroud of snow and ice, reaching from the pinnacles of the Sangre de Cristo Mountains to the summits of the San Juans, obscured geographic features that would have been obvious in any season (Richmond 1990).

Men and animals suffered terribly from the mind-numbing, extremities-freezing cold. Fast-burning sagebrush, the only fuel on the valley floor, provided little campfire comfort. Hunger and thirst crazed the animals. Diary entries suggest that Frémont wanted to continue toward his predetermined route, the Saguache Canyon. Once across the Continental Divide, the expedition should have moved quickly through the desert Basin and Range country. Williams, however, advocated turning north toward Poncha Pass. Frémont rejected this suggestion. Williams then advocated a route that supposedly would cut their travel time by two days. Each day had become critical if they were to pass through the Rockies before Christmas. Godey and others encouraged Frémont to take Williams's advice (Carvalho 1857; Schlaer 2000).

A FATEFUL DETOUR

The party turned southwest along the frozen channel of Carnero Creek, a tributary of Saguache Creek (figure 14.1). Three primary streams in the La Garita Mountains feed Carnero Creek, which breaks onto the valley floor through a basalt formation known as the Hell Gate. Here the force of water roiling through the narrow gap prevented the men from maintaining their path along the frozen riverbed. As if acknowledging doubts and possibly whether the expedition might succeed in crossing the Continental Divide, Frémont had two elk carcasses cached. The expedition diverted its course from Carnero Creek to cross over a hill south of the Hell Gate. After coming off the hill, the men found themselves in Coolbroth Canyon, a

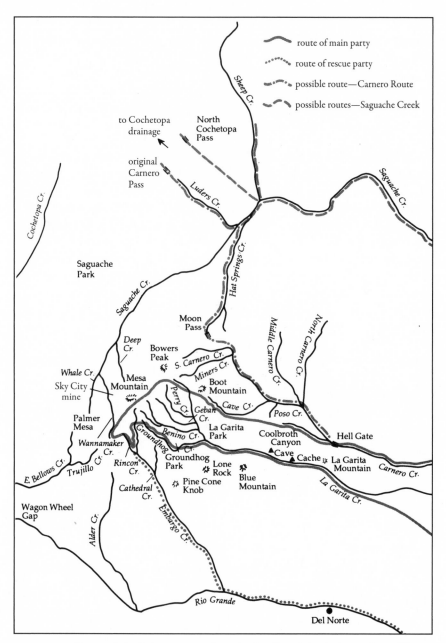

FIGURE 14.1 Route of the Frémont party into the San Juan Mountains and possible routes to Chochetopa Complex (Richmond 1990). *Courtesy,* Colorado Historical Society.

minor tributary of South Carnero Creek. Instead of saving days or miles, Williams's "shortcut" had expended both time and energy that would have been unnecessary had the expedition continued toward the more open route along Saguache Creek (Richmond 1990).

The drainages of Carnero Creek and Coolbroth Canyon have a configuration like a V, similar in appearance at the point of convergence yet miles apart in geophysical differences as they reach toward the summits of La Garita Mountains. Here began the circumstances that dictated the expedition's absolute failure. Had Frémont's men proceeded into South Carnero Canyon, they would have ascended to Moon Pass, dropped onto the Hat Springs Creek tributary of the Saguache, followed Luder's Canyon to Carnero (Old Cochetopa) Pass, and then made their way toward the Lake Fork of the Gunnison and points beyond. Perhaps the winter of 1848–1849 precluded passage through the San Juan Mountains by any route (Hume 1862; Marvin 1933; Richmond 1990).

Not realizing that a route leading to the Cochetopa Complex and the Continental Divide lay just a short distance to their right, the expedition ascended Coolbroth Canyon and passed over the ridge that defined the drainage of Cave Creek. One of Richard Kern's watercolors depicts their desolate surroundings. In the painting, caves pock solid rock cliffs along the north side of the frozen creek. A snow-clad, twin-peaked mountain rises ahead of three men posing by the rock walls. Often in winter, grasses and shrubs protrude from the skimp of snow that normally hugs the double summit of Boot Mountain. Such was not the case that December day in 1848. In attempting to ascend Boot Mountain along the source canyon of Cave Creek, men mounted on mules toiled through chest-deep snow for hours just to pass a few hundred feet. This day brought the first mention of mules succumbing to the traumas of cold, hunger, and fatigue (Richmond 1990).

The men's jubilation at finally reaching their objective, the summit of the twin-peaked mountain presumed to mark the Continental Divide, plunged into disbelief. Before them lay a higher dome surrounded by snow-covered shoulders, ridges, and valleys—Mesa Mountain. Turning to the east, the exhausted explorers viewed the snow-clad Sangre de Cristo Mountains across the forty-mile span of the San Luis Valley. Below them, veering northwest from the Hell Gate, rose the South Carnero Canyon—so near and yet so inaccessible from this elevation.

The expedition dropped onto the headwater of La Garita Creek and made camp on its west slope at the timberline. Scott scouted ahead and reported not only a road but also grass protruding through the snow. Determined to cross the Divide, the men rose at dawn and drove the animals along the ridge that separates John's and Deep Creeks, tributaries of Saguache Creek, from tributaries that flow into La Garita Creek—Perry's, Geban, and East and West Benino Creeks. The promised grass turned out to be the tips of willows covered by deep snow.

That night the expedition members huddled miserably amid sparse cover at the head of West Benino Creek. The next day almost brought total destruction. In trying to pass directly over the summit of Mesa Mountain, even Bill Williams became stupefied with the cold. McGehee's narrative (1891) describes the old man astride his mule, dazed and shivering in his military "monkey jacket." Frantic animals, lunging wildly, disappeared into snowdrifts dozens of feet deep. After hours of toil at trying to break a road, the men turned back and spent another bitter night in their previous camp area.

Above the Benino Creek camp, just beyond the north end of a snowy cornice that sometimes lingers into summer, stands a massive rock. Passing along the same dividing ridge Frémont's expedition followed from La Garita Creek camp, La Garita Stock Driveway makes a sharp corner along the west edge of this rock outcropping. At that very point, butted against the base of the rock, is the outline of a D-shaped rock shelter. Incised into the rock's face about ankle height and windblasted almost to obscurity is the date 1848. Which member of the expedition left this record in witness of his horrible ordeal is unknown. Perhaps he believed he was marking his tomb.

Godey took charge the next day. Using mauls contrived from hacked trees, the men beat a shoulder-wide passage through snow 18 feet (5.5 m) deep. More pack animals succumbed while crossing the south face of the peak. Finally, the men broke over the ridge and dropped into three spruce copses at the head of Wannamaker Creek (Richmond 1990).

NOTHING BUT SNOW

The expedition members spent the better part of a week in the Wannamaker Creek camp. Fierce storms pummeled the mountain daily. The men clustered around burning trees hacked off at snow level. The stumps in the Wannamaker camp measured 16 to 30 feet (4.9 to 9 m) aboveground closest to the ridge and 4 to 5 feet (1.2 to 1.5 m) farther into the swale. Some stumps still stood in the late 1980s, but many had fallen to become part of the forest debris. As the trees burned, snow melted to form soggy, cold, smoky depressions around the men and their tents. Ben Kern named his surroundings "Camp Dismal." Just beyond the camp, the starving animals, having devoured each others' manes and tails, screamed in misery.

Frémont had sent Preuss with a scouting party to determine the possibility of proceeding farther west. The small party followed the ridge above Trujillo Creek toward Palmer Mesa. Everything before them was blanketed in white. With supplies running low, the men exhausted, the animals dying, and blizzards hammering the mountains day after day, Frémont had no choice but to return to the San Luis Valley. As the severe weather broke beneath blue sky and warming sun, the men packed their belongings, supplies, and equipment into large bundles, which they ferried across the ridge dividing Wannamaker Creek from Rincon Creek. Godey's mule, Polly, was driven until she fell in her tracks (Preuss 1958).

A SPATE OF HOPE

The change in weather, as well as their departure from the Wannamaker camp, lifted the men's spirits. The new camp, perched near the edge of a sheer drop above the headwater of Rincon Creek, became "Camp Hope." Historians refer to this site as the "Christmas Camp," for here Frémont and his men passed the holiday. Some made merry with elk stew, minced mule-meat pies, rice, donuts, biscuits, extra rations of sugar and macaroni, and coffee with alcoholic spirits. Frémont isolated himself and delved into his volume of *Blackstone*.

Twenty-one days had passed since the men camped at the base of Mosca Pass, not 100 miles (160 km) to the east. They still had not passed the Continental Divide. Frémont decided that after returning to the floor of the San Luis Valley, the expedition would proceed west by the route he had anticipated taking—the Saguache Valley. To do so, the expedition needed fresh animals and additional supplies. The closest villages, whether north of Taos or in the Ojo Caliente Valley, were over 100 miles away.

On December 26, a four-man relief party, composed of King, Thomas Breckenridge, Creutzfeldt, and Williams, started down Rincon Canyon, toward the valley of the Rio Grande. Frémont allotted sixteen days for the trip to the New Mexican settlements and back. During that time, the rest of the men would move the baggage off Mesa Mountain. After rendezvousing near present-day Del Norte, the full expedition would proceed to California (Colville 1996; Old Spanish Trail Association 2008).

At Camp Hope the men built two crude sleds, or *sledges,* to convey baggage down Rincon Canyon. After losing control of one, which crashed and spewed its contents, the men abandoned the second sled. Its remains, recovered at the Christmas Camp in the 1950s by Ranger Mark Ratliff and a group of Del Norte citizens, are displayed at the Rio Grande County Museum.

On December 27, the men began moving the baggage to a camp 2 miles (3.2 km) downstream from the Christmas Camp. The Kerns noted the day as sunny and warmer, with birds singing. The weaker men "merrily" rode the large bundles filled with supplies downhill. On December 29, Ben Kern lamented that his riding mule, the fattest of the remaining animals, had been butchered. The lower Rincon camp soon earned the name "Camp Disappointment." Godey, who had accompanied the relief party to the Rio Grande so he could guide the rest of the expedition off the mountain, returned with news that this drainage was too steep and rocky to ferry out the baggage.

During the trek across the ridge between the La Garita Creek camp and Mesa Mountain, Frémont had observed a fairly open route leading back toward the San Luis Valley. To intercept it, the men had to haul the heavy bundles up the ridge that divided Rincon Creek and the upper reaches of Embargo Creek. The Kerns consolidated their belongings into a bale that weighed about 200 pounds (90 kg). Assisted by Gregario, Juan, and Manuel, the Kern party pushed and pulled the bundle uphill with "a little coaxing, perseverance & hard labor," according to Kern's diary. The Kerns' mess passed New Year's Eve at the head of East Embargo Creek. Richard Kern sketched and later painted a watercolor depicting their holiday camp near copper-colored cliffs. Edward Kern fixed a treat of minced mule-meat pies. Some men fired their guns to celebrate the arrival of the New Year; others contemplated the events of the past month.

By New Year's Day, the expedition had moved into a camp among the trees at the head of Groundhog Creek. As with other Frémont camps on Mesa Mountain, the Groundhog Creek site sits at the edge of the forest on the west side of the creek, where the men could catch the early-morning sun. Perhaps because he and his brothers had made merry the night before, Ben Kern complained about having to push his bundle up and over the dividing ridge on his hands and knees.

Groundhog Creek offered the best location of the fourth expedition's camps. The protected west slope at the head of the stream rises gently to provide adequate shelter within the spruce forest. The stream flows through a wide valley that descends gradually toward its juncture with La Garita Creek. The Kerns found this camp so pleasant that they decided to await the relief party's return with fresh mules before descending toward the floor of the San Luis Valley.

DESOLATION AND DEATH

While the Kerns lingered at the Groundhog Creek camp, sketching or writing in their diaries, the other expedition members moved baggage and equipment downstream through Groundhog Park into La Garita Park. Time was running out, as were food, health, and stamina. On January 9, 1849, Ralph Proue, the oldest member of the expedition, fell in his tracks and could not be revived. The men passed his lifeless body as they trudged between the Groundhog Creek camp and the next campsite 3 miles (4.8 km) downstream, near the mouth of Little La Garita Creek.

Frémont's mess, composed of himself, Preuss, Godey, Godey's teenage nephew Theodore McNabb, and the freedman Saunders Jackson, had proceeded to a suitable cache site on La Garita Creek. Known as Point of Rocks, a name not uncommon for the tuff outcroppings that mark most streams along the western edge of the San Luis Valley, the site provided a virtually snow-free area at the base of a broad rock face that radiated solar heat. The Point of Rocks site offered easy access from the floor of the San Luis Valley and could be located without difficulty because a landmark fortress-like prominence rose above the creek—La Garita.

Three weeks had passed since King's relief party had departed from Camp Hope. Convinced that King's party had failed to fulfill its objective, Frémont decided that his mess should move as quickly as possible toward the New Mexican settlements in an effort to save the rest of the expedition. He left Lorenzo Vinsonhaler in command and left orders that the remaining men were to cache the baggage and then proceed as quickly as possible south, along the San Luis Valley toward the Conejos (Rabbit) River.

Within a day, Frémont and his companions came upon a lone Indian with a couple of horses near the oxbow of the Rio Grande, south of present-day Alamosa. The Indian guided Frémont's party to the relief party encamped near the San Luis Hills, still dozens of miles from the New Mexican settlements. Somewhere between Del Norte and Monte Vista, Williams had convinced the other members to leave the shelter of the Rio Grande's cottonwood forest. According to Breckenridge, Williams had spotted smoke and feared that Indians camped along the river might retaliate for his earlier treachery. Exposed to the full fury of icy winds blasting off the snowy summits of the Conejos Mountains, King apparently succumbed to exhaustion and hypothermia. An entry in Preuss's diary suggests that King's body had been cannibalized, a charge supported in a letter written by J. H. Hume, one of Breckenridge's relatives (Hume 1862; Preuss 1958).

With the relief party survivors in tow, Frémont and party rushed toward the settlement of Rio Colorado (Questa, New Mexico). Godey quickly began to procuring relief supplies, including horses for meat and transportation. Villagers volunteered to accompany Godey on the race back to the San Luis Valley to rescue the men left in La Garita Mountains. Preuss remained in Red River to await Godey's return. Frémont proceeded to Taos. Carson, who had returned to his home in Taos, provided succor and solace for his longtime friend.

The animosity and quarrels that had developed between the Kerns and veteran expedition members erupted into full animosity after Frémont's departure. Vinsonhaler sent an order for the Kern brothers to move their baggage to the cache site and to be quick about it. While struggling to reach the cache site at Point of Rocks, the Kerns and other members of the expedition encountered excruciating conditions. Typically, January in the San Luis Valley brings severe subzero temperatures intensified by fierce, biting winds. The canyon of La Garita Creek became a virtual arctic wind tunnel. The men could barely stand against the gale. Some sought shelter in a rock overhang near where Little La Garita Creek joins La Garita Creek.

As McGehee cooked his last bit of food, a mixture of sugar and macaroni, Richard Kern accidentally knocked the concoction into the fire. Rawhide laces and feathers from a dead hawk, cooked into a glue, became their only sustenance until the Kerns' mess finally joined the rest of the expedition members. By then, the other men had divided and digested the meat cached at the "elk camp," as well as a deer that had been killed. The Kerns howled foul over receiving a fore-shoulder of venison as their allotment. However, Thomas Martin stated later that he had killed a "goat" (probably a mountain sheep) and delivered it to the Kerns' mess.

Frostbite and hypothermia began to take their toll. Carver, ranting that he knew how to save the expedition, rushed into the night, never to be seen again.

Manuel, suffering from frozen feet, claimed he could go no farther. After begging to be shot, he returned to a wickiup at the cache site. Charges of rascality, cowardice, and unfit principles peppered the Kerns' diaries as human interactions unraveled. Veteran Thomas Martin declared that the expedition was in chaos—each man for himself. To ward off the temptation of cannibalism, several members agreed that if any man fell behind, the others would build a fire and then press onward.

The Kerns and some of the weaker members vowed to remain together and to leave no one behind. Veteran Charles Taplin offered to stay with the weaker group to serve as a hunter if the opportunity arose. Having caught up with Andrews and Henry Rohrer, who had fallen behind in the push to reach the Conejos River, the Kerns group made camp just below the oxbow of the Rio Grande, south of present-day Alamosa. Richard Kern's watercolor *Relief Camp* depicts this site among gnarled cottonwoods, with the snowy summit of Sierra Blanca in the distance. Ben Kern wrote in his diary that hunger pangs had him eyeing bugs suspended in ice.

McGehee, who considered himself the weakest member of the Kerns party, recorded in his diary an incident that would lead to the greatest controversy regarding the disaster, one that would turn member against member and members against their leader. Historians have debated the issue for decades. With Andrews dead and Rohrer dying, McGehee named those who sat near the fire—himself, John Stepperfeldt, and Taplin. That left Cathcart and the three Kern brothers. The published version of McGehee's diary, "Rough Times in Rough Places" (1891), omitted the name of the person who came to the fire and proposed resorting to cannibalism. McGehee stated that in protesting the deed, he suggested waiting three days. On the third day, McGehee lay prone with hunger. Taplin heard a "halloo." It was Godey with bread and meat.

Despite allegations by other expedition members, some historians have presumed that no one in the Kerns camp participated in cannibalism since Godey arrived within McGehee's three-day deadline. However, Edward Kern's second letter to his sister Mary, written from Taos during his recovery from the ordeal, stated: "Our food is good and for this country quite a satisfactory change from what it was a month ago, when hide ropes gun covers Wolf and mair that's horrible and awfu'—What e'en to name wad be unlawfu'" (Hafen and Hafen 1960:228).

In the list of rawhide laces, pinfeathers, decomposed wolf and hide, mice, snails, worms, bugs, a dead horse, and anything else "horrible and awful" mentioned or utilized as food during the exodus, only one thing would have been "unlawful"—human flesh (Carvalho 1857).

REUNION AND RECRIMINATIONS

Despite Godey's rescue attempt, eight members perished in the drive to reach the Conejos River. Godey intercepted the first survivors, Gregorio and Juan, about 20 miles (32 km) below the Rio Grande's oxbow, near the area where Frémont had intercepted the relief party survivors. The rest of the men, strung along the river, directed Godey to those left behind. Godey had to backtrack to locate George Hubbard, who supposedly had remarked upon leaving Hardscrabble: "Friends, I don't want my bones to bleach upon those mountains this winter amidst that snow" (Hafen and Hafen 1960:146; McGehee 1891). Godey found Hubbard sitting by his fire, dead but still warm. After accounting for all the survivors, Godey and his volunteers headed to the cache site in hopes of retrieving some of the baggage, especially personal belongings. Miraculously, Manuel emerged from his wickiup.

Godey and the survivors did not head for the New Mexican settlements immediately, as might be expected. Most needed time to regain their strength. According to Preuss, it had taken Godey's rescue mission two days to reach the first survivors. A week later, William Bacon, Martin, and Vinsonhaler started for the Rio Colorado in New Mexico. Fifteen days after Godey's departure, those three "hungry brothers" reunited with Preuss. Unable to retrieve the baggage, Godey and his New Mexican volunteers rejoined the Kerns group. Traveling down the Rio Grande, they intercepted the survivors still encamped in the San Luis Hills. The next day Godey and all of the stragglers set camp 10 miles (16 km) upstream on the Culebra (Snake) River, a traditional camp area where the trail from Taos, Pike's "Spanish Road," forded the river (Hafen 1957).

According to Richard Kern's diary, every day was now a "fine day." Perhaps that explained why he seemed to tarry as the other survivors proceeded south to rejoin Frémont. After being caught in a sudden blizzard near Costilla Creek, Kern finally joined Preuss at the Rio Colorado settlement. In his diary, Preuss lamented being as "poor as Job" but acknowledged that having one's life was sufficient (1958:153). In all, ten men, one-third of the expedition, had perished—Proue in the mountains, King south of the Rio Grande, and the other eight between the cache site in La Garita Canyon and the mouth of the Conejos River.

On February 12, 1849, two months after entering the eastern San Juan's La Garita Mountains, twenty-four survivors reunited in Taos. That same day Frémont, having reorganized the expedition, departed for Santa Fe and Albuquerque to proceed to California by the Gila River route. Frémont's decision to leave the Kern brothers in Taos ignited a chain of controversies about the expedition that seared personal letters and newspaper editorials throughout the remainder of Frémont's life. In 1862, the *Liberty* (Missouri) *Tribune* proclaimed:

Glory, Glory, Hallelujah!

Major General John C. Frémont, "the Statesman who never made a speech, the General who never won a battle, the Pathfinder, who is always on the wrong trek, the Millionaire who is not worth a continental dam," has, thank God, retired to private life. (R. W. Settle collection)

The men of Frémont's fourth expedition endured extreme, almost unimaginable physical hardships in La Garita Mountains. Only a handful of exploratory adventures rival their ordeal. Few characters in US history remain as swathed in controversy as John Charles Frémont. Few equally tragic stories generate the ongoing, broad-in-scope, intense interest historians and the general public have in the fourth expedition. Writers have filled pages and tomes with details, hypotheses, and analyses covering a wide array of topics relating to the expedition.

For decade upon decade, the silvery hacked stumps left behind in each of the expedition's campsites stood in mute testimony long after human memory of the participants or their descendants might recall or recount the story. As an old man living in Telluride, Thomas Breckenridge accompanied his son over Stony Pass and down the Rio Grande Valley on an ore-hauling trip to the railhead in Alamosa, Colorado. Along the way, Breckenridge pointed out landmarks related to the tragedy that had almost taken his life. His story, published first in *The Rocky Mountain News* (W. C. Ferril, The Sole Survivor, August 30, 1891) and later in *Cosmopolitan* magazine, would be the last firsthand account.

Whether Frémont or Williams with his detours and delays or the fickle forces of nature should be faulted for the expedition's failure has little relevance to understanding the horrific tribulations those thirty-three pioneer explorers and their guide endured in their attempt to cross the Continental Divide within the San Juan Mountains. When one stands at Inscription Rock today and surveys a 360-degree view of the landscape—Mesa Mountain, with its barren summit and ridges dividing the tributaries of the Saguache from those that flow to the Rio Grande; the Wannamaker swale and Rincon Creek's cliffs; Groundhog and La Garita Parks; twin-peaked Boot Mountain; the Sangre de Cristo Mountains in the distance; the wide span of the San Luis Valley; and the broad horizons of the eastern San Juan Mountain's peaks and canyons—the mind's eye must eliminate the towns, the farms and ranches, the highways and gravel roads, and all modern conveniences—even the ATV and hiking trails—to recreate the scene as it was that winter of 1848–1849.

Whiteness. Nothing but snow, 25 to 50 feet (7.6 to 15.2 m) deep. Cold. Temperatures so frigid every breath stabs the lungs. Cheeks, lips, fingers, and feet turning white with frostbite within minutes. Misery. Insufficient clothing or shelter. Hunger pangs gnawing the belly day after day. Chaos. Brays and screams of anguish from suffering mules. Hurricane-force winds shrieking across timber-less summits. Cursing. Weary men forcing pack animals to move onward and upward, beating narrow paths with hand-hewn mauls, pushing and pulling heavy baggage, shivering in damp clothes in flimsy tents. Day after day, night after night. When the mind sees, hears, and feels every aspect of this historic ordeal, the human sufferings, frailties, conflicts, intrigues, and egos that made up Frémont's fourth expedition become as captivating as any ancient epic.

REFERENCES

Bigelow, J. 1856. *Memoir of the Life and Public Services of John Charles Frémont*. New York: Derby and Jackson.

Carvalho, S. N. 1857. *Incidents of Travel and Adventure in the Far West*. New York: Derby and Jackson; reprinted Philadelphia: Jewish Publication Society of America, 1954.

Colville, R. M. 1996. *La Vereda, a Trail through Time*. Alamosa, CO: San Luis Valley Historical Society.

Frémont, J. C. 1845. *Report of the Exploring Expedition to the Rocky Mountains in the Year 1842—and Oregon and North California in the Years 1843–44*. Washington, DC: Gale and Seaton.

————. 1887. *Memoirs of My Life*. Chicago: Belford, Clark.

Graham, B. 2000. *The Crossing*. Eldorado National Forest Interpretive Association, no city given.

Gwyther, G. 1870. A Frontier Post and Country: *Overland Monthly* 5 (December):520.

Hafen, L. R., and A. W. Hafen, eds. 1960. Frémont's Fourth Expedition: A Documentary Account of the Disaster of 1848–49. *The Far West and the Rockies Historical Series, 1820–1875*. Glendale, CA: Arthur H. Clark.

Heap, G. H. 1853. Central Route to the Pacific. Reprinted in L. R. Hafen and A. W. Hafen, eds., *The Far West and the Rockies Historical Series, 1820–1875*. Glendale, CA: Arthur H. Clark, 1957.

Hine, R. V. 1962. *Edward Kern and American Expansionism*. New Haven, CT: Yale University Press; reprinted as *In the Shadow of Frémont: Edward Kern and the Art of Exploration, 1845–1860*. Norman: University of Oklahoma Press, 1982.

Hume, J. H. 1862. Unpublished letter from Joe Hyatt Hume, Slater, Missouri, October 3, 1937. Transcripted editorial from *The Liberty Tribune* (July 11, 1862). R. W. Settle Collection, William Jewell College, Liberty, MO.

Jackson, D., and M. L. Spence, eds. 1970. *The Expeditions of John Charles Frémont*, vol. 1. Urbana: University of Illinois Press.

————. 1973. *The Expeditions of John Charles Frémont*, vol. 2. Urbana: University of Illinois Press.

Marvin, C. F. 1933. *Climatic Summary of the U.S.—Western Colorado*. Washington, DC: US Government Printing Office.

McGehee, M. 1891. Rough Times in Rough Places. *Century Magazine* (March), reprinted in Opening a Land of Destiny, Part I. *Life Magazine* (April 6, 1959):94–104.

Old Spanish Trail Association. 2008. *West Fork of the Old Spanish Trail in the San Luis Valley, Map and Time-Event Chart*. Old Spanish Trail Association, La Vereda del Norte Chapter. Gunnison, CO: B&B Printers.

Preuss, C. 1848. *Map of Oregon and Upper California from the Surveys of John Charles Frémont and Other Authorities, 1848*. Map portfolio; reprinted and edited by D. Jackson. Urbana: University of Illinois Press.

————. 1958. *Exploring with Frémont: The Private Diaries of Charles Preuss*. E. G. Gudde and E. K. Gudde, eds. Norman: University of Oklahoma Press.

Richmond, P. J. 1990. *Trail to Disaster: The Route of John Charles Frémont's Fourth Expedition*. Monographs in Colorado History. Niwot: University Press of Colorado and Colorado Historical Society.

Ruxton, G.F.A. 1973. *Adventures in Mexico and the Rocky Mountains.* Glorieta, NM: Rio Grande Press.

Schlaer, R. 2000. *Sights Once Seen.* Santa Fe: University of New Mexico Press.

Spence, M. L., ed. 1984. *The Expeditions of John Charles Frémont,* vol. 3. Urbana: University of Illinois Press.

Spencer, F. C. 1929. The Scene of Frémont's Disaster in the San Juan Mountains, 1848. *Colorado Magazine* 6 (July):141–146.

Weber, D. J. 1985. *Richard H. Kern: Expeditionary Artist of the Southwest, 1848–1853.* Albuquerque: University of New Mexico Press for the Amon Carter Museum.

CHAPTER FIFTEEN

San Juan Railroading

Duane Smith

" THE DIFFICULT PROBLEM OF OUR PROSPERITY will then be solved, and we will then have to thank the enterprise and pluck of the much-maligned narrow gauge," asserted Silverton's *San Juan Herald* (April 27, 1882). The long-awaited arrival of the Denver and Rio Grande Western Railroad seemed only a short time away. Silverton could hardly wait for the day to arrive.

On Tuesday evening, June 27, excited townspeople distinctly heard the first engine whistle. Then it arrived. On July 13, the *Herald* proudly announced that the bridge across Mineral Creek, the last obstacle, was nearly completed. Said the editor, "We will help the railroad, and the railroad will help us. That's about the size of it." The *Herald's* rival, the *La Plata Miner*, concurred on this issue. The July 15 edition observed, "So far, all that can be done by the outside world has been done, for by this medium it has been opened to us—what now remains is for us to do—to commence to make ourselves and make good our statements."

Indeed, it was time. Silverton, the "queen of silver land," had been waiting for what seemed an eternity—all of four years—for the iron horse to arrive. And well the residents might be nervous. With a railroad, a community and mining district had a future. Without one, major questions loomed about the present and the future.

With a railroad, particularly in the isolated, lonely San Juans, many benefits arrived that would come in no other way. The cost of living went down, year-round transportation became feasible, investors could arrive with ease, ore could be shipped to the smelter more cheaply, and equipment was brought in more quickly. In addition, the railroad brought an ease, comfort, and convenience no other form of travel could provide in the nineteenth century. Furthermore, the railroad's advent put the town on the map and assured a much better, more profitable future than had been the case before it appeared. In this boom-or-die era, the railroad was a key to success.

It had not always been that way. Some thought the iron horse would lead "immortal souls down to hell" (Smith 1999:4). The man who made that statement did not want the railroad to threaten his canal, once a popular way to transport goods. Nonetheless, railroads burst upon the American scene in the 1830s and 1840s. Indeed, America led the world in accepting railroads and their benefits. By the 1850s, the Midwest and Northeast were crisscrossed by rails, and people across the Missouri clamored for railroad connections. The South, meanwhile, lagged far behind. With a wonderful river system, it preferred the steamboat.

Southerners said no to western railroad expansion. They feared that more free states would come into the union as settlers moved west. The Civil War ended that issue once and for all. As the war dragged on, the US Congress approved the building of a transcontinental railroad, which was completed in 1869 at Promontory, Utah. The iron rail now banded the nation together, opened the West to settlers with a speed and comfort not previously known, and doomed the Native Americans, whose land it divided.

The proclamation "Gold! Gold in the Pike's Peak country" startled Americans in late 1858 and 1859. Out they rushed, nearly 100,000, to get "rich without working"—something they quickly found was a myth. Quite a few stayed, however, and in 1861 a new territory—Colorado—was carved out of Kansas, Nebraska, and Utah, with Denver as its principal city. Almost from the first day, Denverites dreamed of railroad connections to end their isolation and provide a cornucopia of blessings. The Civil War intervened, and—much to their horror—the transcontinental route went north of them, touching the territory only at one corner, at isolated Julesburg.

There was another major problem: the mountains. Colorado's mountains were high and rugged, and the passes were too steep for railroads to navigate. Coloradoans tried to find an acceptable route but failed. Times looked bleak for the struggling territory, whose best mining days seemed behind it.

Then, horror of horrors, the Union Pacific Railroad created an economic rival—Cheyenne, Wyoming, complete with railroad connections. The new territory and its "queen of the mountains and plains" stared disaster in the face. These were dark months for Denverites. Denver's movers and shakers were not about to allow themselves to be bypassed by this "wonder of the age," and they rose to the occasion. They quickly decided to build their own railroad to tie into the Union Pacific somewhere near Cheyenne.

Money was the main problem, but that issue was eventually resolved. Congress provided a 900,000-acre (360,000 ha) land grant, and the Union Pacific gave some aid. Out went the grading crews in 1868, and down went the rails. Finally, in June 1870, the Denver Pacific was completed with a traditional ceremony that included driving in a silver spike presented by the city of Georgetown, Colorado's current silver queen.

Denver's railroad cup overflowed that summer, when the Kansas Pacific reached town in August. The two railroads made Denver the railroad hub of the Rockies and ended any threat Cheyenne might have posed. As former territorial governor John Evans, an inveterate railroader, admitted, "Colorado without railroads is comparatively worthless" (Smith 1999:7). He could have said that for every mining and farming town, farmer, and miner.

The year 1870 was important for another reason. A young Civil War veteran, William Jackson Palmer, had come west with the Kansas Pacific (KP), and he had an idea. All the railroads ran from east to west; he wondered, why not build one north to south, with rail lines into the mountains. After failing to convince the KP to build into the rich Arkansas Valley, he mounted a search for funding. He had caught a terminal case of "railroaditis."

It would take millions of dollars to build Palmer's dream, and he did not have the money. Neither did Denverites, who had nearly completely mortgaged themselves to build the Denver Pacific. Palmer went elsewhere: east and, eventually, to England. On the way, he made the acquaintance of Mary "Queen" Mellen, who later became his wife and whose father provided him with connections.

English investors backed him; for that reason, among others, he selected the 3-foot (0.9-m) gauge, one they were familiar with. More important, in the mountains this narrow gauge allowed trains to go around sharper curves and up steeper grades, and it cost less to build than wider-gauge trains. The tradeoff would be smaller cars and engines, with less capacity for passengers and freight. Back came Palmer, and construction started south from Denver.

Palmer did not take land or money from the government, except for a few acres for railroad yards. The young businessman, however, was also a social planner and town builder, and he planned to fit both into railroad development. Colorado Springs was his first planned community. When the Denver and Rio Grande (D&RG) arrived in September 1871, Palmer's town company was ready to sell lots in this carefully laid out temperance community.

The rails continued beyond Colorado Springs until the 1873 crash and depression stopped construction. Meanwhile, coalfields had been tapped, and Palmer looked ahead to his goal of the D&RG eventually reaching as far south as Mexico. Two things stopped him. One was a railroad rival, the Santa Fe; the other was the discovery of Leadville's silver millions.

Both lines had to go through the Royal Gorge to reach booming Leadville, which led to a brief but bloodless fight in and out of court. The fight was resolved, with the D&RG gaining the gorge and the Santa Fe gaining Raton Pass, thus blocking the D&RG from Santa Fe, New Mexico. Palmer went on to Leadville, and his D&RG became a mountain railroad. He looked longingly at other mining districts. Palmer was determined to go southwest into what appeared to be a bonanza for him and his line: the San Juans.

San Juaners were ecstatic. They could hardly wait. On November 29, 1879, the *La Plata Miner* reported excitedly that contracts had been let for an extension of the D&RG from Conejos to the Animas River. Palmer's surveyors were already in the valley charting a route; the arrival of the promised age could not be far away, surely not beyond the next July.

The paper's editor could not contain himself: "In fact, it is impossible to estimate the great advantage in everyway the completion of this road will be to our camp." Only 164 miles (264 km) away was the "plucky road" from the "rich mining districts of the unexplored San Juan." This would, the editor convinced himself—and his readers no doubt agreed—"prove a rich harvest" for the D&RG (all quotations from the November 29, 1879, issue).

Meanwhile, the D&RG became embroiled in an argument with Animas City, in the Animas Valley, during the winter and early spring of 1879–1880. The railroad typically demanded certain terms from any town it approached. These terms might involve land, money, or the buying of stock, but that was the cost of gaining railroad connections. Animas City's haughty city fathers miscalculated and stubbornly said no to the railroad's demands because they believed the D&RG had to come to their town. The railroad did what it usually did: it started its own nearby community.

In September 1880, the new town of Durango, 2 miles (3.2 km) downriver, saw the light of day. Nearly 2,500 people had crowded into the town by Christmas as the 286 Animas City residents watched in dismay. The following August, the D&RG officially reached its "magic metropolis" on the Animas. Durango celebrated with a parade, a baseball game (defeating Silverton 10–3), races, and a "grand hop" in the evening.

A year later, the railroad reached Silverton. It bypassed Animas City, dooming that community. Such was the result of trying to challenge a railroad in the heyday of its power. Other towns in other places with other railroads could testify to the same result.

At the same time, in 1880, track laying commenced between Antonito and Chama over Cumbres Pass in the southeastern San Juans. This line extended westward and reached Durango in July 1881. Now it was possible to ride the rails between Denver and Durango.

Silverton got its railroad connection, and the results were immediately apparent. San Juan County's gold and silver production jumped from $62,000 in 1882 to $550,000 in 1884. By 1889, it was nearing $1 million, even as the price of silver on the world market dropped. Such was the impact of a railroad. Silverton's population grew, as did its economic sphere, which included all the nearby smaller mining camps.

Silverton was not finished railroading. It was about to become one of Colorado's busiest railroad hubs and could claim not one but four railroads running in and out of the community. A Russian émigré named Otto Mears made this possible.

Mears, the "toll road king of the San Juans," had been the prime mover in opening transportation into the region. Now he turned his attention to railroads. North of Silverton were the mines that were pouring gold and silver into the town and the state. Mears was determined to reach them.

His first railroad effort was directed to Red Mountain, which burst onto the scene in the early 1880s with several rich silver mines. It needed a railroad to reach its full potential, and not just the trails that came in from Ouray and Silverton. Chartered in 1887, the Silverton Railroad reached the Red Mountain district two years later, at Ironton. Mears had hoped to go on through Ironton Park and down the Uncompahgre Gorge, but the grade proved too steep (19 percent) and the danger of snow slides too great, so the rail line ended at the Albany Mine, 18 miles (29 km) from where it began. Mears conducted surveys for both a cog railway and an electric line to bridge the gap to Ouray, but he dropped both ideas in the 1890s.

The Silverton Railroad was a scenic route, climbing from 9,000 feet (2,750 m) to over 11,000 feet (3,350 m) and back again to 9,000 feet. It was also an engineering triumph because of the famous Corkscrew Gulch turntable, which allowed the trains to reach the town of Red Mountain and the nearby mines. By the time the railroad reached Red Mountain, however, the district had slipped past its prime— although for a while in the early 1890s it was a very busy line.

The crash and depression of 1893, coupled with the precipitous collapse of the price of silver, ended Red Mountain's days of prosperity, along with those of many other silver districts. The Silverton Railroad ran on but soon stopped running during the particularly troublesome winter months. Freight and passenger traffic dwindled to almost nothing. Finally, in 1922, the Interstate Commerce Commission gave permission to abandon the line.

Mears's second line was the Silverton Northern. It was incorporated in September 1895 and was projected to reach Eureka, Animas Forks and Mineral Point, then to go through the mountains by way of a three-quarter-mile-long (1.2-km) tunnel to Lake City, by way of Henson Creek. It never got beyond Animas Forks.

The rail line reached Eureka in 1896 and, finally, Animas Forks in 1904. The Eureka segment proved profitable for years, but Animas Forks was well past its heyday when the railroad arrived. The grade was so steep that only a few cars could be pulled that far. The Silverton Northern ceased operations in 1942.

The Silverton, Gladstone and Northerly Railroad was constructed in 1899 to the small camp of Gladstone and the mines near it. The purpose of this little line, just 7 miles (11 km) long, was to bring ores and concentrates to Silverton, thereby reducing shipping costs and making it possible to mine low-grade ore economically.

While this line initially made money, 1907 brought hard times. The mining boom had passed, and smelters closed, reducing its freight to nearly nothing until only three trains ran each week. The original owners lost their troubled railroad in 1915, and Mears purchased it, although he could not make it operate profitably. In 1922, with the closing of the Gold King Mine, the line's career ended. It fell under the jurisdiction of the Silverton Northern during the last years of its existence.

The three little lines had lasted for over fifty years of flamboyant and spectacular Rocky Mountain railroading. Their primary goal of turning many marginal mines into big producers did not pan out as expected, despite great hopes. The decline in the price of silver was no help. There was nothing the railroads or the miners could do about that. The silver issue died with William Jennings Bryan's defeat in 1896 by the Republicans and their gold platform. With the coming of improved highways, the automobile and truck finally did in Silverton's railroads as surely as they had finished off other railroads.

Otto Mears was not finished, however. Frustrated by not being able to go down the Uncompahgre Canyon and its 8-mile (13-km) gap between Ironton Park and Ouray, he decided to go around it—although this was not the only reason he built the 172-mile (277-km) Rio Grande Southern Railway (RGS) from Durango to Ridgway. There were rich mines to be tapped, especially at booming, prosperous Telluride and, to a lesser extent, at Rico and places in between. Incorporated in 1895, the RGS was built from both ends.

As with his other routes, Mears was short of money and received financial backing from easterners and others, eventually including the D&RG. The Durango extension, as it was called, tapped the coal camps of Porter and Hesperus in 1890 and sped on. That same year, work started from Ridgway and, in July 1891, reached one of the rail line's primary goals, Telluride. That branch opened the richest mines along the route.

The RGS's biggest problem (climbing from the valley up and over Lizard Head Pass) was overcome by construction of the Ophir Loop. This proved to be one of the most famous engineering accomplishments in Colorado, as well as a tourist attraction. Construction delays on this segment, though, slowed the entire project. Finally, on October 15, 1891, the line was completed—a year later than Mears had planned. That did not matter at Rico, where a silver spike was driven and a grand "silver ball" was held. A more riotous celebration of workers and uninvited locals was held in local saloons and in the red-light district.

For the RGS, the good years ended all too soon, with the 1893 crash. Mears, though, enjoyed the brief prosperity and issued some of his famous passes to ride the railroad. The beautiful buckskin and solid-silver passes and silver watch fobs became famous. He outdid himself in 1892 with his silver-filigree pass.

The RGS went into receivership in August 1893, one of the victims of the depression that lasted throughout Colorado for another five to six years. Silver mining never regained its former excitement or position in Colorado mining. Mears lost control of his "great dream" and, along with it, a large portion of his personal fortune.

Meanwhile, the D&RG had purchased large blocks of RGS bonds prior to the "silver panic," when profits looked good. After all, the RGS ran from one D&RG track to another, from Durango to Ridgway. While it kept the original name, it became only a DR&G feeder line, serving a silver-mining area that was no longer mining silver.

Fortunately for the railroad, Telluride's mines took up some of the slack until World War I. Also, tourism began to play a more important role because of the interest in what in 1906 became Mesa Verde National Park. The "swing around the circuit," or the "rainbow route," from Denver to Durango, to Silverton, to Ironton, via stage to Ouray, and back to Denver became a tourist must. Traveling through beautiful mountain scenery with the chance to visit the rapidly disappearing mining West—and perhaps to take a side trip to see the Utes—the route had much to offer. A side trip on the RGS allowed visitors to see the world's first cultural park, Mesa Verde.

Despite the fact that the RGS tapped lumbering, tourism, mining, passenger, and freight traffic, it fought a losing battle throughout the twentieth century. The RGS had one more brief moment of notoriety when modified car bodies were fitted for use on the railroad to carry passengers, mail, and a bit of small freight. Thus the Galloping Goose was born. The Goose, however, could not save the Rio Grande Southern. The end came in 1951, and all the tracks were torn up over the next few years.

Two more trains reached the San Juans, including the D&RG's extension from Gunnison to Lake City (figure 15.1). The 36-mile (58-km) Lake Fork branch was finished in 1889, at great expense. It never lived up to expectations because it tapped one of the least-important San Juan mining districts. Lake City's mineral production rallied for a few years after the railroad arrived, then dropped rapidly.

The D&RG eventually purchased the line that ran into Creede during that camp's prosperous mining days in the 1890s, when Cy Warman reported that it was day all day in the daytime, and there was no night in Creede. It, too, succumbed to the usual twentieth-century problems that plagued the mining districts, and the railroad was abandoned.

All that remains of the railroad system that once served the San Juans are the Durango to Silverton and Chama to Antonito lines. These are trips into yesteryear. They carry several hundred thousand tourists per year along the spectacular scenic

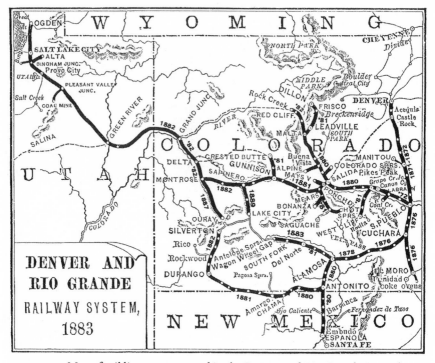

FIGURE 15.1 Map of rail lines as presented in the Denver and Rio Grande Annual Report of the Board of Trustees. *Courtesy,* Colorado Historical Society.

route up the Animas River and its canyon to Silverton and over the spectacular Cumbres Pass, the highest rail pass in North America, into the San Luis Valley.

The railroads opened and developed the San Juans in the nineteenth century as nothing else could have done. They were both praised and damned, desperately needed, and often condescending to the mines and towns they tapped. The glory days ended soon enough, but what a legend they left behind. The German writer and poet Heinrich Heine wrote an appropriate epitaph:

> The ghost of an ancient legend
> That will not let me be.
>
> ("THE LORELEI: A LEGEND FROM
> THE RHINE RIVER VALLEY," 1823)

REFERENCES

Athearn, R. G. 1962. *Rebel of the Rockies: The Denver and Rio Grande Western Railroad.* New Haven, CT: Yale University Press.

Ferrell, M. H. 1973. *Silver San Juan: The Rio Grande Southern Railroad.* Boulder: Pruett.

McCoy, D. A., R. Collman, and R. W. McLeod. 1990s–2005. *The Rio Grande Southern Story*. New York: Sundance Books, 11 vols.

Osterwald, D. B. 1992. *Ticket to Toltec—a Mile by Mile Guide for the Cumbres & Toltec Scenic Railroad*. Lakewood, CO: Western Guideways.

Sloan, R. E., and C. A. Skowronski. 1975. *The Rainbow Route*. New York: Sundance.

Smith, D. A. 1999. *A Legendary Line*. Ouray: Western Reflections.

Thode, J. C. 1989. *George L. Beam and the Denver & Rio Grande*. New York: Sundance.

Part 4

Points of Interest in the Eastern San Juan Mountains

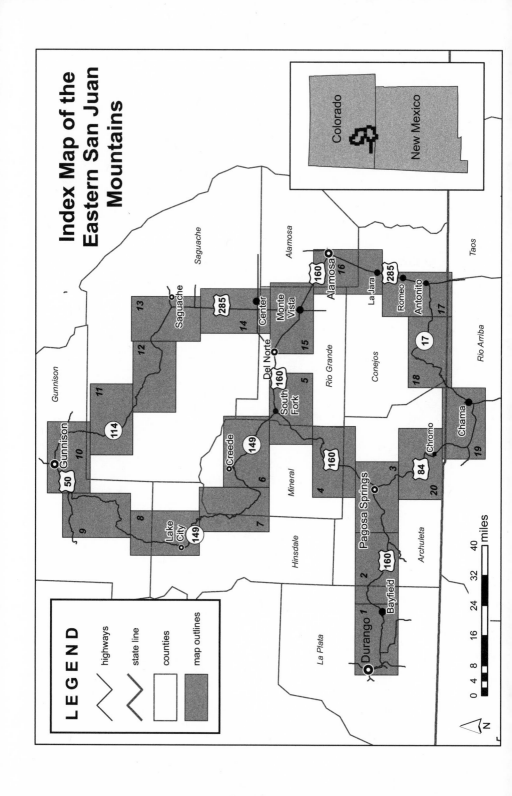

Index Map of the Eastern San Juan Mountains

LEGEND

⋀ highways

⋀ state line

☐ counties

▨ map outlines

Colorado

New Mexico

Gunnison

Saguache

Saguache

Alamosa

285

160

16

Alamosa

La Jara

285

285

Center

Monte Vista

Del Norte

14

15

Romeo

Antonito

17

Rio Grande

Conejos

17

12

13

11

114

160

South Fork

5

Creede

149

160

18

Gunnison

10

50

Gunnison

Mineral

4

6

Chromo

Chama

19

Lake City

149

7

8

9

Hinsdale

Pagosa Springs

3

84

20

Archuleta

Rio Arriba

Taos

La Plata

160

2

Durango 1

Bayfield

miles

0 4 8 16 24 32 40

N

CHAPTER SIXTEEN

Eastern San Juan Mountains
Points of Interest Guide

Rob Blair, Hobey Dixon, Kimberlee Miskell-Gerhardt, Mary Gillam, and Scott White

THIS ROAD LOG IS MEANT FOR THE VISITOR who wishes to further explore the geology, ecology, and human history, as seen from the highways that traverse and circumnavigate the eastern San Juan Mountains. This guide is designed so the traveler can enter a given highway at any point and travel in either direction. On the twenty relief maps herein, the points of interest (POI) are identified as black-filled boxes with numbers that correspond to write-ups in the text. The shaded relief maps come from the US Geological Survey website http://seamless.usgs.gov. Along each highway, green milepost (mp) signs will be found on the right-hand side of the road when mileage increases; if mileage decreases, they appear on the left side of the road. These mileage signs are noted as small white boxes on the shaded relief maps; every fifth box is labeled with the designated mileage. Construction and snowplows may destroy some milepost signs, which may not be replaced accurately.

We recommend that the serious user of this road log refer to US Geological Survey 1:100,000 topographic maps (Montrose, Gunnison, Silverton, Saguache, Del Norte, Antonito, Chama, and Durango quadrangles), which can provide place names, elevations, and greater topographic detail. The information for this road guide was obtained from numerous sources, including a number of previously published geologic road logs (Donnell 1960; Epis 1981; Evanoff 1994; Fassett 1988; Gries and Vandersluis 1989; Lipman and McIntosh 2006). These and other

243

references are included at the end of the chapter. No guide of this sort is exempt from errors; please bring to our attention any oversights and misinformation.

DURANGO TO SOUTH FORK

US 160 (85 miles [137 km])

Maps 16.1–16.5

1. (mp 83.5, Hwy. 160) Beginning of road log (at intersection of US Highways 160 and 550 in Durango). Durango (elevation 6,512 feet [1,985 m]; population 13,922, according to the 2000 census) is the largest municipality in the San Juans and is home to Fort Lewis College and the Durango-Silverton Narrow Gauge Railway. For more detailed information about Durango, please refer to the *Western San Juan Mountains: Their Geology, Ecology, and Human History.*

2. (mp 84.1) Animas River Bridge provides a good view to the west of the Animas–La Plata pump station, located at river level. Mancos Shale slopes are exposed at river level. The Mesa Verde Group lies on the Mancos and is exposed on both sides of the highway. The Mesa Verde Group comprises the Point Lookout Sandstone (lowest), Menefee Shales, and Cliff House Sandstone (highest and most recently deposited), each exposed as south-dipping layers.

3. (mp 86) Moving Mountain landslide is composed of rubble from the Kirtland Formation and can be seen to the west, resting on the Upper Cretaceous Picture Cliffs Sandstone. The landslide occurred December 1931–January 1932. Overlying the Picture Cliffs Sandstone are the Fruitland Coals, followed by the Kirtland Formation and the purple McDermott Formation (seen 0.6 mi [1 km] south). The Animas Formation caps the skyline to the west and south. The highway here passes through the southerly-dipping Hogback Monocline, which marks the structural boundary between the Colorado Plateau and the Southern Rocky Mountains. To the east of the river are the old highway (CO 3) and Carbon Junction, named after the Fruitland Coal outcrop at roadside, just north of the CO 3 junction with US 160. Methane leaks have been detected from these outcrops.

4. (mp 89) Terrace gravels cap the Ewing Mesa surface, uphill to the north, and Florida Mesa, to the south. Both gravel surfaces are overlain with a thin layer of loess. These gravels are mined extensively for aggregate. Erosion of the slopes just north of the highway has exposed a 640 Ka Lava Creek B volcanic ash from the Yellowstone caldera. The ash is a meter-thick, whitish layer a few meters below the terrace surface. The Animas Formation composes the underlying bedrock on both sides of the highway and consists of greenish-gray to tan volcaniclastic mudstones and sandstones of Paleocene age.

5. (mp 103–104) Bayfield. Elevation 6,900 feet (2,104 m); population 1,549 (2000 census). The Los Pinos River, flowing from the north, has incised through the Animas Formation and exposed outcrops on the western side of the channel.

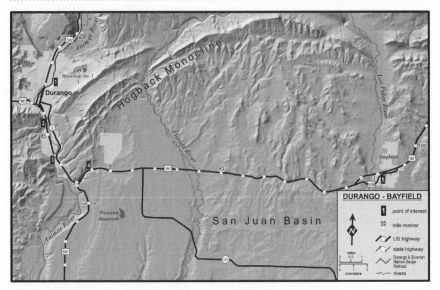

MAP 16.1 Durango (POI 1–5)

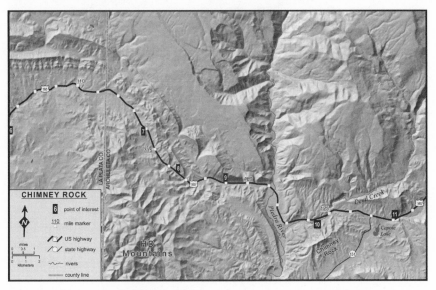

MAP 16.2 Chimney Rock (POI 6–11)

6. (mp 96–115) Animas Formation, of Paleocene age, is exposed in road cuts on both sides of the highway. It is composed of fluvial sandstones and volcaniclastic mudstones. The rolling terrain falls within the piñon-juniper ecosystem, with an understory of Gambel oak.

7. (mp 114) HD Mountains lie 2 miles (3.2 km) to the south and are eroded from the Tertiary Animas and San Jose Formations (sandstone). These mountains are part of the northern margin of the San Juan Basin. The center of the structural and depositional basin is located 50 miles (80 km) south and represents the second-largest gas-production field in the United States, after the Permian Basin field in west Texas. Production has been mostly from the Picture Cliffs Sandstone and Mesa Verde Group and, most recently, from coal-bed methane extracted from the Fruitland Formation. The HD Mountains have been the focus of controversy among energy companies searching for natural gas, environmentalists seeking minimal to no impact, and some landowners who want to be left alone. Most of the Colorado land to the south is part of the Southern Ute Indian Reservation. The HD Mountains support a piñon-juniper ecosystem at lower elevations, which grades into scattered ponderosa pine–Gambel oak on north-facing slopes and at higher elevations.

8. (mp 116–117) Tertiary–Upper Cretaceous section is exposed north of the highway. The road descends a 7 percent gradient for over a mile, from the eroded piedmont surface at Yellow Jacket Summit to Yellow Jacket Creek. From west to east the road crosses the Animas Formation, then the underlying Kirkland, the Fruitland (coals), and the Picture Cliffs Sandstone, entering the Upper Cretaceous Lewis Shale, which opens into a valley. Note the pinkish-orange "clinker" from burned coals in the Fruitland Formation.

9. (mp 118–120) Hogback ridges on both sides of the highway represent steep south-dipping formations: Cliff House Sandstone north of the highway and Picture Cliffs Sandstone to the south. The road follows a strike valley eroded from the Lewis Shale.

10. (mp 124.5) Chimney Rock is the tallest of the prominent spires and ridge to the south and consists of the Picture Cliffs Sandstone. High on the ridge is a lookout tower. Around AD 1000, Ancestral Pueblans constructed a settlement on the ridge, where signal fires could be seen directly from similar settlements to the south and southwest. The signals might have communicated information associated with warnings and seasonal gatherings. This area was designated an Archaeological Area and National Historical Site in 1970. The site is home to over 200 undisturbed structures. It is considered the most isolated "outlier" community connected with the Chaco Canyon culture. Fathers Dominguez and Escalante noted that they passed south of Chimney Rock on August 4, 1776 (Dane 1960).

11. (mp 127–129) Northern boundary of the San Juan Basin is crossed as the highway passes through a notch eroded through the south-dipping Mesa Verde Group. The highway from the west lies on the Lewis Shale, while to the east it lies on the Mancos Shale.

12. (mp 136) Lower Cretaceous Dakota Sandstone is exposed along the north side of the highway, around the curve.

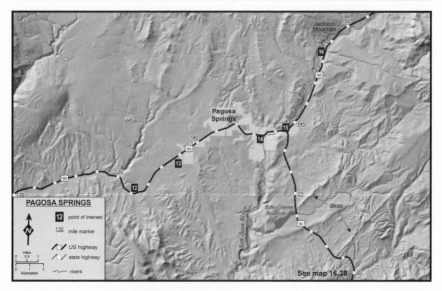

MAP 16.3 Pagosa Springs (POI 12–16)

13. (mp 137–142) Archuleta Anticlinorium is a highland eroded on gently folded Dakota Sandstone and Mancos Shale where the Fairfield Pagosa Village is sited. The anticlinorium consists of northwest/southeast-trending anticlines, synclines, and faults in a zone 75 miles (121 km) long by 6–16 miles (10–26 km) wide (Kelly and Clinton 1960). Along the highway, two distinct fault scarps are visible, one at mp 137.6 and the second at mp 140.8. Both faults produce notable displacement with the east side up, exposing west-facing scarps of Dakota Sandstone. The most easterly of the scarps (Eightmile Mesa Fault) can be traced for 16 miles (Galloway 1980). The Fred Harmon Art Museum and Red Ryder display is located south of the road on the up-thrown block of the fault. Note the stunning view of the skyline to the northeast, which lies within the Weminuche Wilderness Area and is composed of Oligocene volcanic rocks.

14. (mp 143–144) Pagosa Springs. Elevation 7,079 feet (2,159 m); population 1,591 (circa 2000). The town is named after the local hot springs and is derived from the Ute language "Pah," meaning water, and "gosa" as boiling (figure 16.1). The hot springs have likely been frequented by Native Americans, beginning with pre-Puebloan people. Early Europeans observed numerous radiating foot trails from the flowing hot springs adjacent to the San Juan River. The springs' uniqueness and "curative" powers occasionally spawned conflict between competing tribes.

The most recent interesting story stems from a conflict between the Navajos and Utes in 1867. Both tribes claimed rights of occupation, which resulted in skirmishes between the competing groups. After a day of conflict, a final settlement

247

FIGURE 16.1 POI 14. Pagosa Springs spa and hot tubs

strategy was agreed upon: each tribe chose a representative to fight to the death; winner takes all. The Navajos picked their strongest warrior, and the Utes named Colonel Albert Pfeiffer to represent them. Both contestants stripped to the waist and, armed with Bowie knives, tangled. Pfeiffer was smaller and much quicker on his feet. He quickly gained the upper hand and delivered a fatal blow to the Navajo's chest. The Navajo Tribe departed. In 1873 the Brunot Treaty assigned ownership of the springs to the occupying white settlers (according to brochures provided by the Pagosa Springs Chamber of Commerce).

By the early 1500s the Spanish had established an outpost in Santa Fe, and independent Spanish explorers are thought to have passed through the area, seeking gold and silver. An official exploratory expedition led by Don Juan Maria de Rivera in 1765 passed through Pagosa Springs and reached as far north as the Gunnison River. In 1776, Padre Francisco Atanasio Dominguez and Silvestre Velex de Escalante initiated exploration of a route from Santa Fe to California. This path went through Pagosa Springs and became part of the Old Spanish Trail.

The discovery of gold in the San Juans in the 1860s led to increased conflicts between prospectors and Utes. Calls for military intervention resulted in the establishment of Camp Lewis in Pagosa Springs in the fall of 1878, a year and half after

Pagosa Springs had become an official town site. The camp became Fort Lewis in January 1879. One year later, the fort was moved 20 miles (32 km) southwest of Durango, along the La Plata River, where it was deemed more central to local confrontation with the Utes.

Pagosa Springs is located on the southern foothills of the San Juan Mountains, in an area dominated by scattered grassy parks and intervening stands of ponderosa pine and Gambel oak. As with many other towns surrounding the San Juans (e.g., Durango, Cortez, Chama, and Ridgway), Pagosa Springs is built on the Cretaceous Mancos Shale. Nearby wells show that Precambrian gneisses are found between 1,300 and 1,400 feet (~400 m) below the surface. The Jurassic Entrada Sandstone lies unconformably on the Precambrian basement rocks at depth.

The hot waters (30° to 60°C) issue from fractures that penetrate the Precambrian rocks and allow the heated water to rise rapidly from a depth of 6,000 to 7,000 feet (1,800–2,100 m) (Galloway 1980). The water is recycled groundwater and likely took hundreds of years to return to the surface. While in transit, the water was warmed by residual heat left from volcanic activity that occurred more than 20 million years ago. Calcium carbonate dissolved in the water is precipitated at the source of some springs, forming tufa mounds. A prominent mound is located at a parking lot in town, on the north side of the river, just southwest of the San Juan River bridge and across the river from the visitors center.

References (POI 1–14): Condon 1990; Donnell 1960; Evanoff 1994; Fassett 1988; Galloway 1980; Kelly and Clinton 1960.

15. (mp 144.5) Intersection of US 160 (Pagosa Springs to South Fork) and US 84 (Pagosa Springs to Chama, NM). For the route to Chama, see POI #114–104.

16. (mp 148) Jackson Mountain forms the rounded peak 2 miles (3.2 km) northwest of the road (figure 16.2). It consists of an intrusive injected during the Oligocene into the Lewis Shale, forming multiple sills of porphyritic andesite-dacite rock with clusters of small quartz crystals. The andesite is exposed northwest of the highway at mp 150.2. Lipman (2006) calls this the Jackson Mountain Volcano core, and it is the source of exposed andesite several miles north and northeast of the mountain.

17. (mp 151–153) Landslides plague this stretch of the highway. Constant repair is necessary to keep the road fit for traffic as landslide debris creeps toward the San Juan River. Look for the tilted ponderosa pine and fir that form the "drunken forest" between the highway and the river. The slippage has been facilitated by leaking utility pipes that supply water to local farmers and ranchers (figure 16.3). When water-saturated, the Lewis Shale, Picture Cliffs Sandstone, and Fruitland Formation fail easily on the over-steepened slopes. Note the recumbent folds in the coals from the Fruitland Formation exposed northwest of the highway at mp 152.2.

18. (mp 154) Glacial moraine accounts for the hummocky ground and scattered glacial erratics seen to the west. These glacial deposits mark the end moraine

FIGURE 16.2 POI 16. Jackson Mountain, looking north

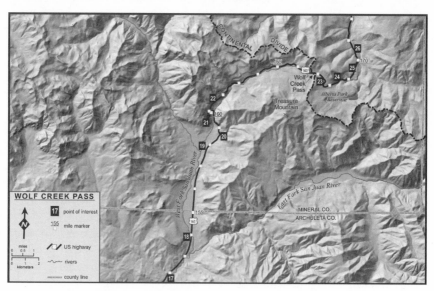

MAP 16.4 Wolf Creek Pass (POI 17–26)

of the West Fork Glacier. Sheep Mountain (12,369 feet; 3,770 m) forms the high peak at due north (figure 16.4).

19. (mp 157–159) The West Fork Valley is a glacially carved valley bounded on all sides by prominent cliffs of Conejos Formation (lower valley walls). The road

FIGURE 16.3 POI 17. Landslide along US Highway 160

traverses a wedge of debris derived from slope wash, mudflows, and landslides moving from east to west toward the river. Periodically, debris flows force closure of the road.

20. (mp 158.7) Treasure Falls, with a vertical fall of 105 feet (32 m), are a short walk from the parking lot. The cliffs and slopes expose coarse volcaniclastic breccias of the Early Oligocene Conejos Formation (35–30 Ma), derived from volcanic debris

251

FIGURE 16.4 POI 18. Sheep Mountain, looking north

flows from the slopes of composite volcanoes that long ago were located north and east of here. A half-mile north of the Treasure Falls turnout is an exposed face of Conejos that contains vugs (small geodes) of white zeolites and other minerals. To see them, park at the Treasure Falls turnout and walk up the highway, because traffic can be hazardous.

21. (mp 160.5) Scenic Overlook of the West Fork (of the San Juan River) Valley. The vista provides good views south into the valley. A cross-section of the Lower Conejos Formation is exposed north of the parking lot. This section consists of volcaniclastic mudstones, sandstones, conglomerates, and breccias derived from composite volcanoes that erupted more than 30 million years ago. To the southwest across the valley is Saddle Mountain (12,033 feet; 3,668 m). A full exposure of the volcanic section is visible here, with Fish Canyon Tuff (at the top) overlying the Pagosa Peak Dacite, Masonic Park Tuff, Treasure Mountain Group, and the Conejos Formation (at the bottom) (figure 16.5). Across the valley to the southeast are cliffs of the Conejos Formation. When the volcanoes erupted, they created hot slurries of fluid mudflows chockfull of coarse debris that swept down their steep slopes. The Conejos Formation is equivalent to the San Juan Formation noted in *The Western San Juan Mountains*. The valley has been eroded through the volcanic strata into the underlying Tertiary Animas Formation and possibly Rio Blanco Formation (Gries and Vandersluis 1989).

22. (mp 161–162) Breccias and loose blocks from the Conejos Formation make this stretch of the highway susceptible to rock fall. Large artificial walls were constructed to minimize damage from periodic collapse of this material. Road outcrops exposed between mp 162.5 and Wolf Creek Pass are dominantly Treasure Mountain Group tuffs and breccias derived from the Platoro Caldera complex,

Fish Canyon Tuff

Pagosa Peak Dacite

Majorie Park Tuff

Treasure Mtn Tuff

Conejos Fm

Conejos Fm

FIGURE 16.5 POI 21. Saddle Mountain, looking west from overlook

located approximately 25 miles (15.5 km) to the southeast. The prominent peak to the east (mp 163–164) is Treasure Mountain (11,908 feet; 3,630 m).

23. (mp 167) Wolf Creek Pass. Elevation: 10,857 feet (3,309 m). The road crosses the Continental Divide, which separates the Rio Grande drainage (east) from the San Juan River drainage (west). The pass receives an average of 45.39 inches (115 cm) of annual precipitation, which includes 435.6 inches (1,101.5 cm) of snowfall (www.wrcc.dri.edu). The original road was opened in 1916. The highway over Wolf Creek Pass requires considerable attention during the winter to minimize catastrophic snow avalanches. Berms, snow sheds, and selective artillery fire are used to control the avalanches.

The Wolf Creek Ski Area is located 1 mile (1.6 km) southeast of the pass and began operation in the fall of 1939. The land east of the ski area has been proposed as the site for the Wolf Creek Village, a planned community that could house 10,000 residents. At the time of writing, this project is embroiled in controversy among developers, land managers, and environmentalists. The Wolf Creek divide is home to wetlands and spruce-fir forest, with alpine tundra at higher elevations. During past glaciations, the divide separated ice flows moving north and east into Pass Creek from glaciers flowing west, down Wolf Creek. The ski area now occupies a cirque basin that fed the Pass Creek Glacier.

The highway crosses stratified volcanic rocks from the Treasure Mountain Group (mp 162.5–169.5). At least seven volcanic units have been mapped, consisting mostly of ash-flow tuffs derived from the Platoro and Summitville Calderas that erupted 28.5–29.5 Ma (Lipman 2006; Lipman and McIntosh, chapter 2, this volume). Overlying the Treasure Mountain Group to the north is the Fish Canyon Tuff, which is derived from La Garita Caldera and dates to 28.0 Ma. Several northwest/southeast-trending faults lie beneath the highway, the ski area, and the proposed village and are collectively called the Pass Creek Fault Zone.

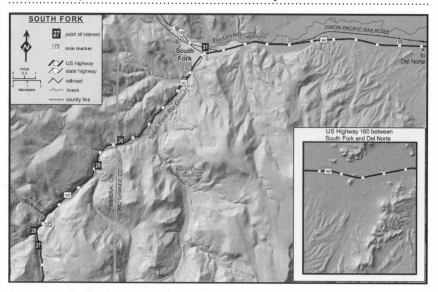

MAP 16.5 South Fork (POI 27–31)

24. (mp 168.8) The snow shed forces snow avalanches to overshoot the highway. Beware of residual snow and ice in the tunnel during winter.

25. (mp 169–170) Treasure Mountain Group and Fish Canyon Tuff outcrop west of the road. The Treasure Mountain Group (upside of highway at mp 169) is mostly a darker reddish-brown tuff and volcaniclastic unit. The Fish Canyon Tuff (low side of highway at mp 170) is a moderately welded tan to pinkish quartz-latite ash-flow sheet.

26. (mp 170.5) The Fox Mountain Plateau to the east is capped by the Hinsdale Basalt (~26 Ma). Lying beneath the basalt are four distinct ash-flow units: the Snowshoe Mountain Tuff (from the Creede Caldera, 26.3 Ma), Wason Park Tuff (South River Caldera, 27.4 Ma), Carpenter Ridge Tuff (Bachelor Caldera, 27.5 Ma), and, in the lower half of the valley, the Fish Canyon Tuff (La Garita Caldera, 28 Ma). The valley of Pass Creek, below the highway, was glaciated 20,000 years ago. The lateral moraines dam the Tucker Lakes, located south and east across the valley. The road to these popular lakes intersects at mp 172.5.

27. (mp 174) Glacially rounded and striated cliffs are exposed north and south of the tunnel.

28. (mp 174.5) Oil seeps have been reported 5 miles (8 km) north, up Hope Creek, north of Big Meadows Reservoir (Gries and Vandersluis 1989:43). The oil seeps along fractures that penetrate the Cretaceous sedimentary source rocks and the overlying volcanic formations.

FIGURE 16.6 POI 30. Fish Canyon Tuff along US Highway 160

29. (mp 178.7) Summitville access road (USFS 380). Summitville, approximately 15 miles (24 km) south, has become an example of how the lack of an environmental database, poor management, lack of proper oversight, and inflexible regulations can lead to expensive consequences for the environment and taxpayers. The abandoned mines and historic town had their heyday in the late nineteenth and early twentieth centuries. The mining-cum-Superfund sites lie at an elevation of 11,500 feet (3,500 m), 2 miles (3.2 km) east of the Continental Divide; thus surface runoff feeds tributaries of the Rio Grande system, including the Alamosa River, Cropsy Creek, and Wightman Fork.

The region experienced three phases of mining activity, beginning with placer gold mining from 1870 to 1873. Between 1873 and 1940, underground mining exploited gold-bearing quartz veins that produced 240,000 troy ounces of gold. In 1984 Galactic Resources Limited, a Canadian company, operated an open-pit mine on the northeastern flank of South Mountain, producing 249,000 ounces of gold. The mine site is located near the west margin of the Platoro-Summitville Caldera complex, above intrusive quartz-latite porphyry surrounded by Summitville andesite. The lode complex has been classified as acid-sulfate epithermal Au-Ag-Cu mineralization associated with advanced argillic (clay) alteration (Pendleton, Posey, and Long 1995). In December 1992, Galactic Resources filed for bankruptcy, necessitating actions to curb additional cyanide spills, prevent further erosion, and reclaim the mine site. As of 2007, the Superfund cleanup had cost taxpayers approximately $185 million. Maintaining the water-treatment plant alone costs $1.5 million per year, and this will likely continue indefinitely.

30. (mp 180–181) Fish Canyon Tuff forms the massive vertically jointed cliffs north of the road (figure 16.6). This uniform-looking tuff is the most voluminous of

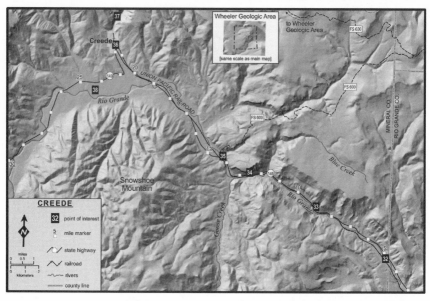

MAP 16.6 Creede (POI 32–38)

all volcanic units in the San Juan Mountains. It covers 5,800 mi² (15,000 km²) and is believed to have an eruptive volume greater than 1,200 mi³ (5,000 km³). It was ejected from La Garita Caldera 28.0 million years ago (see Lipman and McIntosh, chapter 2, this volume).

SOUTH FORK TO GUNNISON
CO 149 (125 miles [200 km])
Maps 16.5–16.10

31. (mp 0) South Fork. Elevation 8,180 feet (2,847 m); population 604 (circa 2000). South Fork is located on 16.5; the road then continues north into 16.6. This town marks the southern point for the Silver Thread Scenic Byway (CO 149) between South Fork and Blue Mesa Reservoir. South Fork was a center for timber production and supported numerous local farms and ranches. The charred forest from the 2002 Million Fire can be seen south of the highway between mp 182 and 186. Note the cliffs of Fish Canyon Tuff west of town.

32. (mp 0–8) The Rio Grande Valley is flanked by cliffs comprising successive volcanic ash-flow tuffs (figure 16.7). The river follows a straight northwest/southeast-trending valley (graben) controlled by the Rio Grande Fault System, which dropped the valley down. Between South Fork and mp 2, the Fish Canyon Tuff (27.6 Ma) is exposed at road level on the southwest side and the Masonic Park

FIGURE 16.7 POI 32. Rio Grande Valley, looking north

Tuff (28.3 Ma) on the northeast side (Lipman, Sawyer, and Hon 1989; Lipman and Steven 1976). Between mp 2 and 8, the base of the valley is all Masonic Park Tuff, while the high ridge to the east is capped by Fish Canyon Tuff. This valley was carved by streams and exposes part of the central volcanic region that includes Lake City, Creede, South Fork, Summitville, and Platoro. The stratigraphy is complicated by the presence of at least 14 volcanic eruptions, many having multiple eruptive phases that resulted in different lithologies. As one should appreciate, geologists have taken decades to untangle the complexities and map more than 180 identified volcanic units.

33. (mp 9) La Garita Caldera wall faces northwest and trends generally northeast to southwest. Outcrops to the northwest are in the collapsed caldera. Roadside exposures northeast of the highway comprise rockfall, landslides, and talus, which obscure much of the bedrock.

34. (mp 11.7–14) Wagon Wheel Gap is a narrow canyon through which early explorers passed. Charles Baker traveled through the gap with prospectors in 1863. He continued over Stony Pass into Bakers Park, which was eventually occupied by the town of Silverton. Upon returning to spend the winter in New Mexico, the group had to abandon a wagon near the gap. Later miners referred to the canyon as "the gap where the wagon wheel was found."

South of mp 13.5 is the Wagon Wheel Gap Hot Spring (4UR Guest Ranch), which in the mid-1970s had a flow rate of 32 gallons (120 liters) per minute and a temperature of 131°F (57°C) (Barrett and Pearl 1976). The Utes frequented this

FIGURE 16.8 POI 34. Columnar jointing in Fisher Dacite

hot spring and called it "Little Medicine." The hot spring suggests rapid upwelling of heated meteoric waters, perhaps associated with the northwest/southeast-trending Rio Grande Fault System (not exposed). The gap's north valley wall is the south edge of the Wagon Wheel Gap lava dome, which is composed of the Fisher Dacite (26.7 Ma). At mp 15, to the east, this dacite displays distinct columnar jointing associated with the cooling of the dome (figure 16.8).

35. (mp 14.2 access) Wheeler Geologic Area (WGA) is located 8 miles (13 km) north of Wagon Wheel Gap and is accessed via USFS 600, a rough 24-mile (39-km) 4-wheel-drive road located at mp 14.2. This area of interest encompasses 1 square mile (2.6 km²) and falls within the La Garita Wilderness (see figure 16.9). The WGA was named after Captain George M. Wheeler, one of four Department of War surveyors (Hayden, King, Wheeler, and Powell) directed to explore the western United States. Wheeler mapped much of Colorado in 1874. The geologic area is noted for its numerous eroded spires and hoodoos (figure 16.9) and was nominated as a National Monument in 1908 by Theodore Roosevelt. In 1950 the National Park Service transferred the monument to the US Forest Service, and in 1969 the acreage was formally reclassified as a "geologic area" because of its small size, poor access, and isolation. More than half of the WGA is covered with colluvium and landslide deposits, yet within this square mile are poorly welded and

Cebolla Creek Tuff

Rat Creek Tuff (dacite)

Rat Creek Tuff (rhyolite)

FIGURE 16.9 POI 35. Wheeler Geologic Area

welded tuffs (volcanic ash) that form spectacular goblins and eroded channels. The greatest erosion occurs on the poorly welded tuffs, which lack cohesion between grains. The ash falls came mostly from a complex eruptive cycle associated with the San Juan Caldera complex (27–26.9 Ma). In the northeast quadrant of the geological area are exposures of the Fish Canyon Tuff (27.6 Ma), which was ejected from La Garita Caldera (Lipman 2006).

36. (mp 21–22) Creede. Elevation 8,852 feet (2,698 m); population 377 (circa 2000). Creede is the Mineral County seat and the only incorporated town in the county. Named after prospector Nicholas C. Creede, it became established in 1891, after Silverton, Telluride, and Lake City were already booming. The valley was originally occupied by camps that took on their own names: Jimtown, Creedmoor, Amethyst, and North Creede. Most of Creede burned down in June 1892, but it quickly recovered and reestablished its wild frontier reputation.

37. Silver mines. Just north of Creede are the three most successful silver mines in the Creede Mining District: the Commodore, Amethyst, and Last Chance Mines. The Commodore (figure 16.10) operated from 1891 until 1976 from the Amethyst Vein. The discovery site on the Amethyst Vein was assayed at about 1,500 ounces per ton of silver.

38. (mp 15–32) Ancient Lake Creede occupied the curved, flat valley floor located south and east of Creede. The valley, or moat, circles Snowshoe Mountain,

FIGURE 16.10 POI 37. Commodore Mine and approximate trace of Amethyst Vein, north of Creede

a resurgent volcanic dome rising up from the center of the collapsed Creede caldera (26.8 Ma). This caldera filled with water during the Late Oligocene to form a lake that encircled the dome. The moat is filled with more than 1,500 feet (450 m) of fine, light yellowish-tan lake sediment called the Creede Formation (figure 16.11). Perhaps half of the moat sediment has been removed during the last 15 Ma, during an erosional cycle noted by Steven, Hon, and Lanphere (1995). Most of the sediment in the lake was deposited within less than 1 million years after eruption and consists of volcaniclastic sediment, air-fall ash deposits, and carbonates, including travertine. Fossils of pine cones, twigs, leaves, bees, flies, and mosquitoes found in the Creede Formation suggest that the climate was similar to, if not slightly warmer and drier than, today (Axelrod 1987). Good exposures of the Creede Formation can be seen at Airport Curve (mp 21.9) and between mp 23 and 28, to the northwest (figure 16.11).

The circular valley displays several distinct terraces representing depositional fill phases associated with glacial epochs. When glaciers were at their maximum, they produced voluminous quantities of sediment. Cosmogenic dating of glacial moraines forming the hummocky topography at mp 32 (Benson et al. 2005) indicates that the last retreat of glaciers began around 20 Ka (figure 16.11). The San

FIGURE 16.11 POI 38. Lacustrine sediments from Lake Creede

Juan Mountains have experienced at least seven major cycles of glaciation (see chapter 3).

39. (mp 41) The scenic overlook looks southwest into Box Canyon and the Upper Rio Grande drainage. To the west is Crooked Creek. San Juan City was located in the valley and served as the departure point for Stony Pass and passage into the western San Juans (Silverton and Ouray). The highway follows a broad northwest/southeast-trending valley for 7 miles (11 km) between mp 32 and 50: the Clear Creek Graben, a fault-bound trough. The northeastern side of the graben is defined by the Bristol Head Fault, and the unnamed southwestern side is defined by a fault that parallels the Rio Grande. The Clear Creek Graben has the same northwest-southeast trend as the Rio Grande Graben (northwest of South Fork) and is offset to the southwest by 10 miles (16 km) by the Creede Caldera. These two fault zones are likely influenced by ancient Precambrian structures that trend northwest to southeast (Baars 1992). The Clear Creek Graben can be traced for nearly 20 miles (32 km) and trends more northerly at the north end. Clear Creek (all forks), Crooked Creek, House Creek, Road Creek, and the Upper Rio Grande drain toward the northeast in entrenched canyons until they reach the Clear Creek Graben, in which they turn to the southeast and coalesce into the Rio Grande.

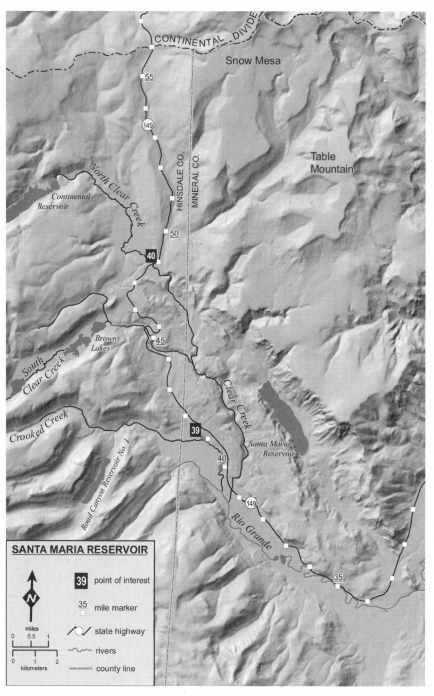

MAP 16.7 Santa Maria Reservoir (POI 39–40)

head scarp

earth flow

FIGURE 16.12 POI 41. Slumgullion Earth Flow

40. (mp 48.9) The Black Mountain Folsom Archeological Site is located about 3 miles (4.8 km) northwest along the North Clear Creek road, which leads to the Continental Reservoir. Just east of the reservoir is a documented Folsom site, suggesting that this region was visited by early Native Americans at 10.9–10.2 Ka. For more information, contact the Rio Grande National Forest or the San Juan Archeological Network.

41. (mp 66) The Slumgullion Earth Flow is seen just north of the road. The earthflow was listed on the National Registry of Natural Landmarks in 1965 and was designated a Colorado Natural Area in 1983. It is an active slope failure from Mesa Seco at 11,600 feet (3,535 m), descending to 9,000 feet (2,743 m) (figure 16.12). The debris extends 4.5 miles (7.2 km) down the valley into the Lake Fork of the Gunnison River, where it spreads laterally and forms a dam that creates Lake San Cristobal. The earthflow is more than 1,000 years old and exhibits intermittent movement, as indicated by the "drunken" forest found on the slide surface. Recent investigations (Coe et al. 2003) suggest a "bathtub model" of failure, in which modern movement is confined to the top of a bathtub-shaped, clay-rich basal layer that formed shortly after initiation of the earthflow. The upper 2 miles (3.5 km) of the earthflow moves between 1.6 and 20 feet (0.5 and 6 m) per year (Coe et al. 2000). The earthflow is associated with acid-sulfate hydrothermally altered volcanic rocks from the Uncompahgre–San Juan Caldera complex.

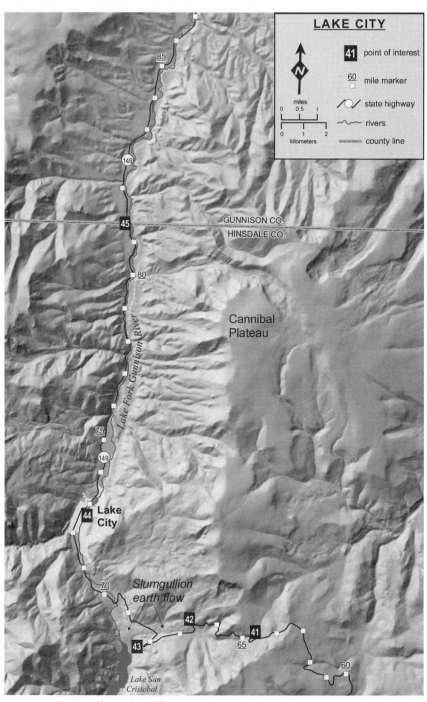

MAP 16.8 Lake City (POI 41–45)

42. (mp 67) Viewpoint. Lake San Cristobal exists because of the landslide dam created by the Slumgullion Earth Flow, which blocks the Lake Fork of the Gunnison River. The Lake Fork flows through a formerly glaciated valley. The remnants of end moraines have been mapped (Atwood and Mather 1932) at the mouth of Slaughterhouse Gulch in Lake City and about 2 miles (3.2 km) north of town, at Sparling Gulch. The Lake Fork of the Gunnison River follows an arc that traces the southern edge of the Lake City Caldera, which lies west of town.

43. (mp 68.5) Alferd Packer Monument commemorates the February–March 1874 "massacre" that resulted in the conviction of Alferd Packer, who was accused of killing and eating his five comrades, after being stranded in a fierce snowstorm. The alleged massacre occurred along the riverbank, months before the first cabin was constructed (by Enos Hotchkiss) at what would become Lake City. During the trial (1883–1886) Larry Dolan, a Lake City saloon keeper, blurted out [paraphrased]: "You man-eating son of [a] bitch. There was [*sic*] seven Democrats in Hinsdale County, and you ate five of them." Packer served sixteen years in prison before he was paroled in 1901. Recent forensic evidence (Bailey 2003) from clothing of the slain men, reexamination of bones, and careful inspection of a pistol from the site of the crime all support Packer's story. Packer claimed he shot Shannon Bell in self-defense as the hunger-crazed, hatchet-wielding Bell attacked him. Bell had already killed the others. Packer admitted to snacking on the dead bodies to survive the winter storm and to eluding authorities, thereby generating negative sentiment against him. Some still believe Packer was guilty as originally charged.

44. (mp 72–73) Lake City. Elevation 8,671 feet (2,643 m); population 375 (circa 2000). Lake City is the Hinsdale County seat and was first occupied by prospectors in 1871. The railroad arrived in 1889 and left in 1933, thus defining the heydays of mining activity. Lake City lies on the northeast corner of the Lake City caldera, which erupted at 23 Ma and was the source of the Sunshine Peak Tuff.

45. (mp 81.4) Hinsdale-Gunnison county line. The Lake Fork Valley descends 20 miles (32 km) to the north (mp 73–93), with little deviation. The valley walls are composed of Conejos-equivalent breccias that came from the Lake Fork volcano, which was located to the east and filled an old paleovalley of the Lake Fork that apparently followed a course similar to that of the existing river (Hon and Lipman 1989). Note ribs of eroded breccia that form tent rocks between mp 81 and 82, to the east of the highway (figure 16.13). At least two distinct glacial outwash terraces can be observed along the Lake Fork Valley between Lake City and Gateview at mp 93.

46. (mp 95.5–98) Fish Canyon Tuff is exposed in road cuts. The highway crosses the axis of a paleovalley that drained northwest and is now filled with Fish Canyon Tuff (Hon and Lipman 1989). The paleovalley developed on 35–30 Ma volcanic breccia and lava flows (Conejos Formation equivalents) and merges here with the northwest/southeast-trending Cimarron Fault Zone.

FIGURE 16.13 POI 45. Hoodoos formed in Conejos Formation

47. (mp 99.8) Road to Powderhorn and Upper Cebolla Creek. Proterozoic Precambrian bedrock is exposed at mp 99.5 and a patch of Sapinero Mesa Tuff at mp 99. A carbonatite mine is located at Iron Hill, about 4 miles (6.4 km) southeast of this intersection (figure 16.14). Carbonatite is an uncommon igneous rock that contains calcite or dolomite crystals. It is of Ordovician age (574–543 Ma) and is associated with Early-Paleozoic tectonic rifting (Fenton and Faure 1970; McMillan and McLemore 2004; Olson et al. 1977). Iron Hill is located on the east side of Cebolla Creek and along the south edge of the northwest/southeast-trending Cimarron Fault Zone (Steven and Hail 1989; Tweto et al. 1976). The northeast block is up, and the overlying volcanic rocks have been stripped down to a pre-volcanic erosion surface of Precambrian crystalline rock. North of the fault, both the Lake Fork and Cebolla Creek flow in canyons entrenched into the underlying Precambrian rocks.

48. (mp 106.7) At Nine-Mile Hill, the Hinsdale Formation (basalt) caps ridges on both sides of the road. The lava flows have been dated between 19 and 16 Ma (Steven and Hail 1989). Between mp 108 and 109, Fish Canyon Tuff rests on Precambrian crystalline rocks. The exposed Precambrian surface is a modified

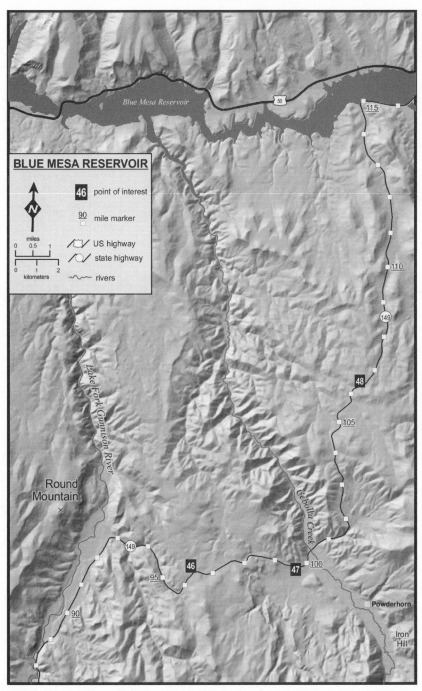

MAP 16.9 Blue Mesa Reservoir (POI 46–48)

FIGURE 16.14 POI 47. Iron Hill carbonatite mining area

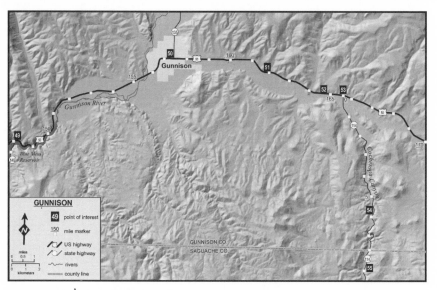

MAP 16.10 Gunnison (POI 49–55)

relict erosion surface and exhibits rounded concordant summits initially stripped of overlying younger rocks during the Early Cambrian. Later sedimentary cover was removed during the Paleocene and Eocene. Erosion continues today.

49. (mp 117.5) Intersection with US 50 (at mp 148) and Blue Mesa Reservoir. The road log picks up at Gunnison, to the east. Precambrian rocks are exposed and are overlain by thinning edges of volcanic breccias and ash flows.

GUNNISON TO SAGUACHE
US 50 (8 miles [12.9 km]), CO 114 (62 miles [100 km])
Maps 16.10–16.13

50. (mp 156–158) Gunnison. Elevation 7,703 feet (2,348 m); population 5,409 (circa 2000). In 1853, John Gunnison was exploring for an east-west route for the railroad and passed through the valley that is now the home of Gunnison and Western State College. The town is located at the geographic center of the Upper Gunnison drainage basin, which has an area of nearly 4,000 mi^2 (10,100 km^2). The valley is used mostly for cattle ranching and supports crops of hay and sorghum. Sagebrush-steppe is prevalent in the broad open valleys between 7,500 and 9,000 feet (2,290 and 2,743 m). The surrounding ranges, both north and south, consist of mixed-conifer forests dominated by Douglas fir broken by occasional pockets of aspen (J. Sowell2008, from cpluhna.nau.edu). At higher elevations, Engelmann spruce and subalpine fir dominate. Precambrian gneisses (1.7 Ga) are exposed south of Gunnison at river level and are overlain by Tertiary ash flows and volcanic breccias.

51. (mp 161–163, US 50) West Elk Breccia is exposed in outcrops along the road. The breccia was erupted at 35–30 Ma and consists of andesite debris-flow deposits derived from composite volcanoes that existed to the north and are equivalent to the Conejos and San Juan Formations found to the south in the San Juan Mountains. These flows fill local erosional channels cut into the underlying Mesozoic sedimentary rocks. The thin rim rocks seen to the north are Early Cretaceous Dakota Sandstone, which overlies the Jurassic Morrison Formation.

52. (mp 163.8) Junction Creek Sandstone (Jurassic age) is seen in road cuts overlying Precambrian crystalline rocks.

53. (mp 165.6, US 50) Intersection with CO 114. This marks the zero milepost for CO 114, which proceeds south. Precambrian gneisses are exposed at road level.

54. (mp 5.9, CO 114) Uranium mine dumps and pits are seen scattered across hillsides in the Precambrian metamorphic rocks. The prospecting occurred during the 1950s uranium boom and, in places, reworked earlier gold mines along quartz veins.

55. (mp 8.8) Pillow basalts are exposed in Precambrian metavolcanic rocks (1.72 Ga) at highway level on the east side of the road, south of the Cochetopa Hideaway entrance (figure 16.15).

56. (mp 13.2 and 14.5) Igneous contacts between the dark-reddish Cochetopa Granite (1.65 Ga) and the dark-gray Cochetopa Metagabbro (1.72 Ga) are exposed to the west. The former was forcefully intruded in the latter. To the north, the Cochetopa Canyon broadens, reflecting the weaker nature of the metamorphic rocks. The relatively resistant granite dominates to the south, where a narrow steep-sided canyon cuts through it.

FIGURE 16.15 POI 55. Precambrian pillow basalts

57. (mp 15–18) Cochetopa Canyon. This narrow canyon creates an ecological "tunnel" along Cochetopa Creek. The riparian habitat supports a mix of willows, blue spruce, and narrow-leafed cottonwood. Birders should keep an eye out for the American dipper, white-throated swift, and rock wren. Pink pegmatite dikes (mp 16.2) intrude the Cochetopa Metagabbro and are associated with the Cochetopa Granite (figure 16.16).

58. (mp 18–19) Cochetopa Granite (1.65 Ga) intrudes metavolcanic and metasedimentary formations (1.72 Ga). Beyond the canyon walls, the granite is overlain by Junction Creek Sandstone, Morrison Formation, and an overlying volcanic blanket of Fish Canyon Tuff and welded Carpenter Ridge Tuff.

59. (mp 19–26) Cochetopa Park, Dome, and Caldera (figure 16.17) are seen to the south. Cochetopa is Ute for "pass of the buffalo." This area is a corridor for lynx, elk, and black bear. The treeless area is a shrub-grassland habitat that supports mountain bluegrass, nodding brome, Thurber fescue, big sagebrush, and snowberry. Pocket gophers (Thomomys talpoides) are common. In the Dome Lakes State Wildlife Area to the south, breeding ducks, including Cinnamon Teal and Gadwall, have been seen. This shrub-grassland is also the type locality for the Cochetopa Series, a unique soil type defined by a Mollic epipedon A/B horizons and an Argillic Bt horizon.

FIGURE 16.16 POI 57. Pegmatite dikes intrude the Cochetopa Metagabbro

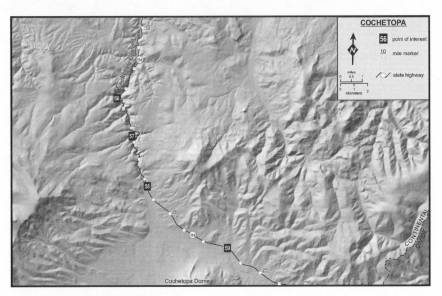

MAP 16.11 Cochetopa (POI 56–59)

The central topographic dome consists of intruded rhyolite lava flows that punched through the weakened roof of the caldera floor. The surrounding moat (low grassland area) is associated with the collapsed caldera and is filled with tuffs

FIGURE 16.17 POI 59. Cochetopa Dome and surrounding Caldera, looking south

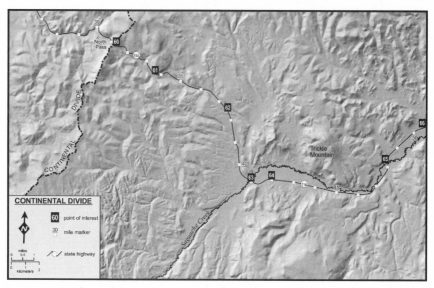

MAP 16.12 Continental Divide (POI 60–66)

and sandstones (Lipman 2006). The Cochetopa Caldera never erupted directly. Instead, the caldera collapsed in response to the horizontal draining of its shallow magma pool to the south into the Nelson Mountain caldera, which erupted at 26.9 Ma. Upon venting, part of the Nelson Mountain Tuff flowed northward onto the newly formed Cochetopa Caldera depression. The fact that the Nelson Mountain Caldera is small and underfit and yet produced a voluminous ash flow of > 500 km³ supports this interpretation (Lipman 2006). No eruptive ash flow has been associated with the Cochetopa Caldera. The white tuffaceous sediments exposed along the highway contain pumice fragments and are derived from an eruptive source external to the Cochetopa Caldera.

FIGURE 16.18 POI 61. Columnar basalts

60. (mp 31.2) North Pass allows paved crossing of the Continental Divide at 10,149 feet (3,093 m). The pass is located near the western margin of the North Pass Caldera, the source of the Saguache Creek Tuff (32.2 Ma).

61. (mp 33.4) Columnar joints from a dacitic lava dome (32.2 Ma) are exposed north of the road (figure 16.18). The thick dacites in this area are younger than the Saguache Creek Tuff and fill the North Pass Caldera along with the younger Carpenter Ridge and Fish Canyon Tuffs.

62. (mp 37.5) Carpenter Ridge Tuff–Fish Canyon Tuff contact is visible to the north, across East Pass Creek (figure 16.19). The densely welded Carpenter Ridge Tuff, from the Bachelor Caldera (27.5 Ma), caps the less-welded Fish Canyon Tuff, from La Garita Caldera (28.0 Ma). The Fish Canyon weathers to form rounded outcrops. In exposures south of the road, the Carpenter Ridge displays two phases: one, a glassy, densely welded zone that interfingers with the other, which contains vuggy lenses and represents vapor-phase crystallization. The top of the Fish Canyon is barely exposed at road level along the south side of the highway.

63. (mp 40.8) The Sheep Creek Fault trends north up along the valley, as can be seen from the bridge over Saguache Creek. The west side is down. The grassy east-dipping slopes of Carpenter Ridge Tuff (27.5 Ma), on the west side of the fault, reflect drag folding (bending of layers adjacent to the fault).

64. (mp 42) The North Pass Caldera (eastern rim) is to the north. The Fish Canyon Tuff (28.0 Ma), seen on the west side (the inside) of the eastern rim, filled

FIGURE 16.19 POI 62. Carpenter Ridge Tuff overlying the Fish Canyon Tuff

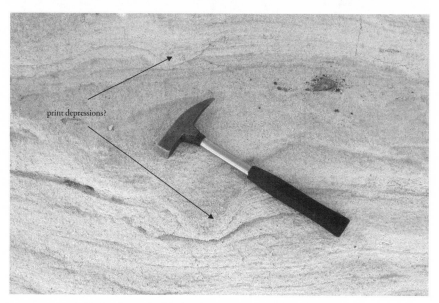

FIGURE 16.20 POI 65. Cross-bedded sediments and possible footprints

the North Pass Caldera approximately 4.2 million years after the Saguache Creek Tuff was erupted from it. These Fish Canyon Tuffs are braced up against Conejos andesites (> 32.2 Ma), forming the lower bold cliffs. The contact indicates the eastern rim of the caldera.

65. (mp 48.3) Cross-bedded sediments are exposed in the road cut just north of the highway. Overlying these eolian and fluvial sediments is the basal Saguache Creek Tuff (32.2 Ma), noted for its black, welded vitrophyre. This dark glassy unit

FIGURE 16.21 POI 67. Houghland Hill, capped by 21.8 Ma basalts from the Hinsdale Formation

erupted from the North Pass Caldera west of here. Reworked white tuffaceous sediments are found between many of the ash-flow units in this area and reflect different ages of a similar depositional environment. At this outcrop, curious sags in the sediments at contact layers have been interpreted as possible rhino-like footprint depressions (Emmett Evanoff, unpublished notes) (figure 16.20).

66. (mp 50–51) Hills to the north with dip slopes of volcaniclastic sediments (Conejos Formation) are considered outlying remnants of an old composite volcano. Fish Canyon Tuff forms the flat-lying, stratified layers.

67. (mp 52–55) The Bonanza Tuff (33 Ma) is exposed in several outcrops at road level. The eruptive source is from the north, and lithology varies from a basal dacite to an upper crystal-poor rhyolite. This tuff blankets paleo-topography eroded on the Conejos lavas and breccias, resulting in discontinuous outcrops of the Bonanza Tuff. To the south, Houghland Hill (figure 16.21) displays a thick stratigraphic section of major ash-flow sheets beginning at the base, with Bonanza, Fish Canyon, and Carpenter Ridge Tuffs overlain by a basaltic lava cap of Miocene Hinsdale Formation (21.8 Ma).

SAGUACHE TO DEL NORTE

US 285 (23 miles [37 km]), CO 112 (13 miles [21 km])
Maps 16.13–16.15

68. (mp 61.5, CO 114; 86, US 285) Saguache. Elevation 7,697 feet (2,346 m); population 578 (circa 2000). End of CO 114 and intersection with US 285, which connects to Monte Vista and Del Norte to the south. Saguache comes from the Ute word "sagwach," meaning blue or green. The San Luis Valley, to the east, is a structural basin that holds more than 16,000 feet (4,900 m) of Miocene-Pliocene

FIGURE 16.22 POI 69. Volcanic outcrop south of Saguache

Santa Fe Group sediments. From Late-Cretaceous time through Early Tertiary (70–40 Ma), this area was part of a structural high called the San Luis Highland, which included what are now the south end of the Sawatch Range, the Sangre de Cristos, and possibly the Brazos Uplift, southeast of Chama, NM. Sediments from this highland shed westward into the San Juan Sag, the Chama Basin, and the San Juan Basin (Gries and Vandersluis 1989).

69. (mp 84) Volcanic rocks are encountered in two road cuts near mp 84. The northern exposure is a lava flow of unknown age. The southern road cut displays volcanic strata inclined to the south (figure 16.22). The basal layer is a volcanic breccia (Conejos Formation) overlain by a thin white tuff layer and basaltic lava flows (Hinsdale?). The flows in places exhibit an *aa* texture. The ridges may indicate inverted topography where, at the time of eruption, the existing flows filled paleodrainages and hardened to resist erosion that eventually removed the surrounding country rock. Another hypothesis suggests that more widely spread flows were followed by tilting, erosion, and deposition of Late-Tertiary sediments, which buried part of the eroded flows. Between 37 and 35 Ma, composite volcanoes developed north, west, and southwest of here, defining the early San Juan Volcanic Field with debris flows and breccias, which are lumped together in the Conejos Formation. This formation is believed to account for 60–70 percent of the

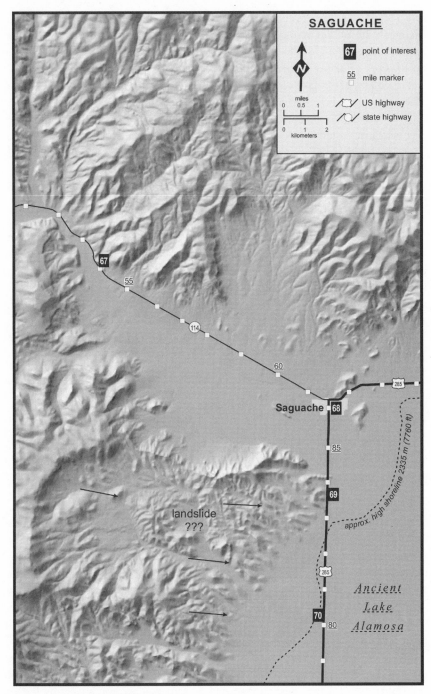

MAP 16.13 Saguache (POI 67–70)

total volume of erupted volcanic material in the San Juan Mountains (Lipman and McIntosh 2008). Beginning about 25 million years ago and possibly earlier, the San Luis structural basin formed. It is a half graben, bounded on the east by the north/south-trending Sangre de Cristo Fault, which produced more than 20,000 feet (6,000 m) of displacement (Gries and Vandersluis 1989). The Sangre de Cristo Mountains form the eastern skyline.

70. (mp 80.2) The 38th parallel is marked by a sign commemorating Korean War veterans. By coincidence, this sign also marks the approximate high shoreline (7,660 ft; 2,335 m) of ancient Lake Alamosa, which filled the San Luis Valley at 440 Ka.

71. (mp 76–77) Russell Lakes State Wildlife Area occupies 3,000 acres (1,200 hectares) to the east of the road and contains mountain shrublands and wetlands. It is a haven for waterfowl and can be accessed using boardwalks and trails. The ponds contain bull rushes and cattails and are surrounded by greasewood and rabbitbrush communities. Rabbitbrush is a native shrub indicative of aerated soils, while black greasewood is associated with wet, alkali-rich soils. The white crusting of salt is a mixture of sodium carbonate, bicarbonate, and sulfate. Sightings of osprey, mergansers, cormorants, snowy egrets, herons, and gulls are frequent. In the spring and fall, sandhill cranes migrate north and south, respectively, through the San Luis Valley. La Garita Mountains (volcanic) lie to the west, and the up-faulted Sangre de Cristo Mountains dominate to the east.

72. (mp 69) Penitente Canyon lies 8 miles (12.9 km) west and is a popular rock-climbing area. It is located a few miles southwest of the community of La Garita and was originally a ritual site of the Catholic sect known as Los Hermanos Penitentes. This brotherhood, known for self-flagellation, painted the Virgin Mary on one of the cliff faces. The rock is composed of moderately welded Fish Canyon Tuff (28 Ma). The Penitente Canyon trail is lined with chokecherries, rabbitbrush, currants, and wild clematis. It is also a haven for finches, mourning doves, and robins.

The northern variation of the Spanish Trail passes by the canyon entrance and through La Garita, on its way to Saguache. North of Penitente Canyon is Carnero Creek, which provided the access to the San Juans for Captain John Charles Frémont's expedition during the winter of 1848–1849. The group was caught by early winter storms, and ten men died (Richmond 1990). The survivors eventually retreated (most by way of La Garita Creek, south of Penitente Canyon) to Taos (see Richmond, chapter 14, this volume).

73. (mp 63, US 285; mp 13.1, CO 112; junction of highways) Central-pivot irrigation circles can be seen in all directions. The San Luis Valley utilizes central-pivot irrigation methods because those use less water than other irrigation delivery systems. The valley supports more than 2,550 of these circles and half circles (S. Vandiver, Rio Grande Water Conservation District, personal communication, December 2007). Each consists of sprinkler heads mounted on a rotating feeder

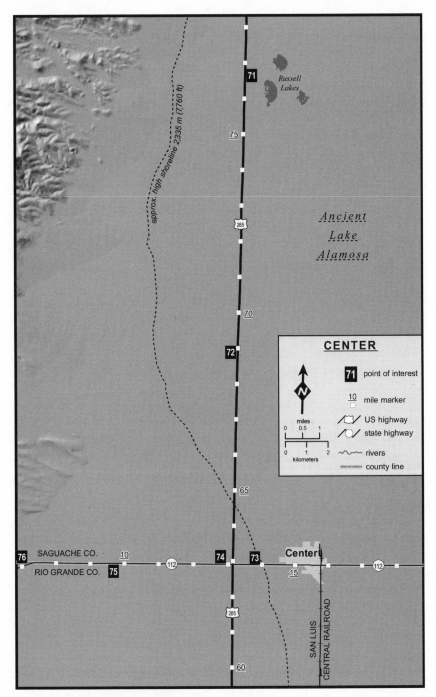

MAP 16.14 Center (POI 71–76)

pipe from a single well head. It can take three days for the feeder pipe to make one rotation around a 160-acre (65-hectare) quarter section. Electric motors, computers, and gauges maintain the integrity of rotating pipe. This part of the San Luis Valley receives less than 7 inches (18 cm) of rain per year, but with added irrigation it can support cool-season crops, such as alfalfa, malting barley (for the Coors Brewery), wheat, potatoes, and lettuce. The town of Center lies 2.5 miles (4 km) to the east and has a population of 2,400. It is the hub of potato production in the valley. In a recent production year (circa 2000), 23.8 million hundredweights (1,080 kilotonnes) of potatoes were produced from 76,800 acres (31,080 hectares).

74. (mp 13, CO 112) The Great Sand Dunes, seen 35 miles (65 km) to the east, is an accumulation of windblown sand at the base of the Sangre de Cristo Mountains. Much of the dunes dates from the Late Pleistocene (18 Ka) and was likely constructed episodically over the past 130 Ka or longer. The sand is derived from two sources. First, sand eroded from the Sangre de Cristo Mountains by streams and overland flow is deposited along the Medano and adjacent creeks. Second, fine sand from the Rio Grande floodplain (alluvial and glacial outwash) is picked up by the prevailing southwesterly winds and deposited at the base of the Sangre de Cristo Mountains. Together, these sands are mixed, reworked, and deposited at the base of Mosca and Medano Passes, where the air is forced upward by the topography before passing through a wind gap in the Sangre de Cristo Mountains. The resulting sand sea displays many dune types, such as barchans, transverse, longitudinal, climbing, star, and parabolic. The highest dunes tower 600 to 700 feet (183 to 213 m) above the valley floor.

75. (mp 7–13, CO 112) River gravels deposited by the Rio Grande in this area form a broad-based, gentle-gradient alluvial fan that covers more than 400 mi^2 (1,050 km^2). River cobbles, or "potato rocks," uncovered from the fan are often piled up at section corners. The cobbles also armor the bottoms of irrigation ditches after the fines have been washed away.

76. (mp 7, CO 112) Elephant Rocks, to the west, shows up as a "boulder field" strewn across a gently southeasterly sloping surface. The rocks are the result of spheroidal weathering of jointed Fish Canyon Tuff, producing hundreds of rounded outcrops, or tors, that resemble a herd of elephants to those with a good imagination. The piñon-juniper forest here is good habitat for rock wrens, green-tailed towhees, and piñon jays.

77. (mp 0.6, CO 112) The Summer Coon volcanic center lies 6 miles (10 km) due north of Del Norte, is 7.5 miles (12 km) in diameter, and can be seen as an eroded, forested dome on the skyline. Summer Coon (active at 35.9–33 Ma) is the best-preserved composite volcano that predates the late-stage ash-flow eruptions (< 30 Ma) in the San Juans. It represents one of perhaps 20 stratovolcanoes that produced the intermediate breccias and andesitic lavas that compose the Conejos Formation. Erosion of the volcano has exposed over 700 radial dikes emanating

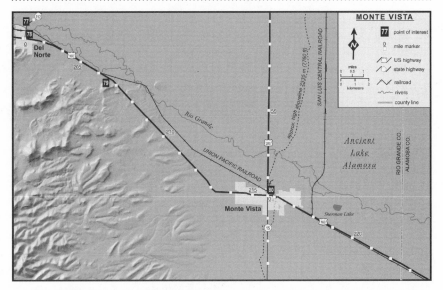

MAP 16.15 Monte Vista (POI 77–80)

from a central stock (Evanoff 1994). Summer Coon is a corruption of the names Sumner and Kuhn, two early ranchers in the area.

Just west of Del Norte are the eroded remnants of a rhyodacite lava dome considered to be an independent eruptive center that predates Summer Coon (Parker, Grau, and Thomas 1991).

DEL NORTE TO ALAMOSA, US 160 (31 MILES [50 KM]); ALAMOSA TO ANTONITO, US 285 (28 MILES [45 KM])

Maps 16.15–16.17

78. (mp 0, CO 112; mp 202, US 160) Del Norte. Elevation 7,874 feet (2,400 m); population 1,705 (circa 2000). The town site was settled by Spanish farmers in 1859 and became incorporated in 1872. The town is the eastern gateway to the San Juan Mountains. The Pinos Creek Road (mp 201.5), west of town, leads to Summitville. This road follows a cliff of Conejos Formation that exposes a basal-flow breccia (James 1971). The highest mountain visible to the south is Bennett Peak (13,203 feet; 4,024 m). A well-developed glacial cirque is carved into the northeast side and cradles two tarns called the San Francisco Lakes. The San Francisco Glacier terminus is just below 10,800 feet (3,290 m). The Bennett Peak summit is composed of the Hinsdale Formation (basalt) that overlies ignimbrite sheets of Treasure Mountain.

79. (mp 206.5) Hinsdale Formation (Oligocene-Pliocene basalt) forms the black rim rock of the mesa and adjacent hills to the south. The basalt flowed on a paleo-surface that dips several degrees to the east and is a combination of the original slope angle and subsequent tilting that accompanied the sagging of the San Luis Valley and evolution of the Rio Grande Rift.

80. (mp 214–217) Monte Vista. Elevation 7,663 feet (2,336 m); population 4,529 (circa 2000). The town was a water stop called Lariat in the summer of 1881, when the railroad was extended from Alamosa to Del Norte. After several name changes, it became Monte Vista (Spanish for "mountain view") in 1886, when the city was incorporated. The city is an important center for agriculture; principal cash crops include potatoes and malting barley (typical annual county production of 2.1 million bushels on 20,000 acres [8,100 hectares]). Other crops, all under irrigation, include spring wheat (780,000 bushels from 7,700 acres [3,100 hectares]), alfalfa, and grass hay (109,000 tons on 33,400 acres [13,500 hectares]). Ranchers raise cattle, hogs, and sheep. Hogs and sheep are fattened locally, while calves are shipped as "feeders" out of the valley. Three local wildlife refuges occupy more than 100,000 acres (40,500 hectares) of wetlands and nutrient-rich habitat for tens of thousands of resident and migratory birds, including loons, pelicans, herons, egrets, swans, hawks, eagles, falcons, sandpipers, and owls. In the fall and spring, as many as 20,000 migrating sandhill cranes are present on the valley floor. The annual Crane Festival, held every March, celebrates the birds and provides educational and photographic opportunities for birders and other wildlife lovers.

81. (mp 231–233, US 160; mp 34, US 285) Alamosa. Elevation 7,544 feet (2,299 m); population 7,960 (circa 2000). The road log follows US 285 south to Antonito. Alamosa was built on the west bank of the Rio Grande and sits at the intersection of US 160 and 285. It is centrally located in the San Luis Valley and is the home of Adams State College, founded in 1921. The town once served as a railroad hub, although the glory days have long passed. Peat deposits from a historical wetland were mined east of town in the 1970s and 1980s. The peat dates at 13.4–11.3 Ka, with further organic deposition occurring 6.7–4.3 Ka. Tufa deposits at the site were dated at 6.7–3.9 Ka (Schumann and Machette 2007).

Fossil teeth and tusk fragments indicate that Mastodon once frequented the area. The San Luis Valley held a vast lake between 3.5 Ma and 440 Ka (see figure 5.5). During this time, lake sediments of the Alamosa Formation were deposited (Machette, Marchetti, and Thompson 2007). The maximum lake extent was 93 by 30 miles (150 by 48 km), and it was more than 200 feet (60 m) deep. Evidence of the lake includes fine clay sediment concentrated in the central part of the basin and eroded shoreline platforms, gravel bars, and spits. The outflow channel was carved into the Servilleta Basalts, which flowed into the southern San Luis Basin 5–4 Ma. The overflow event occurred at approximately 440 Ka, during the maximum lake height associated with rapid deglaciation. Lake sediment cores reveal numerous expansions and shrinkages of the lake over a 3-million-year period. Once the overflow channel was established around 440 Ka, the Rio Grande drainage "instantly"

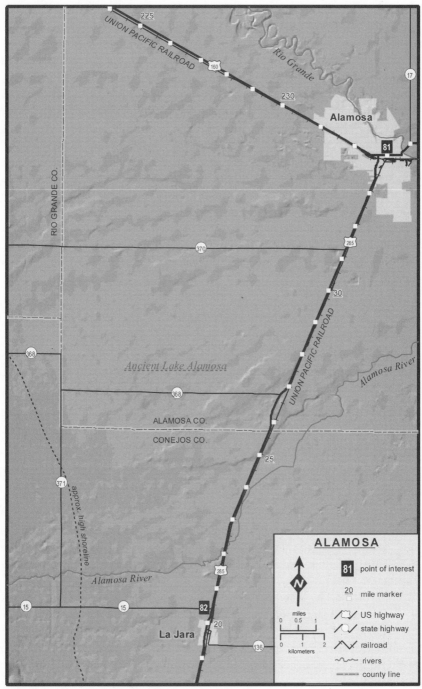

MAP 16.16 Alamosa (POI 81–82)

increased its watershed by 8,500 mi² (22,000 km²). The clay sediment deposited during the life of the lake created confined aquifers that produce artesian wells when penetrated by drill holes (Machette, Marchetti, and Thompson 2007).

82. (mp 20.6, US 285) Hot Creek State Wildlife Area is located 12 miles (19 km) west of here in the foothills of the San Juan Mountains. Hot Creek is fed by several warm springs that elevate its temperature to 54°F (12 °C). The park supports mixed vegetation of desert shrub, piñon-juniper, and ponderosa pine. It is a known habitat for bighorn sheep and antelope.

83. (mp 13) Romeo. Elevation 7,736 feet (2,358 m); population 375 (circa 2000). Jack Dempsey hails from the town of Manassa, located 3 miles (4.8 km) east of Romeo. Dempsey held the world heavyweight boxing title from 1919 to 1926 and was nicknamed the Manassa Mauler because of his ferocious boxing style. He secured the title on July 4, 1919, by knocking Jess Willard down seven times in the first round and pummeling him to such an extent that Willard could not answer the bell for the fourth round. Willard suffered a broken jaw, broken ribs, lost teeth, and partial loss of hearing in one ear. Twelve miles (19 km) to the west is the La Jara State Wildlife Area.

ANTONITO TO CHAMA
CO–NM 17 (49 miles [79 km])
Maps 16.17–16.19

84. (mp 6, US 285) Antonito. Elevation 7,888 feet (2,404 m); population 873 (circa 2000). Antonito is built on a large alluvial fan consisting of reworked glacial outwash deposited by the Conejos River, the largest tributary of the Rio Grande. The town is the eastern terminus of the Cumbres and Toltec Scenic Railroad, which links Chama, New Mexico, the western terminus. This 64-mile (103-km) track heads southwest into New Mexico and follows Rio de los Pinos westward, crossing Cumbres Pass after jogging back into Colorado. It is considered the highest and longest active narrow-gauge railway in North America (Burroughs and Butler 1971).

85. (mp 36–38) San Antonio Mountain (10,935 feet; 2,593 m), seen to the south, is a dome-shaped shield volcano composed of dacite dated at 3.1 Ma (figure 16.23). The volcano is associated with the structural evolution of the Rio Grande Rift and the San Luis Valley.

The volcano Los Mogotes forms the high conical hills seen due west and is best viewed by those driving toward Chama. This is a shield volcano with basalt flows (Hinsdale Formation, 5.3–4.7 Ma) that cover nearly 100 mi² (250 km²) to the north (Lipman 1975). The high mesas to the south are also capped by Hinsdale Basalt. To the south, US 285 is built on basalt flows of the Servilleta Formation (4.5–3.6 Ma).

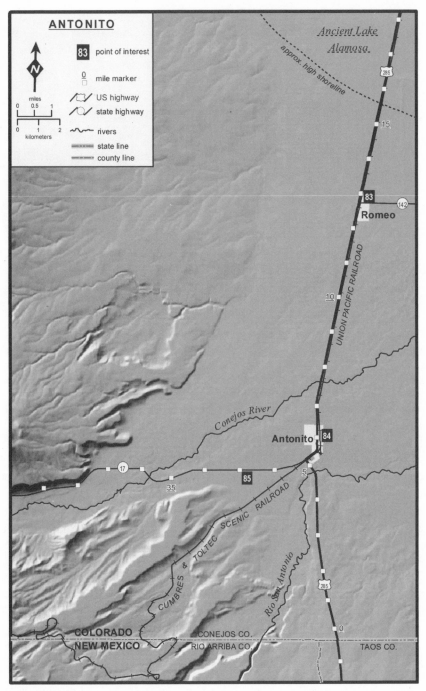

MAP 16.17 Antonito (POI 83–85)

FIGURE 16.23 POI 85. San Antonio Mountain, a shield volcano

86. (mp 28.8–30) Los Pinos Formation is exposed north of the road and consists of white-bedded sandstones and conglomerates derived from the weathering and erosion of volcanic ash flows and breccias of the Conejos Formation. It is overlain by the resistant, multilayered, and alkaline Cisneros Basalt (5.3–4.7 Ma), which is associated with the volcano Los Mogotes. Los Pinos Formation is > 25 Ma and, in this area, about 600 feet (183 m) thick. It is the most dominant sedimentary formation found parallel to the eastern edge of the San Juan Mountains south of Monte Vista.

87. (mp 26.8) The volcano Los Mogotes is visible due east. Piñon-juniper dominates the south-facing slopes seen north of the road.

88. (mp 18–25) Treasure Mountain Tuff (29.5–28.5 Ma) is exposed high on both sides of the canyon as multilayered cliffs that overlie breccias of the Conejos Formation exposed in the lower 200–400 feet (60–120 m) of the valley. Treasure Mountain Tuff was vented from the Platoro Caldera complex, located about 20 miles (32 km) to the west and northwest. Along this stretch of the canyon, look for rockslides, slumps, and debris flows. In particular, note the large landslide masses that moved across the valley from southwest to northeast, between mp 18 and 19, and forced the river to hug the north valley wall.

89. (mp 17.2) Platoro, a historic mining town, is located about 20 miles (32 km) northwest, on USFS 250. The road also allows access to Summitville. One-and-

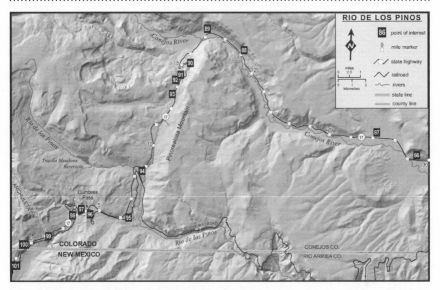

MAP 16.18 Rio de los Pinos (POI 86–101)

a-half miles (2.4 km) along USFS 250, Precambrian granites are exposed beneath the Conejos Formation.

90. (mp 14.4) Conejos Valley overlook provides a view northwest toward Platoro. Lower slopes in the canyon are Conejos Formation, overlain by cliffs of Treasure Mountain Tuff and a high rimrock of Hinsdale Basalt (26 Ma). The riparian life zone along the river represents only 1 percent of the landscape, yet it provides habitat for approximately 96 percent of the wildlife habitat for fish and bird species (from the roadside sign).

91. (mp 13.5–16) Landslide deposits and talus abound in this section of the highway (figure 16.24). These deposits eroded from the Conejos and lower units of the Treasure Mountain Tuff.

92. (12–13.5) Glacial deposits left from glaciers that came from the west and permafrost from the last glaciation account for much of the chaotic rubble and hummocky topography seen west of road.

93. (mp 9–12) La Manga Pass (10,230 feet; 3,118 m) (high point at mp 11.2) displays exposures of light-colored Los Pinos Formation in road cuts. The highway marks the trace of the north/south-trending Cumbres Fault, with the west side down (Lipman 1975). The high ridgecrest to the east is Pinorealosa Mountain, named by early Spanish settlers to describe a "land of many spruces." In 1879, a fire devastated part of this spruce forest, which, at these altitudes, is still recovering. Pinorealosa Mountain is eroded from the Treasure Mountain Tuff and has two high points capped by Masonic Park Tuff.

FIGURE 16.24 POI 91. Landslide deposits

94. (mp 8.6) Rio de los Pinos flows in the broad valley to the south. The valley follows the trace of the north/south-trending Cumbres Fault, which continues over La Manga Pass and across the Conejos Valley to the north (Lipman 1975). In this area, the fault dropped the west side down about 800 feet (240 m). The hummocky lower slopes on the west side of the Pinos Valley are landslide deposits. The higher peak to the southwest is Mount Neff, which is composed of the Treasure Mountain Tuff, with a thin cap of welded Masonic Park Tuff at the summit.

95. (mp 5–7.5) View of Pinorealosa Mountain to the east. It consists of Treasure Mountain Tuff overlying the Conejos Formation. Landslide deposits underlie the highway and allow slow transport of huge scattered blocks of Conejos Formation. Looking north, the road crosses (mp 8.3–9) Treasure Mountain Tuff, beneath a prominent cliff of welded Masonic Park Tuff.

96. (mp 4) Cumbres Pass, at 10,022 ft (3,455 m), is not only the high point for the railroad but also represents the highest rail pass in the United States. In

FIGURE 16.25 POI 98. Windy Point

1848 the US Army engaged the Ute and Jicarilla Apache Indians in battle. Old Bill Williams, a military scout earlier adopted as a Ute tribesman, was wounded.

97. (mp 3.5) Conejos Formation overlook and outcrop view to the west. The Conejos has been mapped as far south as the Picuris Range, near Taos (Lucas et al. 2005). Much of the Conejos overlies the Rio Blanco Formation, which is composed of coarse sediment shed from the pre-volcanic San Juan Dome. The San Juan Volcanic Field can be classified into three stages: (1) an early stage of more intermediate andesite and dacite lavas and volcaniclastic deposits, collectively called the Conejos Formation, which was associated with composite volcanoes that existed 35–30 Ma and composes more than 60 percent of total volume of the volcanic field; (2) a multiple ash flow (ignimbrite) phase of air-fall deposits (30–26 Ma); and (3) a late stage of bimodal trachybasalt and silicic rhyolite lava flows (26–22 Ma). In total, the San Juan Volcanic Field encompasses more than 9,650 mi² (25,000 km²) and erupted volume in excess of 9,600 mi³ (40,000 km³).

98. (mp 2.3) Windy Point can be seen to the northeast, above the highway switchbacks, where the rail line hugs the exposed base of craggy Conejos breccias (figure 16.25). The Masonic Park Tuff caps the skyline at the summit of Neff Mountain, with the Treasure Mountain Tuff beneath.

99. (mp 1–2.3) Wolf Creek Valley is to the south. To the east, in the far side of the valley, the Late Eocene Blanco Basin red beds lie unconformably on a steeply dipping Mesozoic sequence consisting of the Jurassic Morrison Formation (figure 16.26). In Wolf Creek Canyon, exposures of the underlying Wanakah Formation

FIGURE 16.26 POI 99. Angular unconformity with Tertiary Blanco Formation over Jurassic Morrison Formation

and Entrada Sandstone can be found. The Blanco Basin Formation predates the earliest volcanism, which took place around 37–35 Ma. It is derived from the pre-existing and now-exposed sedimentary and Precambrian outcrops and is mostly an arkosic sandstone and conglomerate. To the west, the Blanco Basin Formation rests unconformably on the Burro Canyon and Dakota Sandstones. The angular unconformity indicates uplift (Laramide) and erosion before the Oligocene cycle of volcanism.

100. (mp 0, CO 17; mp 9.6, NM 17) State line between Colorado and New Mexico.

101. (mp 8.5) Stratigraphic section to the north, from top to bottom, includes Tertiary Conejos Formation, Blanco Basin (red beds), and Cretaceous Mancos Shale, surrounded by landslide debris on lower slopes (figure 16.27). To the south, the road crosses glacial moraine with glacial erratics scattered across the surface.

102. (mp 5) Burro Canyon and Dakota Sandstones, of Lower Cretaceous age, dip southwesterly from the west side of the highway. Road is on a glacial moraine.

103. (mp 4.2) Fault in Dakota Sandstone to the west of the road. Note red beds of the Morrison Formation in road cuts to the east.

Conejos Fm

Blanco Basin

Mancos Shale

FIGURE 16.27 POI 101. Exposed stratigraphic section

CHAMA TO PAGOSA SPRINGS

US 84 (49 mi [79 km])

Maps 16.19–16.20

104. (mp 0–2, NM 17; 161, US 84) Chama. Elevation 7,875 feet (2,400 m); population 1,173 (circa 2006). The town is named for the Rio Chama, which might have come from the Tewa Pueblo word "tzama," meaning "here they have wrestled," or Chama might have been used to describe the red muddy waters of the river (Osterwald 1992). Between 1876 and 1882, the Denver and Rio Grande Western Railroad laid 500 miles (800 km) of narrow-gauge track from Denver over Cumbres Pass to Chama and then to Durango. The railroad stopped carrying passengers in 1951 and ceased all transport in 1968. The only section of this route that has been operating since 1971 is the Cumbres and Toltec Scenic Railroad, a 64-mile (103-km) journey between Chama and Antonito.

The town is built on Wisconsin-age terraces (60–17 Ka) adjacent to the Rio Chama. Immediately north of town, on both sides of the river, are higher terraces (100 feet, or 31 m, above river) that are most likely of Durango age (360–250 Ka; ages based on work from Durango; M. Gillam, personal communication, 2009). The bedrock beneath Chama and the surrounding countryside to the south is Cretaceous Mancos Shale. To the east of town are outcrops of Dakota Sandstone dipping gently to the southwest. The distinct knob just west of town is Rabbit Peak, which is capped by the Mesa Verde Formation. The Mancos Shale is notorious

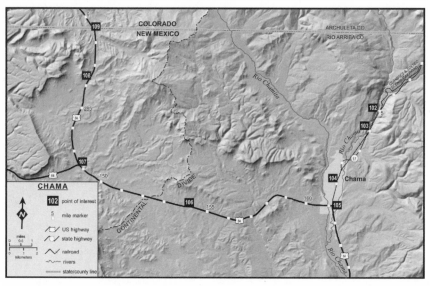

MAP 16.19 Chama (POI 102–109)

for its association with landslides. This region is no different. Fifty percent of the area north between Chama and the Colorado border is mapped as landslide (Muehlberger 1967).

105. (mp 0, NM 17) Junction of US 64 (mp 161.1) and US 84 (mp 293.6).

106. (mp 150–158) Mesa Verde Formation overlies the Mancos Shale and forms the cliffs north of the road.

107. Junction US 64 (mp 148.7) and US 84 (mp 281.2). The points of interest guide follows US 84 between Chama and Pagosa Springs; US 64 heads to Farmington, NM, and follows the old rail route between Chama and Dulce.

108. (mp 285.5) Mesa Verde Formation (Upper Cretaceous) forms a prominent cliff outcrop to the west: Eagle Point (figure 16.28). Three miles (4.8 km) to the east is Chromo Mountain, which is composed of Conejos volcaniclastic rocks overlying the Lewis Shale.

109. (mp 287.8 and mp 0) State line. About 5 miles (8 km) west of here, the state boundary makes a half-mile jog northward to join the 37th parallel. The 1868 survey was supposed to place the entire CO-NM boundary at the 37th parallel; however, because of a surveying error while moving east to west, the line angles slightly south. Once the error was caught, the remaining line (to the west) was shifted north. In 1902–1903, the US Congress authorized a second survey to correct the matter, but protests from Colorado and arguments from New Mexico resulted in a US Supreme Court ruling (1925) that the original survey stands (Osterwald 1992).

FIGURE 16.28 POI 108. Mesa Verde Formation and Eagle Point

FIGURE 16.29 POI 108 and POI 110, looking northeast at Chalk Mountains

110. (mp 6) The Chromo Oil Field, nearby, follows the axis of the Chromo Anticline, which trends northwest to southwest. Production comes from fractures in Mancos Shale (Gries and Vandersluis 1989). Six miles (10 km) to the east-northeast is Navajo Peak (11,323 feet; 3,452 m), which displays a prominent south-facing cliff of Conejos Formation. This mountain marks the southern point of the Chalk Mountains. Six miles east of Navajo Peak is Banded Peak (12,778 feet; 3,896 m), which has a base of Conejos and overlying ash-flow tuffs of Treasure Mountain and Masonic Park and is capped by the resistant Summitville Andesite.

111. (mp 7) Hogback composed of Mesa Verde Formation is crossed by the highway. The hogback marks the northern limb of the Chromo Anticline, which strikes southeastward. The steep northeasterly dip results from drag folding along the Chromo Fault, which trends northwest to southeast (northeast side down) (Gries and Vandersluis 1989). Confar Hill climbs on the Lewis Shale. To the east is V-Rock eroded from Tertiary intrusive rock (figure 16.30).

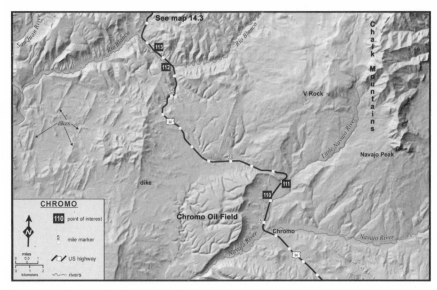

MAP 16.20 Chromo (POI 110–113)

FIGURE 16.30 POI 111. V-rock, eroded from Tertiary intrusive rock

FIGURE 16.31 POI 112. Intrusive dike exposed to west

112. (mp 16.4) Dike from bridge over the Rio Blanco is visible to the west and forms a prominent wall that cuts through the Mancos Shale (figure 16.31).

113. (mp 17.5) Dike is visible in the road cut to the northeast. Dozens of these northeast-trending Tertiary dikes are visible from satellite imagery, suggesting crustal extension across this region, perhaps coinciding with the evolution of the Rio Grande Rift and the San Luis Valley.

114. (mp 28) Intersection of US 84 and US 160. Pagosa Springs is to the west and Wolf Creek Pass and South Fork are to the east. End of road log.

REFERENCES

Atwood, W. W., and K. F. Mather. 1932. *Physiography and Quaternary Geology of the San Juan Mountains, Colorado*. Professional Paper 166. Washington, DC: US Geological Survey.

Axelrod, D. I. 1987. *The Late Oligocene Creede Flora*. Publications in Geological Science 130. Berkeley: University of California.

Baars, D. L. 1992. *The American Alps*. Albuquerque: University of New Mexico Press.

Bailey, D. P. 2003. Solving the American West's Greatest Mystery: Was Alferd Packer Innocent of Murder? *Pathways Magazine* 1–6.

Barrett, J. K., and R. H. Pearl. 1976. *Hydrogeological Data of Thermal Springs and Wells in Colorado*. Information Series 6. Denver: Colorado Geological Survey.

Benson, L., R. Madole, G. Landis, and J. Gosse. 2005. New Data for Late Pleistocene Pinedale Glaciation from Southwestern Colorado. *Quaternary Science Reviews* 24:49–65.

Burroughs, R. L., and A. P. Butler. 1971. Third Day Rail Log Antonito, Colorado to Chama, New Mexico. In H. L. James, ed., *Guidebook of the San Luis Basin, Colorado*. New Mexico Geological Society, no city given, 43–100.

Coe, J. A., W. L. Ellis, J. W. Godt, W. Z. Savage, J. E. Savage, J. A. Michael, J. D. Kibler, P. S. Powers, D. J. Lidke, and S. Debray. 2003. Seasonal Movement of the Slumgullion Landslide Determined from Global Positioning System Surveys and Field Instrumentation, July 1998–March 2002. *Engineering Geology* 68(1–2):67–101.

Coe, J. A., J. W. Godt, W. L. Ellis, W. Z. Savage, P. S. Powers, D. J. Varnes, and P. Tachker. 2000. *Preliminary Interpretation of Seasonal Movement of the Slumgullion Landslide as Determined from GPS Observation July 1998–July 1999*. US Geological Survey Open-File Report 00-102 (on-line edition).

Condon, S. J. 1990. Geologic and Structure Contour Map of the Southern Ute Indian Reservation and Adjacent Areas, Southwest Colorado and Northwest New Mexico. US Geological Survey Map I–1958.

Dane, C. H. 1960. Early Explorations of Rio Arriba County, New Mexico, and Adjacent Parts of Southern Colorado. In E. C. Beaumont and C. B. Read, eds., *New Mexico Geological Society Guidebook, 11th Field Conference*, 113–127. Socorro: New Mexico Geological Society.

Donnell, J. R., ed. 1960. *Geological Road Logs of Colorado*. Denver: Rocky Mountain Association of Geologists.

Epis, R. C. 1981. Supplemental Road Log No. 2, Gunnison to Saguache, Colorado. In R. C. Epis and J. F. Callender, eds., *Western Slope Colorado 1981*. New Mexico Geological Society, no city given, 64–74.

Evanoff, E. 1994. Late Paleogene Geology and Paleoenvironments of Central Colorado. *Rocky Mountain GSA Field Guidebook* 6–8 (May):1–97.

Fassett, J. E. 1988. Second Day Road Log from Durango, Colorado around Northeast Rim of San Juan Basin via Bayfield, Chimney Rock, Arboles, Allison, and Ignacio, Colorado and Back to Durango. In J. E. Fassett, ed., *Geology and Coal-Bed Methane Resources of the Northern San Juan Basin, Colorado and New Mexico*. Denver: Rocky Mountain Association of Geologists, 337–351.

Fenton, M. D., and G. Faure. 1970. Rb-Sr Whole-Rock Age Determinations of the Iron Hill and McClure Mountain Carbonatite-Alkalic Complexes, Colorado. *Mountain Geologist* 7:269–275.

Galloway, M. J. 1980. *Hydrologeologic and Geothermal Investigation of Pagosa Springs, Colorado*. Special Publication 10. Denver: Colorado Geological Survey.

Gries, R. R., and G. D. Vandersluis. 1989. *Laramide and Cenozoic Geology Road Log*. Denver: Rocky Mountain Association of Geologists.

Hon, K., and P. W. Lipman. 1989. Western San Juan Caldera Complex. In C. E. Chapin and J. Zidek, eds., *Field Excursions to Volcanic Terranes in the Western United States*, vol. I: *Southern Rocky Mountain Region*. Memoir 46. Santa Fe: New Mexico Bureau of Mines and Mineral Resources, 350–380.

James, H. L., ed. 1971. *Guidebook of the San Luis Basin, Colorado*. Twenty-Second Field Conference, September 30–October 2. Socorro: New Mexico Geological Society. (logs Chama to Antonito, both rail and road; Del Norte to South Fork; Monte Vista to Saguache; Summer Coon volcano)

Kelly, V. C., and N. J. Clinton. 1960. Fracture Systems and Tectonic Elements of the Colorado Plateau. *University of New Mexico Publications in Geology* 6:1–104.

Lipman, P. 1975. Geologic Map of the Lower Conejos River Canyon Area, Southeastern San Juan Mountains. Miscellaneous Investigations Map I–901. Reston, VA: US Geological Survey.

———. 2006. Geologic Map of the Central San Juan Caldera Cluster, Southwestern Colorado [1:50,000]. Geologic Investigation Series I-2799. Reston, VA: US Geological Survey.

Lipman, P. W., and W. C. McIntosh. 2006. Field Trip to Northeastern San Juan Mountains: North Pass and Cochetopa Park Calderas. Boulder: Geological Society of America, Rocky Mountain Section, May 15–16.

———. 2008. Eruptive and Noneruptive Calderas, Northeastern San Juan Mountains, Colorado: Where Did the Ignimbrites Come From? *Geological Society of America Bulletin* 120(7–8):771–795; doi: 10.1130/B26330.1.

Lipman, P. W., D. A. Sawyer, and K. Hon. 1989. Central San Juan Caldera Cluster, Field Guide 3: South Fork to Lake City. In C. E. Chapin and J. Zidek, eds., *Field Excursions to Volcanic Terranes in the Western United States,* vol. I: *Southern Rocky Mountain Region.* Memoir 16. Socorro: New Mexico Bureau of Mines and Mineral Resources, 330–350.

Lipman, P. W., and T. A. Steven. 1976. Geologic Map of the South Fork Area, Eastern San Juan Mountains, Southwestern Colorado. Miscellaneous Investigations Series Map I–966. Reston, VA: US Geological Survey.

Lucas, S. G., A. B. Heckert, K. E. Zeigler, D. E. Owen, A. P. Hunt, B. S. Brister, L. S. Crumpler, and J. A. Spielman. 2005. The San Juan Volcanic Field and Cretaceous Depositional Systems: Third-Day Road Log from Chama to Cumbres Pass, Los Brazos, and Heron Lake. In S. G. Lucas, K. E. Zeigler, V. W. Lueth, and D. E. Owen, eds., *Geology of the Chama Basin.* Fifty-sixth Annual Fall Field Conference Guidebook. Socorro: New Mexico Geological Society, 58–69.

Machette, M. N., D. W. Marchetti, and R. A. Thompson. 2007. Ancient Lake Alamosa and the Pliocene to Middle Pleistocene Evolution of the Rio Grande. In M. N. Machette, M.-M. Coates, and M. L. Johnson, eds., *2007 Rocky Mountain Section Friends of the Pleistocene Field Trip—Quaternary Geology of the San Luis Basin of Colorado and New Mexico, September 7–9, 2007.* Open File Report 2007-1193. Reston, VA: US Geological Survey, 157–167.

McMillan, N. J., and V. T. McLemore. 2004. Cambrian-Ordovician Magmatism and Extension in New Mexico and Colorado. *New Mexico Bureau of Mines and Mineral Resources Bulletin* 160:1–11.

Muehlberger, W. R. 1967. *Geology of Chama Quadrangle, New Mexico.* Bulletin 89. Socorro: State Bureau of Mines and Mineral Resources, New Mexico Institute of Mining and Technology.

Olson, J. C., R. F. Marvin, R. L. Parker, and H. H. Mehnhert. 1977. Age and Tectonic Setting of Lower Paleozoic Alkalic and Mafic Rocks, Carbonatites, and Thorium Veins in South-Central Colorado. *Journal of Research,* US Geological Survey 5:673–687.

Osterwald, D. B. 1992. *Ticket to Toltec—a Mile by Mile Guide for the Cumbres & Toltec Scenic Railroad.* Lakewood, CO: Western Guideways.

Parker, D. F., D. A. Grau, and C. D. Thomas. 1991. Open-System Magmatic Evolution of the Summer Coon and Del Norte Volcanoes, Conejos Formation, San Juan Mountains

Volcanic Field, Colorado. In J. Betsy and J. Zidek, eds., *Field Guide to Geologic Excursions in New Mexico and Adjacent Areas of Texas and Colorado*. Bulletin 137. Socorro: New Mexico Bureau of Mines and Mineral Resources, 21–28.

Pendleton, J. A., H. H. Posey, and M. B. Long. 1995. Characterizing Summitville and Its Impacts: Setting the Scene. In H. H. Posey, J. A. Pendleton, and D. Van Zyl, eds., *Proceedings: Summitville Forum '95*. Special Publication 38. Denver: Colorado Geological Survey, 1–12.

Richmond, P. J. 1990. *Trail to Disaster*. Boulder: Colorado Historical Society and University Press of Colorado.

Schumann, R. R., and M. N. Machette. 2007. Late Pleistocene to Early Holocene Paleoecology of the Mr. Peat Wetland Deposit, Alamosa County, Colorado. In M. N. Machette, M.-M. Coates, and M. L. Johnson, eds., *Rocky Mountain Section Friends of the Pleistocene Field Trip—Quaternary Geology of the San Luis Basin of Colorado and New Mexico, September 7–9, 2007*. Open-File Report 2007–1193. Reston, VA: US Geological Survey, 147–156.

Steven, T. A., and W. J. Hail Jr. 1989. Geologic Map of the Montrose 30' × 60' Quadrangle, Southwestern Colorado. Miscellaneous Investigations Series Map I–1939. Reston, VA: US Geological Survey.

Steven, T. A., K. Hon, and M. A. Lanphere. 1995. Neogene Geomorphic Evolution of the Central San Juan Mountains near Creede, Colorado. Miscellaneous Investigation Series, Map I–2504. US Geological Survey.

Tweto, O., T. A. Steven, W. J. Hail Jr., and R. H. Moench (comps.). 1976. Preliminary Geologic Map Montrose 1° × 2° Quadrangle, Southwestern Colorado. Map MF-761. US Geological Survey.

Glossary

This Glossary includes some of the terms commonly used in the text. It is not a complete listing of all technical terms. For more in-depth definitions, please consult technical dictionaries or general texts in geology and biology. For example, some of the geologic definitions came from the third edition of *Dictionary of Geologic Terms* by Robert Bates and Julia Jackson (American Geological Institute, 1984).

ablation: process by which snow and ice are lost from a glacier (wastage).

acid-sulfate alteration: wall-rock alteration as a result of leaching of alkali metals by strongly acid hydrothermal fluids and characterized by the presence of alunite in the alteration assemblage.

adsorption: adhesion of gas molecules or gas molecules in solution to a solid surface by weak molecular bonding.

algal mounds: small hill-shaped limestone structures formed by the remains of calcium-carbonate-producing algae.

alluvial: associated with or formed by running water.

alpine: associated with high mountains (e.g., the European Alps); more generally, associated with areas above timberline.

andesite: a dark-colored, fine-grained rock of volcanic origin containing sodic plagioclase feldspar and one or more of the mafic minerals (e.g., biotite, hornblende, pyroxene).

angular unconformity: an unconformity in which younger sediments rest on the eroded surface of tilted or folded older rocks.

anisotropy: a property that is different in one direction from the way it is in another direction relative to a reference point within a body, such as an aquifer.

annual: a plant that completes its entire life cycle and dies within a single year.

anticline: a fold, sides down and center up, wherein the core contains older rocks.

aquifer: a water-bearing stratum capable of transmitting water to a well.

aquifer test (pumping test): a series of methods used to quantify the hydraulic properties of aquifers.

arboreal/arborescent: associated with, living in, or having the form of a tree.

Archeozoic: the earlier part of Precambrian time.

arete: a sharp-edged ridge or spur, commonly found above the snowline in rugged mountains, formed by glacial action in the heads of valleys and the backward growth of adjoining cirques.

argillic: pertaining to clay minerals, as in argillic wall-rock alteration.

asexual reproduction: reproduction that does not involve sex; there is only one parent, and the offspring are genetically identical to the parent.

association: a group of plant species that tend to occur together in places that have similar environmental conditions.

asthenosphere: the layer of the Earth's crust below the lithosphere; soft, probably partially molten, the place where magmas are generated, and the site of convection.

basalt: a fine-grained, dark, mafic igneous rock composed largely of plagioclase feldspar and pyroxene.

base flow: water in a stream channel that comes from groundwater discharge.

basement: the oldest rocks in a given area; a complex of igneous and metamorphic rocks that underlies all of the sedimentary formations.

basin: (1) a depressed area with no surface outlet; (2) the drainage area of a stream; or (3) a low area in the earth's crust where sediments have accumulated.

batholith: a large mass formed from the cooling of magmatic rock at depth, with extensive surface exposure and no known floor.

biota: all of the various forms of plants, animals, and microorganisms that live in an area.

botryoidal: a texture in which a mineral aggregate has a spherical shape or surface composed of radiating crystals.

breccia: rock formed by angular broken fragments of older rocks, held together by a mineral cement or a fine-grained matrix.

caldera: a large basin-shaped, generally circular volcanic depression commonly formed by collapse of the magma chamber in the mouth of a volcano after venting.

carbonaceous: a rock or sediment rich in carbon or organic matter; coaly.

carbonate: 1. a mineral containing the anionic structure of CO_3^{-2}, e.g., calcite and aragonite ($CaCO_3$); 2. a sediment formed of the carbonates of calcium, magnesium, and/or iron (e.g., limestone and dolomite).

carnotite: a strongly radioactive, canary-yellow to greenish-yellow secondary mineral; an ore for uranium and vanadium.

cirque: a steep-walled semicircular hollow high in a mountain valley, formed by erosive glacial action.

clastic: a sedimentary rock formed from mineral particles (clasts) that were mechanically transported.

coevolution: adaptation or other form of evolution that takes place in two closely associated species in response to the effects of one on the other.

Colorado lineament: a proposed belt of Precambrian faults or zones of weakness in the basement, about 100 mi (160 km) wide, extending northeastward from the Grand Canyon to the Rocky Mountain front near the Colorado-Wyoming border.

columnar jointing: the breaking of an igneous rock into parallel prismatic columns by cracks produced by thermal contraction when cooling.

community (biotic community): an assemblage of organisms of many different species that all live and interact with one another within a particular area.

confined aquifer: a water-bearing stratum bound by impermeable layers above and below the layer and often pressurized.

cordillera: mountain ranges in an extensive series.

creep: 1. the slow, imperceptible downslope movement of rock and soil as a result of gravity; 2. slow deformation of solid rock resulting from a constant stress over a long period of time.

crustose lichen: a lichen that is closely appressed to its substrate (such as a rock surface) such that it cannot be removed from it; it may be immersed in its substrate.

cryptobiotic soil: a complex grouping of lichens, fungi, algae, mosses, and perhaps other organisms that grow on the surface of soils and stabilize the soil, protect it from erosion, and in some cases fix N_2 from the atmosphere.

cuesta: 1. a ridge with one steep and one gentle face formed by the outcrop and slower erosion of a resistant, gently dipping bed; 2. a type of landform similar to a mesa but with a sloping top and one side higher than the opposite side.

cyanobacteria: simple single-celled, filamentous or colonial aquatic organisms that have prokaryotic cell structure and are photosynthetic; some have the ability to fix molecular nitrogen from the atmosphere into compounds that can be used by plants and animals.

debris flow: a moving mass of rock fragments, soil, and mud, with more than half of the particles larger than sand size. Rates vary from less than 3 ft/yr (1 m/yr) to over 90 mi/hr (150 km/hr).

dendrochronology: the technique of determining age or date by using the patterns found in tree rings.

dessication: the process of drying out.

dike: a roughly tabular body of igneous rock that cuts across the structure of adjacent rocks or cuts massive rocks.

diorite: a plutonic rock with composition intermediate between granite and gabbro; the intrusive equivalent of andesite.

ecosystem: a biotic community and the physical environment within which it lives.

emesis: vomiting.

endogenetic: a term applied to processes that originate within the earth.

ephemeral pond: a pond that has water in early summer or after rain and that dries out during periods without rain.

eukaryotic: a descriptor for cells that have membrane-bound organelles (such as a nucleus or chloroplasts).

evapotranspiration: the combined effects of evaporation and transpiration.

exogenetic: processes originating at or near the surface of the earth.

extrusive: igneous rock that has been erupted onto the earth's surface (e.g., lava flows and volcanic ash).

fault: a fracture along which displacement has occurred.

fauna: a list of the species of animals living in an area.

felsic: an adjective derived from feldspar + silica, applied to an igneous rock that has abundant light-colored minerals; also applied to those minerals (quartz, feldspar, feldspathoids, muscovite).

flora: a list of the species of plants living in an area (the adjective is "floristic").

fluvial: of, or having to do with, rivers.

foliation: a planar set of minerals or banding of mineral concentrations found in a metamorphic rock, or any planar arrangement of textural or structural features in a rock.

foliose lichen: a lichen with a leafy growth form.

footwall: the rock beneath an inclined fault, ore body, or mine working.

fruticose lichen: a lichen with a shrubby or branched growth form.

gabbro: a black, coarse-grained, intrusive igneous rock composed of calcic feldspars and pyroxene; the intrusive equivalent of basalt.

gangue: the non-economic material accompanying valuable minerals in an ore deposit.

geomorphology: the study of surface landforms and their interpretation of the basis of underlying geology and climate.

gneiss: a foliated metamorphic rock that commonly displays bands of light, granular minerals, and bands of dark, flaky, or elongate minerals.

graben: an elongate downdropped crustal block bounded by faults on its long sides.

granite: a coarse-grained, intrusive igneous rock composed of quartz, orthoclase feldspar, sodic plagioclase feldspar, and micas; the intrusive equivalent of rhyolite.

granodiorite: a group of coarse-grained plutonic rocks intermediate in composition between quartz diorite and quartz monzonite.

gravity anomaly: the difference between the observed value of gravity at a point and the theoretically calculated value. Excess observed gravity gives a positive reading, which is indicative of relatively dense material below surface.

groundwater: water that flows and is stored in the earth's subsurface.

groundwater outflow: water that leaves the watershed in the subsurface within the groundwater-flow system.

habitat: a place where an organism lives (i.e., where it finds all of its requirements for life).

halophyte: a plant that can tolerate high concentrations of salt in the soil.

hanging valley: a tributary glacial or stream valley whose mouth is high above the floor of the main valley.

hanging wall: the overlying rock above an inclined fault, ore body, or mine working.

heap-leach: a method of extracting metals from ore wherein the crushed ore is placed on an imperious pad and a leach liquor (typically cyanide) is percolated through it, dissolving the metals and transporting them to collecting ponds.

herbaceous plant: an herb, i.e., a plant that has no wood tissue.

herpetofaunal center: a region that has high diversity (i.e., many different species) of reptiles and amphibians.

hogback: a ridge formed by the slower erosion of hard strata, having two steep, equally inclined slopes.

horn: a high pyramidal peak with steep sides formed at the intersection of several cirques.

horst: an elongate, upthrown crustal block bounded by faults on its long sides.

hydraulic conductivity: a measure of the ability for rock to transmit fluid from one location to another.

hydraulic gradient: the rate of change of hydraulic pressure per unit distance along the direction in which groundwater flows in an aquifer.

hydraulic head: pressure generated in an aquifer.

hydro-geochemistry: the chemistry of ground and surface waters in which their interactions with geological materials are reflected.

hydrogeology: the science of subsurface waters, their interactions with surface waters, and the geological controls on the occurrence of these waters.

hydrograph: a graph of total stream discharge versus time.

hydrothermal: of or pertaining to hot water or the action of hot water.

hydrothermal alteration: chemical and physical changes in bodies of rock associated with the passage of hot fluids and vapors through rock bodies.

hypha: a fine, microscopic filament of a fungus; the entire fungus is made of uncountable numbers of hyphae.

igneous: a rock or mineral solidified from molten material or magma, or the processes related to the formation of these rocks (from the Latin word *ignis*, meaning fire).

input: water that enters a watershed, primarily as precipitation of rain or snow, and whose measurable amount can be used to estimate a water balance.

intergranular: the space between mineral- or rock-fragment grains that make up a rock.

intermediate: an igneous rock that is transitional between basic and silicic (or between mafic and felsic), with a silica content of 54 percent to 65 percent.

intrusion: the emplacement of magma into preexisting rock or the rock so formed.

intrusive rock: igneous rock formed by emplacement into existing rock.

joint: a large, relatively planar fracture in a rock across which there is no relative displacement of the two sides.

laccolith: a roughly mushroom-shaped igneous intrusion that has domed the overlying rocks.

lacrimation: excessive tear production.

leeward: on the side of an object away from the oncoming wind; protected from the wind.

linear regression: analysis of the relationship between two variables, X and Y, finding the best straight line through the data.

lithify: to change to stone, or to consolidate a loose sediment into a solid rock.

lithosphere: the Earth's outer, rigid shell, above the asthenosphere and containing the crust, continents, and plates.

lode: a mineral deposit in which the ore is in the form of veins, disseminations, or breccia fillings.

lysimeter: a device that measures soil moisture.

mafic: an igneous rock made up mainly of dark, iron- (ferric) and magnesian-rich minerals, or such minerals individually.

magma: molten rock material generated within the Earth from which igneous rocks are derived (adj.: magmatic).

marine: of, belonging to, or caused by the sea.

mass wasting: a general term for the downslope movement of soil and rock material from cliffs or slopes, as a result of gravity.

metamorphic rock: a rock whose original mineralogy, texture, or composition has been changed as a result of pressure, temperature, or gain or loss of chemical components.

metamorphism: the processes of mineral, chemical, and structural changes to solid rocks as a result of pressures and heat at depth.

metasomatism: the process of practically simultaneous capillary solution and deposition by which a new mineral may grow in the body of an old mineral, usually with little disturbance of the textural or structural features.

meteoric water: water that comes from the atmosphere.

microclimate: climatic conditions (e.g., temperature and moisture availability) within a very small area, perhaps as small as a square meter; compare with "regional climate," which affects very large areas and is the kind of climate measured by the National Weather Service.

microorganisms: very small organisms, usually visible only with a microscope, such as bacteria, fungi, and algae.

monocline: an S-shaped fold that connects two relatively horizontal parts of the same stratum at different elevations.

moraine: a deposit of glacial till left at the margin of an ice sheet.

mycelium: a macroscopically visible network of fine, microscopic filaments (hyphae) constituting the "body" of a fungus.

mycobiont: the fungus partner in a lichen.

mycorrhiza: a symbiotic association between the fine roots of a plant and hyphae of a fungus benefiting host and obligatory for most of the world's plants.

nunatak: an isolated knob of bedrock that projects prominently above the surface of a glacier and is surrounded by glacial ice.

nurse plant: a larger plant that creates a more favorable environment (e.g., shade for a smaller plant of the same or a different species).

NWS: National Weather Service.

oolite: a rock, usually limestone, made up of many small rounded bodies that resemble fish eggs, formed of calcium carbonate in concentric layers around a nucleus such as a sand grain.

orogeny: the processes of mountain building by which large areas are folded, faulted, metamorphosed, and subjected to plutonism.

orographic effect: the induced, generally upward movement of an air mass that encounters mountains.

output: water that leaves a watershed.

overland flow: surface water that does not infiltrate into the groundwater-flow system, particularly when the ground is saturated, and ends up directly in a stream channel.

PDO: Pacific Decadal Oscillation.

perennial: a plant that can live longer than a single year.

permeability: the degree of connectivity of intergranular and fracture spaces and their ability to transmit water in an aquifer.

Phanerozoic: the part of geologic time represented by rocks in which the evidence of life is abundant—Cambrian and later time.

phenocryst: a large crystal surrounded by a finer matrix in an igneous rock.

photosynthesis: the chemical process by which green plants convert solar energy into food energy.

phycobiont: the alga partner in a lichen.

phyllite: a metamorphosed rock, intermediate in grade between slate and mica schist.

placer: typically a stream or river deposit of sand, gravel, or both that contains particles of ore minerals such as gold, cassiterite, or ilmentite.

plate tectonics: a theory of global tectonics in which the lithosphere is divided into a number of rigid plates that move horizontally and interact with each other at their boundaries.

playa: the flat floor of a closed basin in an arid region.

pluton: a large igneous intrusion, formed at depth in the crust.

plutonic: pertaining to igneous activity at depth.

porosity: the volume of intergranular space and fracture openings relative to the total volume of the rock or aquifer.

porphyry: an igneous rock that contains conspicuous phenocrysts.

PRISM: Parameter-Elevation Regressions on Independent Slopes Model.

prokaryotic: a descriptor for cells that do not have membrane-bound organelles (such as a nucleus).

Proterozoic: the more recent of the two great divisions of Precambrian time.

PVC: polyvinyl chloride.

pyroclastic rock: a rock formed from fragments of volcanic rock scatted by volcanic explosion.

pyroclastic texture: the unsorted, angular texture of the fragments in a pyroclastic rock.

quartz latite: a porphyritic extrusive rock composed primarily of plagioclase, alkali feldspar, biotite (or hornblende), and quartz. Now called *rhyodacite*.

quartz monzonite: a granitic rock in which quartz comprises 10 percent to 50 percent of the felsic components, as in a granite, but with a much lower alkali feldspar and higher plagioclase content than granite.

quartzite: (1) a very hard, clean, white metamorphic rock formed from a clean quartz sandstone; (2) a clean quartz sandstone so well cemented that it resembles (1).

recharge: water that percolates into the ground and feeds the groundwater system.

rhizoids: microscopic hairs that absorb water and nutrients from the substrate into a lichen and that anchor the lichen to the substrate.

rhyolite: the fine-grained volcanic equivalent of granite, light-brown to gray, typically porphyritic.

ring fault: a steep-sided fault pattern that is cylindrical in outline and commonly associated with caldera subsidence.

salt anticline: an anticlinal structure formed by the squeezing up of salt in a plastic state, rupturing and deforming the overlying rocks.

saxicolous: living on rocks.

schist: a metamorphic rock characterized by strong foliation.

sedimentary rock: a rock formed by the accumulation and cementation of mineral grains transported by wind, water, or ice or chemically precipitated at the site of deposition.

sexual reproduction: reproduction that involves sex (i.e., the offspring obtain genetic material from two parents and are genetically different from either parent).

shaft: a vertical or inclined tunnel driven as part of mine workings.

silicic: a silica-rich igneous rock or magma, usually containing quartz and feldspars.

sill: a horizontal tabular intrusion that parallels the planar structure or strata of the surrounding rock.

skarn: limestones and dolomites that have been altered into lime-bearing silicates by the introduction of large amounts of Si, Al, Fe, and Mg.

slate: the metamorphic equivalent of shale; it is compact, fine grained, hard, and splits into slabs and thin plates along cleavage planes.

SNOTEL: Snow Telemetry.

solar insolation: the area on which solar radiation strikes.

solifluction: a process in which soil that is saturated with water slowly creeps or flows downhill.

species: a group of organisms that all share similar physical characteristics and are capable of interbreeding.

spreading center: a zone of tectonic plate divergence, commonly ocean ridges, where partially molten mantel material upwells and new lithosphere is created.

static water level: the elevation to which water will rise in a well naturally when there is no interference.

stock: an intrusion with the characteristics of a batholith but less than 39 mi^2 (100 km^2) in surface area.

stope: an underground excavation created by the extraction of ore.

storage: groundwater held in pores and fracture spaces in an aquifer.

stratovolcano: a volcano consisting of alternating layers of lava and pyroclastic rock, with abundant dikes and sills.

stratum: a layer of sedimentary rock, separable from layers above and below a bed (plural: strata).

structural: pertaining to rock deformation or features that result from it.

subalpine: just below the alpine zone (i.e., just below timberline).

subduction: the process of one lithospheric plate descending beneath another.

sublimation: a physical process in which a solid is transformed into a gas without passing through the liquid state (e.g., ice turning into water vapor without melting first).

substrate: the kind of rock out of which a soil has developed or on which a plant is growing.

symbiosis: a very close ecological relationship between two different species of plants, animals, or microorganisms.

talus: a deposit of large, angular fragments of physically weathered bedrock at the base of a cliff or steep slope.

tectonics: the study of movements and deformation of the crust on a large scale.

tectonism: a general term for all movement of the crust by tectonic processes, including the formation of ocean basins, continents, plateaus, and mountain ranges.

till: an unconsolidated sediment that contains all sizes of fragments from clay to boulders deposited by glacial action, usually unbedded.

timberline: the highest elevation at which trees or forests grow.

transform fault: a strike-slip boundary along which displacement suddenly stops or changes form or along which plates slide past each other.

transpiration: water loss to the atmosphere from plants that take up surface to near-surface waters.

tuff: a rock composed of pyroclastic debris and fine ash. If particles are melted slightly together from their own heat, it is a "welded tuff."

unconfined aquifer: an aquifer capable of transmitting water in all directions.

unconformity: a surface that separates two strata where there is a break or a gap in the geologic record representing an interval of time during which deposition stopped or erosion occurred.

upwarp: the uplift of a broad region without extensive faulting.

uraninite: a strongly radioactive metallic mineral, the chief ore of uranium.

vegetation zone: a range of elevation in the mountains in which one finds a similar set of plant species and similar climatic conditions.

volcaniclastic: pertaining to a clastic rock that contains volcanic material, without regard to its origin.

volcanism: the processes by which magma and associated gases rise into the crust and are extruded onto the Earth's surface and into the atmosphere.

water table: the level to which water will rise in an unconfined aquifer.

well yield: a minimum amount of groundwater supplied to a well by pumping from an aquifer.

windward: on the side of an object facing toward the oncoming wind without any protection from the wind.

Contributors

PHILIP M. BETHKE, geologist (retired), US Geological Survey, Reston, VA

ROB BLAIR, professor of geosciences (emeritus), Fort Lewis College, Durango, CO

GEORGE BRACKSIECK, former publisher and editor of *Rock & Ice Magazine*, Boulder, CO

JONATHAN SAUL CAINE, research geologist, US Geological Survey, Denver, CO

RODNEY A. CHIMNER, professor of wetland ecology, Michigan Technological University, Houghton, MI

DAVID COOPER, ecologist, Colorado State University, Fort Collins, CO

HOBEY DIXON, professor of biology (emeritus), Adams State College, Alamosa, CO

KIMBERLEE MISKELL-GERHARDT, consulting geologist, Durango, CO

MARY GILLAM, consulting geologist, Durango, CO

DAVID A. GONZALES, professor of geosciences, Fort Lewis College, Durango, CO

ANDREW GULLIFORD, professor of anthropology, Fort Lewis College, Durango, CO

MIKE JAPHET, wildlife biologist, Colorado Division of Wildlife, Durango, CO

KARL E. KARLSTROM, professor of geosciences, University of New Mexico, Albuquerque, NM

DEBORAH KENDALL, professor of biology, Fort Lewis College, Durango, CO

JULIE E. KORB, professor of biology, Fort Lewis College, Durango, CO

J. PAGE LINDSEY, professor of biology (emeritus), Fort Lewis College, Durango, CO

PETER W. LIPMAN, volcanologist (emeritus scientist), US Geological Survey, Menlo Park, CA

WILLIAM C. MCINTOSH, geologist, New Mexico Bureau of Geology and Mineral Resources, Socorro, NM

JAMES R. MILLER, professor of oceanography, Rutgers University, New Brunswick, NJ

KOREN NYDICK, research ecologist, US National Park Service, Kings National Park, CA

IMTIAZ RANGWALA, climatologist, NOAA, Boulder, CO

PATRICIA JOY RICHMOND, San Luis Valley historian and author, Dolores, CO

BILL RITTER JR., 41st governor of Colorado (2007–2010), Denver

DUANE SMITH, professor of history, Fort Lewis College, Durango, CO

SCOTT WAIT, wildlife biologist, Colorado Division of Wildlife, Durango

SCOTT WHITE, professor of geosciences, Fort Lewis College, Durango, CO

ANNA B. WILSON, geologist, US Geological Survey, Denver, CO

ROSALIND Y. WU, forest ecologist, US Forest Service, Durango, CO

Index

Page numbers in italics indicate illustrations.

Abies spp. 141, 164, 165, *plate 34*
Acidification, water, 116–17
Acid leaching, Summitville District, 52–53
Adams State College, 282
Adularia-sericite, 41, 46, 48, 56
Agaricus spp., 144
Agriculture, San Luis Valley, 278, 280, 282
Air pollution, lichen monitoring, 139, 147
Alamosa, 282, *283*
Alamosa, Lake, 66, 278, 282, *plate 13*
Alamosa Formation, 282
Alamosa River, mining contamination of, 52, 53–54, 255
Alamosa River granitic intrusions, 23
Albany Mine, 235
Alces alces, 131, 141
Algae, 138; in lakes, 120–21
Alluvial fans, Animas Valley, 74
Alpha-Corsair Vein System, 49
Alpha Mine, 49
Alunite, 65
Amanita spp., 142, 143
American Fluorspar Company, 50

Ames Power Plant, 114
Amethyst Mine (Creede), 41, 43, 44, 259
Amethyst Vein, 41, 259, *260*; mining on, 43–45, 48
Amphibians, 124
Ancestral Puebloans (Anasazi), 190, 246
Ancestral Rocky Mountains, 7, 12, *plates 3, 4*
Andrews, Amos, 216, 218
Andrews, Elijah, 216, 226
Animas City, 234
Animas City moraine, 69
Animas Forks, 208, 235
Animas Formation, 8, 62, 244, 245, 246, 252
Animas River, 66, 145, 197, 244; glacial cycles on, 69–70; incision of, 71–72
Animas River Bridge, 244
Animas Valley, 74, 234
Animas Valley Glacier, 72
Animas Watershed, 89, *plate 20*
Antonito, 234, 237, 284, *285*
Ants, 180
Aphids, 175, 181–2
Aquifers, *plate 24*; hydraulic properties, 89–93

Archuleta Anticlinorium, 8, 247
Armillaria: *mellea*, 167; *ostoyae,* 141
Arphia conspersa, 175
Arrastra Gulch, 206
Arthropods, 174, 177. *See also by type*
ASARCO, 57
Asclepias spp., 176
Ash flows: Oligocene, 63–64; Tertiary, 23
Aspen; aspen stands, 140, 164, *plate 33*; fire
  regimes and, 161–63; fungi associated with,
  141, 144, 145
Aspen anomaly, 12
Atwood, Wallace, 61–62
Avalanches, debris, 74

Bachelor, 43–44
Bachelor Caldera, 28, 35, 43, 254, 273
Bachelor Mine, 41, 44
Bacon, William, 227
Baker, Charles, 205, 257
Baker's Park, 205, 206, 257
Baldy Cinco, *plate 8*
Banded Peak, 293
Barite: in Creede Mining District, 44, 47, 49; in
  Summitville District, 53
Barlow and Sanderson Stage Company, 207
Batrachochytrium dendrobatidis, 123
Bayfield, 155, 244
Bayfield-Vallecito, 186
Beale, Edward F., 215
Bear Creek Fire, 120
Bears, grizzly, 203
Bees, 178
Beetles, 177, 178, 180, 181; bark (*Dendroctonus,
  Procryphalus, Trypophloeus* spp.), 160, 163,
  167, 169; engraver (*Ips, Scolytus* spp.), 151,
  157, 167
Bell, Shannon, 265
Benino Creek camp, 222
Bennett Peak, 281
Benthic zone, 122
Benton, Thomas Hart, 214; Frémont's expedi-
  tions, 215, 216
Betula glandulosa, 132
Bighorn (Mountain) sheep (*Ovis canadensis*),
  141, 185, 186, 193, *plate 35*; die-offs, 191–92;
  distribution of, 190–91
Big Meadows Reservoir, 254
Biological diversity, of fens, 130–31
Biological productivity, of lakes, 120–22
Birch, bog (*Betula glandulosa*), 132

Birds, 141
Black Mountain Folsom Archeological Site, 263
Black Mountain Tuffs, 23
Black Pitch, 44
Blanco Basin formation, 62, 289, *290, 291*
Blue Creek Tuff, 26, 28
Blue Mesa Reservoir, 256, *267,* 268; fish stock-
  ing in, 124–25
Bogs, 129. *See also* Fens
Boletes: king (*Boletus edulis*), 144–45, *plate 28*;
  orange-capped (*Leccinum insigne*), 145
Bolyphantes index, 177
Bonanza (Bonanza City), 55
Bonanza Caldera, 21, 22; mineralization, 39,
  55–57
Bonanza Group, 57
Bonanza Mine, 56
Bonanza Tuff, 22, 275
Boot Mountain, 221
Borers (*Agrilus, Saperda* spp.), 163
Box Canyon, 261
Brazos Uplift, 276
Breckenridge, Thomas, 223, 225, 228
Brewerton, George, 219
Bristol Head Fault, 261
Bromus tectorum, 157
Brunot, Felix, 206
Brunot Treaty, 206, 248
Buckbean (*Menyanthes trifoliata*), 131, *plate 26*
Budworm, spruce (*Choristoneura occidentalis*),
  167
Bugs, 175; green stink (*Acrosternum hilare*), 176,
  181, 182
Bulldog Mountain Fault, 46–47
Bulldog Mountain Mine, 46, 47, 48
Bulldog Mountain Vein System, 43, 45, 46–47
Burro Canyon, 290
Butterflies, 175, 178; monarch (*Danaus plexip-
  pus*), 176

Caddisflies, 180
Caldera complexes/clusters, 10, 17, *plate 8*; cen-
  tral, 24–26, 33–35. *See also by name*
Calderas, *plates 9–11*; and mineralization,
  39–41, 50–51; Tertiary, 19–21, 22, 23–24,
  26–31. *See also by name*
Camp Dismal, 222
Camp Disappointment, 223
Camp Hope, 223
Camp Lewis, 205, 248–49
Canker, Cytospora, 163

Cannibalism: on Frémont expedition, 225, 226; on Packer expedition, 207–8, 265
Cannibal Plateau, 64
Cantherellus cibarius, 140, 141, 145
Carabidae, 177
Carbonatite, 266
Carbon dioxide, 130
Carbon Junction, 244
Carex spp., 131, 132, *plate 26*
Carnero Creek, 219, 278
Carpenter Ridge Tuff, 28, 35, 254, 270, 273, *274*
Carson, 209
Carson, Kit, 205, 213, 217, 219
Cat Creek granitic intrusions, 23
Catfaces, 152
Cathcart, Andrew, 216, 226
Cattle, 210; and fire regimes, 158–59; grazing impacts, 133, 158–59
Cave Creek, 221
CDOW. *See* Colorado Division of Wildlife
CDPHE. *See* Colorado Department of Public Health and Environment
Cebolla Creek, 266
Cebolla Creek caldera, *plate 8*
Cebolla Creek Tuff, 26, 29, *30*, 31
Cenozoic, 61; paleogeography, *plates 5, 6*; volcanism, 8–9
Center, *279, 280*
Center for Snow and Avalanche Studies, dust and snowpacks, 105–6
Central Caldera cluster/complex, 24–26, 33–35
Cervus elaphus, 14, 185, 188–90, *plate 35*; and peatlands, 131, 133
CF&I. *See* Colorado Fuel and Iron Corporation
Chalk Mountain, *293*
Chama, 234, 237, 291
Chama Basin, 276
Chanterelles (*Cantherellus cibarius*), 140, 141
Chattanooga Fen, 132
Cheatgrass (*Bromus tectorum*), 157
Chimney Rock Archeological Area, 155, *245*, 246
Chiquito Peak Tuff, 23, 26
Chlorite, Creede Mining District, 46
Choristoneura occidentalis, 167
Christmas Camp, 223
Chromo, *294*
Chromo Anticline, 293
Chromo Fault, 293
Chromo Mountain, 292
Chromo Oil Field, 293

Cicadas, 175, 182
Cimarron Fault Zone, 265, 266
Cisneros Basalt, 286
Class I Federal Areas, 117
Clean Air Act, 117
Clear Creek, 261
Clear Creek Graben, 35, 261
Clear Lake, 121
Cliff House Sandstone, 244, 246
Climate, 99; long-term, 100–102; recent, 104–7; seasonal, 103–4
Climate change, 130, 170
Clytocibe, 143
Cochetopa, *271*
Cochetopa Caldera, 31, 35, 270, *272, plate 8*
Cochetopa Canyon, 269, 270
Cochetopa Complex, 220(fig.), 221
Cochetopa Creek, 270
Cochetopa Dome, 29, 270, *271–72, plate 8*
Cochetopa Graben, 35
Cochetopa Granite, 269, 270
Cochetopa Hills, 22
Cochetopa Metagabbro, 269, 270, *271*
Cochetopa Park, 29, 31, 270
Cochetopa Pass, 218–19
Cochetopa Series, 270
Cocomongo Mine, 56
Colorado Department of Public Health and Environment (CDPHE), 57
Colorado Division of Wildlife (CDOW), 187, 197–98
Colorado Fuel and Iron Corporation (Colorado Fuel and Iron Company) (CF&I), 50
Colorado Mineral Belt, 6, 9
Colorado Natural Area, Slumgullion Earth Flow, 263
Colorado–New Mexico State line, 292
Colorado Plateau, *4*, 8, 244; dust storms from, 105–6
Commodore Mine, 41, 44, 259, *260*
Conejos Formation, 19, 21; highway views of, 250, 251–52, 265, *266*, 275, 276, 281, 286, 287, 288, 289, 290, *291*, 293
Conejos River, 54, 66, 186–87, 284
Conejos Valley, 287, 288
Confar Hill, 293
Conglomerates, 63
Conservation movement, 210
Continental Divide, 84, 253, *272*, 273
Continental Reservoir, 263
Coolbroth Canyon, 219, 221

Coprine toxins, 144
Coprinus: atramentarius, 144; *comatus,* 140, 144
Corkscrew Gulch turntable, 235
Cornwall Mountain, 23
Corsair Mine, 49
Cortinarius sp., 144
Cottongrass, Altai (*Eriophorum altaicum*), 131
Crane Festival, 282
Crater Creek granitic intrusions, 23
Creede, 207, 208–9, 237, *256*, 257, 259
Creede, Lake, 64, 259–60, *261*
Creede, Nicholas C., 41, 48, 55, 209, 259
Creede Caldera, 28, 31, *32*, 35, 64, 254, 261;
    mineralization, 39, 40–41, 43, 49
Creede Formation, 31, 260
Creede Graben, 43
Creede Mining District, 17, 28, 33, 35, 41–42,
    51, 259; Amethyst Vein in, 43–45; Bulldog
    Mountain Vein System in, 46–47; Equity–
    North Amethyst Vein System in, 47–48; OH
    Vein in, 45–46; replacement ores in, 49–50;
    Solomon–Holy Moses Vein System in, 48–49
Creede Scientific Drilling Project, 17
Creede sowbelly, 44
Creedite, 50
Cretaceous era, *plate 6*; Laramide orogeny, 7–8,
    62, *plate 5*
Creutzfeldt, Frederick, 218, 223
Crooked Creek, 261
Cropsy Creek, 255
Cross, Whitman, 61
Crustaceans, 124
Crustal stretching, 11
Cryptobiotic soil, 140–41
Cryptoporus volvatus, 141
Cumbres and Toltec Scenic Railroad, 284, 291
Cumbres Fault, 287, 288
Cumbres Pass, 234, 284, 288–89, 291
Cutler Formation, 7, 62
Cutthroat trout (*Salmo clarki*), 185; Colorado
    River (*S. c. pleuriticus*), 124, 125, 195, 196,
    197–98, *plate 35*; greenback (*S. c. stomias*),
    195; Rio Grande (*S. c. virginalis*), 195, 196,
    197, 198, *plate 35*
Cyanobacteria, 138
Cystoderma amianthinum, 138

Dakota Sandstone, 246, 247, 269, 290, 291
Danaus plexippus, 176
Daphnia, 124
Debris fans, 73–74

Deep Creek Graben, 35
Deer, 133, 141, 186
Del Norte, 206, 207, 223, 281
Dempsey, Jack, 284
Dendroctonus, 160, 167; *rupifinis,* 169
Denver, water use and, 95
Denver and Rio Grande (D&RG) Railroad, 231,
    233–34, 237–38, *238*, 291
Denver Pacific Railroad, 233
Devil Creek Fire, 159
Dolan, Larry, 265
Dolores River, 66, 89, 196
Dome Lakes State Wildlife Area, 270
Dominguez, Francisco Atanasio, 246, 248
Douglas fir (*Pseudotsuga menziesii*), 141, 164
Drought, 151, 157, 180; and crown fires,
    168–70
Durango, 69, 236, 244, *245*; D&RG railroad in,
    234, 237
Durango Fish Hatchery, 125
Durango-Silverton Railroad, 144
Dust, and snowpack, 105–6

Eagle Mine, 55
Eagle Point, 292, *293*
East Embargo Creek, 224
East Willow Creek, 48
Ecosystems, aquatic, 113
Eightmile Mesa Fault, 247
Elephant Rocks, 280
Elephant's head (*Pedicularis groenlandica*), 131,
    plates 26, 27
Elk (*Cervus elaphus*), 14, 185, 188–90, *plate 35*;
    and peatlands, 131, 133
El Niño Southern Oscillation (ENSO), 103
Empress Josephine Mine, 56
Engineer Pass, 208
English, David, 4
ENSO. *See* El Niño Southern Oscillation
Entrada Sandstone, 249, 290
Eocene, 8; erosion during, 62–63
EPA. *See* U.S. Environmental Protection Agency
Equity Fault, 47
Equity Mine (Creede), 43, 47
Equity–North Amethyst Vein System, 45, 47–48
Eriophorum altaicum, 131
Erosion, 12, 120; glacial, 68–69; Miocene-
    Pliocene, 65–67, *plate 12*; Oligocene-
    Miocene, 64–65; Paleocene-Eocene, 62–63
Escalante, Silvestre Velex de, 246, 248
Euphotic zone, 118

Eureka, 74
European-American settlement, and fire suppression, 158–59, 165
Euxoa auxiliaris, 178, *179*
Evans, John, 233
Evapotranspiration, 87, 89
Ewing Mesa, 244

Fairfield Pagosa Village, 247
Fault systems, 33, 35, 247, 261, 273, 287, 288, 293; and groundwater flow, 94–95; and hot springs, 93–94, 258
Fens, 129, *plates 26, 27*; biological diversity of, 130–31; impacts and threats to, 132–33; types of, 131–32
Fir (*Abies* spp.), 141, 144, 164, 165, *plate 34. See also* Douglas fir
Fire exclusion, 158, 165, 166
Fire-hazard mitigation, ponderosa pine forests, 160–61
Fire histories, 152, 153, 154, *156*
Fire regimes, 152, 154; aspen, 161–63; mixed-conifer forest, 163–67; piñon-juniper woodland, 155–57; ponderosa pine forest, 157–61; spruce-fir forest, 167–70
Fires, 45, 120, 126, 256; and insect ecology, 174, 181–82; crown, 168–70; Holocene, 73–74; suppression of, 158–59
Fish, 120; in food webs, 121–22; mercury in, 125–26; stocking of, 123–25, 197–98
Fish Canyon Tuff, 22, 26–27, 28, 35, 259, *plate 8*; along Colorado 114, 270, 273–74, 275; along Colorado 149, 257, 265, 266; along Highway 160, 252, 253, 254, 255–56; along Highway 285, 278, 280
Fisher Dacite, *258*
Fitzpatrick, Thomas "Brokenhand," 217
Flat-slab model, 9–10
Flies, 176; spiny tachnina (*Paradejeania rutiliodes*), 178, *180*
Florida Mesa, 69, 244
Fluorite, 46
Fluorspar deposits, at Wagon Wheel Gap, 41, 50
Fly agaric (*Amanita muscaria*), 138, 143, *plate 28*
Folsom site, 263
Food webs, in lakes, 120–23
Forest history, 152
Fort Lewis, 205, 249
Fort Lewis College, 244
Four Corners region, 5, 10, 189; power plant pollution, 126, 147

4UR Guest Ranch, 257
Fox Mountain, 64
Fox Mountain Plateau, 254
Frémont, Jesse Benton, 214, 216
Frémont, John C., 205; career of, 213–14; expeditions of, 214–15; fourth expedition of, 216–28, 278
Freshwater ecosystems, 178–79; fire impacts on, 182–83; insects in, 179
Fruitland Coal, 244, 246
Fruitland Formation, 246, 249
Fuller Lake, 121
Fungi, 141, 163; distribution of, 139–40; edible, 144–45; and lichen, 137, 138; poisonous 142–44. *See also* Lichens; Mushrooms
Fur trappers, Taos, 205

Galactic Resources Limited, Summitville District, 52, 255
Galena, Creede Mining District, 46, 47, 48
Galerina autumnalis, 143
Galloping Goose, 237
Galls, 176
Gas production, 246
Gastrointestinal toxins, 144
Geological Survey of the Territories (Hayden), 61
Geomorphology, 62
Geranium, Richardson's (*Geranium richardsonii*), 132
Gladstone, 236
Glaciation, glaciers, 12, 62, 253, 281, *plate 14*; cycles of, 260–61, *plate 15*; moraines, 249–50, 254, 290; outwash from, 265, 284; Quaternary, 67–71; rock, 72–73; till, 71–72
Godey, Alexis, Frémont expedition, 217, 218, 222, 223, 224, 225, 226, 227
Gold, 48, 52, 56, 206, 248–49
Gold King Mine, 236
Gondwana, 7, *plate 3*
Goose Creek, 50
Grabens, 35, 43
Grasshoppers, 176, 182; speckled-wing (*Arphia conspersa*), 175
Gravity anomalies, San Juan, 33
Grazing, 133, 210, *plate 32*; fire and, 158–59
Great Depression, 45
Great Sand Dunes National Park, 204, 218, 280
Gregorio, 216, 224, 226
Grenadier Range, 7
Groundhog Creek, 224
Groundhog Park, 224

Groundwater, 79, 80, 84, 87, 132, 133, *plate 22*; geological controls of, 94–95; hot springs, 93–94, 249, 258; hydraulic properties, 89–93, *plate 24*
Gunnison, 237, *268*, 269
Gunnison, John, 205, 269
Gunnison Basin, 95
Gunnison River, cutthroat trout, 196, 197
Gyromitra spp., 145; *montana,* 143
Gyromitrin, 143

Happy Thought Mine, 44
Hare, snowshoe, 188
Harmon Art Museum, Fred, 247
Hatcheries, 197
Hayden, F. V., *Geological Survey of the Territories,* 61
Hayden Survey, 206
HD Mountains, 246
Hebeloma crustuliniforme, 144
Hermanos Penitente, Los, 278
Hesperus, 236
Highland Mary Lake, 119, *120*; zooplankton in, *122*
Highway 17 (CO-NM), points of interest on, 284–90
Highway 50 (US), 269
Highway 84 (US), 159; points of interest on, 291–95
Highway 144 (CO), point of interest on, 269–75
Highway 149 (CO), points of interest on, 256–68
Highway 160 (US), 159, 161; points of interest on, 244–56, 281–82
Highway 285 (US), points of interest on, 275–81, 282–84
Hinsdale Basalt, 284, 287
Hinsdale County, 265
Hindsdale County Courthouse, 208
Hinsdale Formation, 266, 281–82, 284
Hinsdale trachybasalts, 64
Hispanics, 203, 204, 210, 248
Historical range of variability (HRV), 152; aspen stands, 162–64; mixed-conifer forests, 165, 166; ponderosa pine forests, 159, 160; spruce-fir forests, 169
Hogback, 293
Hogback Monocline, 8, 244, 246
Holocene: fires and debris fans, 73–74; rock glaciers, 72–73

Holy Moses Mine, 41, 48, 209
Homestake Mill, 48
Homestake Mining Company, 46, 48
Homestead Act, 210
Hoodoos, *266*
Hope, Lake, 114, 118
Hope Creek, 254
Horseshoe Mountain member, 23
Hotchkiss, Enos, 265
Hot Creek State Wildlife Area, 284
Hot springs, 12, 257–58, *plate 23*; groundwater flow and, 93–94, 249
Houghland Hill, *275*
House Creek, 261
HRV. *See* Historical range of variability
Hubbard, George, 227
Huerto Andesite, 28, 36
Human impacts, 74–75
Hummingbirds, 141
Humphreys Mill (Creede), 44
Hydrogeology, *plate 20*; aquifers, 89–93, *plate 24*; San Juan Mountains, 79–80; water budget in, 86–87
Hydrologic cycle, 86, 88–89, *plates 16, 21*
Hydrothermal activity, 21
Hyperabyssal stocks, 22

Ibotenic acid, 143
Ignimbrite flare-up, Oligocene, 10–11
Ignimbrites, *plates 9, 10*; Cenozoic, 17–19; Oligocene, 63–64; Tertiary, 19–21, 23, 24, 26–27
Indian agencies, 205
Inky caps (*Coprinus* spp.), 144
Inocybe, 143
Inonotus andersonii, 141
Insects, 141, 173, 174; carnivorous, 176–77; drought and fire impacts on, 181–82; freshwater, 179–80; herbivorous, 175–76; high-elevation, 177–78. *See also by type*
Invertebrates, 124; in lakes, 122–23
Ips confusus, 151, 157
Iron Hill, 266, *268*
Ironton, 235
Irrigation system, San Luis Valley, 278, 280
Irving Formation, 64
Isotoma hiemalis, 177

Jackson, Saunders, 216, 224
Jackson Mountain, 249, *250*
Jackson Mountain Volcano core, 249

Jarosa Peak, 64
Jarosite, 65
Jasper, 54, 55
Jicarilla Apache, 189, 289
Juan, 216, 224, 226
Junction Creek Sandstone, 269, 270
Juniper (*Juniperus* spp.), 155, 164

Kansas Pacific Railroad, 233
Karst topography, 114
Kearny, Stephan Watts, 215
Kenny Flats, *plate 32*
Kerber Creek, 56, 57
Kern, Ben, 216, 222–27 *passim*
Kern, Edward, 216, 224, 225, 226, 227
Kern, Richard, 216, 217, 221–27 *passim*
King, Henry, 218, 223, 225, 227
Kirtland Formation, 244, 246
Kite Lake, 117

Laccoliths, Cenozoic, 9
Lactarius spp., 141
La Garita, 278
La Garita Caldera, 27–28, 253, 254, 256, 257, 273; faulting in, 33, 35–36
La Garita Creek, 204; Frémont expedition on, 221, 224, 225
La Garita Mountains, 27–28, 35, 278; Frémont expedition in, 213, 219–26, 228
La Garita Park, 224
La Garita Stock Driveway, 210, 222
La Garita Wilderness Area, 117, 258
La Jara Canyon unit, 23
La Jara State Wildlife Area, 284
Lake City, 69, 207, 208, 237, 257, *264*, 265
Lake City Caldera, 265
Lake Fork (Gunnison) River, 66, 89, 263, 265
Lake Fork Valley, 265
Lake Fork volcano, 265
Lakes, *plate 25*; biological productivity of, 120–22; characteristics of, 114–15; chemistry of, 115–18; ecosystems, 113–14; fish stocking in, 123–25; invertebrates in, 122–23; mercury pollution in, 125–26; water quality in, 118–20
La Manga Pass, 287, 288
Land grants, Spanish, 204
Landslides, 4, 74, 244, 249, *251, 263*, 265, 287, *288*; and Mancos Shale, 291–92
La Plata Mountains, 186
Laramide orogeny, 7–8, 9–10, 12, 62, *plates 1, 5*

Laramide Rocky Mountains, 8, 9, *plate 5*
Lariat, 282
Last Chance Mine (Creede), 41, 43, 44, 209, 259
Laurasia, 6–7, *plate 3*
Laurentia, *plate 2*; volcanism, 5–6
Lava Creek B volcanic ash, 244
Lava domes, Tertiary, 23
Leccinum insigne, 145
Lentic systems, 179
Lewis Shale, 246, 249, 293
Lichenometry, 146–47
Lichens, 137, *plate 30*; and cryptobiotic soil, 140–41; dating techniques, 146–47; types of, 138–39
Litaneutria minor, 176
Little Giant Mine, 206
Little Molas Lake, 114
Livestock, 210; and fire regimes, 158–59
Los Mogotes volcano, 284, 286
Los Pinos Formation, 286, 287
Los Pinos Graben, 35
Los Pinos Indian Agency, 206, 208
Los Pinos River, 204, 244
Lotic systems, 179; insects in, 179–80
Lumber industry, 210, 256, *plate 32*
Lynx (*Lynx canadensis*), 185, 186, *plate 35*; reintroduction of, 187–88

McClelland Mountain, 28
McDermott Formation, 244
McGehee, Micajah, 216, 222, 225, 226
McNabb, Theodore, 224
Macomb, John H., 205
McPhee Reservoir, 114, 124, 126
Macrophytes, 122
Mafic dikes, Cenozoic, 11
Magmatism, 6, 21, 63; Laramide orogeny, 9–10. *See also* Volcanism
Mammoth-Revenue Vein, 54
Manassa, 284
Mancos River, 145
Mancos Shale, 244, 247, 249, 290, 293, 295; and landslides, 291–92
Mantises (*Litaneutria; Stagmomantis* spp.), 176
Mantle degassing, 12
Mantle uplift, 12–13
Manuel, 216, 224, 226, 227
Marcy, R. B., 219
Marine deposits, Proterozoic, 6
Martin, Thomas, 225, 226, 227

Masonic Park, 293
Masonic Park Tuff, 26, 252, 256–57, 287, 288, 289
Mastodon, 282
Mears, Otto, 206, 208; railroads, 235, 236–37
Medano Creek, 280
Médano Pass, 218
Menefee Shales, 244
Menyanthes trifoliata, 131, *plate 26*
Mercury, in lakes, 125–26
Mesa Mountain, 221, *plate 8*
Mesa Seco, 263
Mesa Verde Formation, 291, 292, *293*
Mesa Verde Group, 244, 246
Mesa Verde National Park, 73, 181, 237
Methane, 130, 246
Mexican War, 205, 214–15
Milankovitch cycles, 66, 68, 71
Milkweed (*Asclepias* spp.), 176
Million Fire, 256
Mineralization, 6, 21, 23, 39, 117; adularia-sericite, 41; in Bonanza Caldera, 55–57; in Creede Mining District, 33, 43, 44–45, 49; in Platoro-Summitville Caldera Complex, 50–51; Stunner, 54–55; in Summitville District, 51–52
Mineral Mountain, *plate 8*
Mines, mining, 39, 133, 255, 265, 266; Bonanza District, 55–57; Creede district, 41–51, 208–9, 259; pollution from, 53–54, 57, 196–97; road building and, 206–7; Summitville District, 52–53, 255; uranium, 269
Mining Act (1872), 206
Miocene, 64, 65
Miramonte Reservoir, 117
Missionary Ridge, 74, 161
Missionary Ridge Fire, 73–74, 120, 145, 151, 159, 170
Missouri, as trade center, 215
Mites, 177
Mixed-conifer forests, *plate 34*; fire regimes, 163–70
Molas Pass, 143
Moles, 141
Monon Hill, ore bodies, 49–50
Monte Vista, 282
Moose (*Alces alces*), 131, 141
Moraines, 69, 249–50, 254, 290
Morels (*Morchella* spp.); false, 143, 145
Morrison Formation, 62, 269, 270, 289, *290*
Mosca Pass, 218

Mosquitos, 177
Mosses (*Sphagnum* spp.), 131, 132
Moths, 175, 178; miller (*Euxoa auxiliaris*), 178, *179*
Mountain belts: history of, 3–5; Precambrian, 6, 12
Mountain building, 5, 13
Mountain goat, 141
Mountain lion. *See* Puma
Mountain Studies Institute (MSI), lake survey, 115, 117
Mountain Village, 133
Mount Hope Tuff, 26
Mount Zirkel Wilderness, 147
Moving Mountain landslide, 244
MSI. *See* Mountain Studies Institute
Muscarine poisoning, 143
Muscimol, 143
Mushrooms, 138, 139–40, *plates 28, 29*; collecting, 145–46; edible, 144–45; honey (*Armillaria ostoyae*), 141; poisonous, 142–44
Mutualism, lichen, 137–38
Mycelium, 145–46

Nacimiento Formation, 8
Nacimiento Uplift, 8
Narraguinnep Reservoir, mercury pollution in, 126
National Registry of Natural Landmarks, Slumgullion Earth Flow, 263
Native Americans, 189, 190. *See also* Jicarilla Apache; Navajos; Utes
Navajo Peak, 293
Navajos, 189; and Pagosa Springs, 247–48
Navajo Volcanic Field, 11
Needle Mountains, 5, 6, 62, 68
Needle Mountains block, 8
Needle Mountains Uplift, 8, 11, 62
Neff, Mount, 288, 289
Nelson Mountain Caldera, 272, *plate 8*
Nelson Mountain Tuff, 29, *30*, 31, 35, 272
Nelson-Wooster-Humphreys Tunnel, 44
Nickel Plate exploration tunnel, 46
Nicolett, Joseph N., 214
Nine-Mile Hill, 266–67
Nival insects, 177
North, Oliver, 197
North American Plate, 7, 19
North Mosca area, fen, *plate 27*
North Pass, 273, *plate 8*
North Pass Caldera, 22, 36, 273–74, 275

North Pass cycle, 22
Nutrient cycles, 121

Oak, Gambel (*Quercus gambelii*), 141, 157–58, 163
OH Vein, 43, 45–46, 48
Oil seeps, 254
Ojito Creek Tuffs, 23
Ojo Alamo Sandstone, 8
Old man's beard (*Usnea* sp.), 146, *147*
Old Spanish Trail, 204, 248, 278
Oligocene: Southern Rocky Mountains uplift, 10–11; volcanism, 63–64
Ophir Loop, 236
Orellanine, 144
Orographic effect, 84
Ouray, Chief, 204, 206
Ouray (city), 208, 235
Overland flow, 88
Ovis canadensis, 141, 185, 186, 193, *plate 35*; die-offs, 191–92; distribution of, 190–91

Pacific Decadal Oscillation (PDO), 103
Packer, Alferd G., 207–8, 265
Packer Monument, Alferd, 265
Pagosa Hot Springs, 205
Pagosa Peak Dacite, 26, 27, 35, 252
Pagosa Springs, 205, 206, 207, *247*, 247–49; lynx, 186, 187
Paleocanyons, Oligocene-Miocene, 64
Paleocene, 8, 62
Paleogeography, *plates 2–6*
Paleo-Indian sites, 204, 263
Paleozoic rocks, 8, 61; plate tectonics, 12, 266, *plate 3*
Palmer, Mary "Queen" Mellen, 233
Palmer, William Jackson, 233
Pangea, *plate 3*; formation, 6–7
Paradox Basin, 7
Pass Creek–Elwood Creek Fault Zone, 51, 54
Pass Creek Fault Zone, 35, 253
Pass Creek Glacier, 253
Pathogens, 141, 177
PDO. *See* Pacific Decadal Oscillation
Peatlands, 282; biological diversity, 130–31. *See also* Bogs; Fens
Peat moss, arctic (*Sphagnum balticum*), 131, 132
Pedicularis groenlandica, 131, *plates 26, 27*
Penitente Canyon, 204, 278
Permafrost, rock glaciers and, 73
Pfeiffer, Albert H., 205, 248

Phenacomys intermedius, 185
Phoenix Mine, 48
Pholiota squarrosa, 144
Photosynthesis, in lakes, 118–19
Picea: *engelmannii*, 138, 141, 164, 167; *pungens*, 164
Picture Cliffs Sandstone, 244, 246, 249
Picuris Range, 289
Piedra area, 141, 164
Piedra River, 159, 204
Piedra Watershed, 89
Pine (*Pinus*): fire regimes, 152, 155–61, 164; piñon (*P. edulis*), 151, 180; ponderosa (*P. ponderosa*), 138, 141, 165, *plates 31, 32*
Piñon-juniper woodlands, fire regimes, 155–57
Pinorealosa Mountain, 287, 288
Pinos Creek Road, 281
Pinos Valley, 288
Pinus. See Pine
Placer mining, 52
Plants, in peatlands, 130–31
Plate tectonics, 3; Paleozoic, 12, 266, *plate 3*
Platoro, 23, 257, 286
Platoro center, 21
Platoro Caldera complex, 23–24, *24*, 26, 35, 252–53, 255, 284; mineralization, 39, 50–51
Pleistocene, 66; cutthroat trout, 196; glaciation, 67–71; mammals, 186, 190, 282
Pliocene, 66; glaciation during, 67–68
Point Lookout Sandstone, 244
Point of Rocks, Frémont expedition, 224, 225
Polk, James, 214
Pollinators, 178
Pollution, 5, 102; from Summitville mines, 53–54; water pollution, 196–97
Poncha Pass, 204
Ponderosa pine forests, *plates 31, 32*; fire regimes, 157–61
Porphyry Peak, 56
Porter, 236
Power plants, pollution from, 126, 147
Precambrian era, 5; mountain belts, 6, 12
Precambrian rocks, 61, 266, 268, 269, *270*, 287, 290, *plate 1*
Precipitation, 84, 86–87, 103, *plate 19*
Preuss, Charles, 216, 218, 222, 224, 225, 227
Procryphalus macronatus, 163
Pronghorn antelope, 141
Prospect Basin, fen in, *plate 26*
Prospectors, 203, 205
Proterozoic, *plate 2*; volcanism, 5–6

Proue, Ralph, 216, 224, 227
Pseudotsuga menziesii, 141, 164
Puebloan tribes, 189, 190
Puma (*Puma concolor*), 185, 186, 193–95, *plate 35*
Puzzle System, 46–47
P Vein, 43, 45–46, 48

Quaternary, 62; glaciation, 67–71
Quercus gambelii, 141, 157–58, 163, 176
Questa (NM), 225, 227

Rabbit Peak, 291
Railroads, 210, 231–32, 288, 291; construction of, 233–38
Ra Jadero Tuff, 23
Ranching, 210, 282
Rat Creek Caldera, *plate 8*
Rat Creek Tuff, 29, *30*, 31, *259*
Ratliff, Mark, 223
Rawley Andesite, 22
Rawley Mine, 56, 57
Razor Creek Dome, 29, *plate 8*
Red Mountain, 235
Red Ryder display, 247
Reservoirs: ecosystems of, 113–14; fish stocking in, 124–25
Rhizocarpon geographicum, 146, *plate 30*
Rhodochrosite, in Creede Mining District, 44, 47
Rico, railroad to, 236
Ridgway, 69, 236
Ridgway Conglomerate, 62
Rifting, Cenozoic, 11–12
Rincon Creek, 223
Rincon de la Osa, 203
Ring faults, 21
Rio Blanco, 295
Rio Blanco Formation, 252, 289
Rio Chama, 291
Rio Colorado (Questa), 225, 227
Rio de los Pinos, 284, 288
Rio Grande, 69, 204, 255, 280; ancestral, 64, 65, 66; headwaters, 84, 196, 203
Rio Grande Fault System, 258
Rio Grande Graben, 35, 51, 261
Rio Grande National Forest, 263
Rio Grande Pyramid, 203
Rio Grande Rift Zone, 11–12, *18*, 19, 33, 64, 282, 284, 295, *plate 6*; bimodal volcanism, 35–36
Rio Grande Southern Railroad (RGS), 236–37

Rio Grande Valley, 256–57, *257*
Rio Grande Valley Glacier, 72
Rio Grande Watershed, 66–67, 72, *plate 18*
Rivera Juan María de, 204, 248
Road Creek, 261
Roads: impacts on fens, 132–33; mining boom and, 206–7, 208. *See also various highways by number*
Rock Creek member, 23
Rock dust, in alpine lakes, 118–19
Rock glaciers, 72–73
Rock slides, 74
Rohrer, Henry, 226
Romeo, 284
Roosevelt, Theodore, 210, 258
Root rot, 141, 167
Rose, Corydon, 208
Russell Lakes State Wildlife Area, 278
Russula spp., 141; *emetica,* 144
Ruxton, George, 216

SAD. *See* Sudden aspen decline
Saddle Mountain, 252, *253*
Saguache, 207, 208, 275–76, *277*
Saguache Creek Tuff, 22, 273, 274–75
Saguache Park, *plate 8*
Saline lake, in Creede Caldera, 40–41
Salix spp., 131
Salmo clarki, 185, *plate 35*; *pleuriticus*, 124, 125, 195, 196, 197–98; *stomias*, 195; *virginalis*, 195, 196, 197, 198
Salmon, kokanee, 124–25
San Antonio Mountain, 284, *286*
San Cristobal, Lake, 263, 265
Sand Creek / Mosca area, 132
San Francisco Glacier, 281
San Francisco Lakes, 281
Sangre de Criso Fault, 278
Sangre de Cristo Range, 8, 66, 276, 278, 280
San Jose Formation, 8, 246
San Juan anomaly, 33
San Juan Archeological Network, 263
San Juan Basin, 8, 62, 95, 246, 276
San Juan Caldera complex, 259, 263
San Juan City, 261
San Juan Dome, 289
San Juan River, 66, 197, 204; at Pagosa Springs, 247, 249
San Juan Sag, 62, 276
San Juan Uplift (Needle Mountains Uplift), 8, 62, 68

San Juan Volcanic Field, 5, 9, 10, 11, 12, 51, 63, 81, 289, *plates 1, 6*
San Luis Basin, 66, 95, 278, 282
San Luis Caldera complex, 28, 29–31, 47, *plate 8*
San Luis Highland, 276
San Luis Hills, 225
San Luis Peak block, 35, *plate 8*
San Luis Valley, 35, 66, 79, 210, 275–76, 282, 284, 295; as catchment basin, 64–65; cutthroat trout in, 196, 197; Frémont expedition in, 218–19, 224, 225; human use of, 203–5; irrigation system, 278, 280
San Miguel Basin, 95
San Miguel River, ancestral, 66
San Miguel Watershed, 89, *plate 18*
Santa Fe, 204, 248
Santa Fe Group, 276
Santa Fe Formation, 65
Santa Fe Railroad, 233–34
Santa Maria Reservoir, *262*
Sapinero Mesa Tuff, 266
Sawatch Range, 8, 276
Sawatch trend, 22
Sawtooth Mountain, 29, *plate 8*
Scolytus spp., 167
Scott, Glen, 62
Seaways: Cretaceous, 62, *plate 5*; Paleozoic, *plate 4*
Sedges (*Carex* spp.), 131, 132
Sediment cores, Quaternary, 68
Sediment flows, 74
Sericite, in Creede Mining District, 41
Servilleta Basalt, 66, 282
Servilleta Formation, 284
Shaggy mane (*Coprinus comatus*), 140, 144
Shaw, Edith, 210
Shaw, Ernest, 210–11
Sheep, 204, 210; grazing impacts, 158–59
Sheep Creek Fault, 273
Sheep Mountain, 250, *252*
Shepherds, 203
Sherman Silver Purchase Act, 43
Shiprock, 11
Shirley, 56
Shrews, 141
Silver, 204; in Bonanza district, 56; crash, 209, 236, 237; in Creede district, 43, 47, 48, 49
Silver Lake, 117
Silver Thread Scenic Byway, 256
Silverton, 205; lynx in, 186, 187; railroad, 231, 234–35, 237

Silverton, Gladstone and Northerly Railroad, 236
Silverton Northern Railroad, 235
Silverton Railroad, 235
Slaughterhouse Gulch, 265
Slumgullion Earth Flow, 4, *263*, 265
Snowdon Lake, rock dust in, 118–19
Snowdon Mountain block, 7
Snowdown Peak rock glacier, 73
Snowmelt: and lake chemistry, 115; and temperature, 104–5
Snowpack: and dust storms, 105–6, 115; and groundwater recharge, 87, 95
Snowsheds, on Wolf Creek Pass, 254
Snowshoe Mountain, 259–60
Snowshoe Mountain Dome, 35, 64, 259–60
Snowshoe Mountain Tuff, 26, 31, 254
Solomon-Holy Moses Fault System, 43
Solomon-Holy Moses Vein System, 48–49
Southern Rocky Mountains, 3, *4*, 8, 65, 244; mid-Tertiary volcanism, 17, 19; rifting and, 11–12; uplift of, 10–11, 66
Southern Rocky Mountain Volcanic Field, 17–19
Southern Ute Indian Reservation, 246
South Fork, 207, *254*, 256, 257
South Fork Tuffs, 23
South Mountain–Platoro Fault Zone, 54
South River Caldera, 29, 254
South San Juan Wilderness Area, 117, 167; Colorado River cutthroat trout, 197–98
Spanish. *See* Hispanics
Sparassis crispa, 141
Sparling Gulch, 265
Sphagnum spp., 131, 132
Sphalerite, in Creede Mining District, 46, 47, 48
Spiders, snow-adapted, 177
Spring Creek moraine, 69
Springs, *plate 23. See also* Hot springs
Springtails (Collembola), nival (*Isotoma hiemalis*), 177
Spring turnover, 119
Spruce (*Picea*): blue (*P. pungens*), 164; Engelmann (*P. engelmannii*), 138, 141, 164, 167; fungi associated with, 144, 145
Spruce-fir forests, 138; fire regimes, 167–70
Squirrel Creek, 57
Squirrel Gulch mine, 56
Squirrels, and lichens, 141
Stagmomantis limbata, 176
Staphylinidae, 177

Stepperfeldt, John, 226
Stewart Peak, *plate 8*
Stock driveways, 210
Stockton, Robert F., 215
Stoneflies, 180
Stony Pass, 206–7, 257, 261
Stratovolcanoes, 21, 63, 280–81
Stream flow, 87; watersheds and, 88–89
Structures, central caldera complex, 33–35
Stunner, 54–55
Subduction, flat-slab, 9–10
Sudden aspen decline (SAD), 163
Sulfide ores, in Creede Mining District, 46
Summer Coon volcano, 64, 280–81
Summitville, 255, 257, 281; mineralization, 23, 39, 51–52
Summitville Andesite, 293
Summitville Boulder, 53
Summitville Caldera, 253, 255; mineralization, 50–51, 54
Summitville Consolidated Mining Company, 52
Summitville District, 255; mineralization, 23, 39, 51–52; pollution from, 52–53
Sunfish, green, 124
Sunlight Lake, *plate 25*
Sunnyside, 41; mineralization in, 49–50
Sunset Ranch, 210
Sunshine Peak Tuff, 265
Superfund Site, Summitville District as, 54, 255
Symbiosis, lichen and, 137–38

Taos: Frémont expedition at, 225, 227; trappers and, 204, 205
Taplin, Charles, 226
Tectonic basins, Proterozoic, 6
Telluride, 69, 133; railroad to, 236, 237
Telluride Conglomerate, 11, 62
Tellurium City, 207
Temperatures: changes in, 107–8; datasets, 109–10; long-term trends in, 99–103; recent trends in, 104–7; seasonal trends, 103–4
Tertiary, 61; caldera complexes, 26–31; ignimbrite calderas, 19–21; volcanism, 17–19, 22–26
Timber industry, 210, 256
Toads, boreal, 122–23
Topography, and watersheds, 81–82, *plate 18*
Tourism, railroads and, 237–38
Toxins, mushroom, 142–44
Trail Ridge (Pagosa Ranger District), *plate 34*

Trappers, Taos, 205
Treasure Falls, 251–52
Treasure Mountain, 253, 281, 293
Treasure Mountain Tuff, 286, 287, 288, 289
Treasure Mountain Group, 23, 26, 252, 253, 254
Trophic state, in lakes, 120–23
Trout, stocking of, 123–24, 125, 197. *See also* Cutthroat trout
Trout Lake, 114, 121
Trout Mountain, 64
Trypophloeus populi, 163
Tucker Lakes, 254
Tuffs, Tertiary, 22, 23, 26–31. *See also by name*
Turkey Springs, 160
Twilight Gneiss, 6

Uncompaghre Basin, 95
Uncompaghre Gorge, Upper, 69
Uncompaghre–San Juan Caldera complex, 263
Uncompaghre Watershed, 89, *plate 18*
Unconformities, Eocene erosional, 62–63
Union Pacific Railroad, 232–33
U.S. Environmental Protection Agency (EPA), 57; Superfund site, 54, 255
U.S. Forest Service, 57, 117, 210–11, 258
Upper Animas Gorge, 69
Upper Animas Watershed, research wells on, 91, 93
Upper Pine River Watershed, wildfires in, 126
Upper Rio Grande Valley, glaciation of, 72
Upper Rio Grande Watershed, 187, 261
Upper San Juan Basin, cutthroat trout in, 196
Upper Uncompahgre Gorge, 69
Uranium mining, 269
Usnea sp., 146, *147*
Ute Fire, 159
Ute Ridge Tuffs, 26
Utes, 189, 203, 204, 206, 289; and Pagosa Springs, 247–49; and Wagon Wheel Gap hot springs, 257–58

Vallecito Reservoir, 114, 161; fish stocked in, 123, 124, 125; wildfires, 120, 126
Valsa sordida, 163
Vegetation zones, *112*
Verpa spp, 145
Vinsonhaler, Lorenzo, 224, 227
Volcanic rocks, permeability of, 93, 95
Volcanism, 61, 62, 81; Cenozoic, 8–9, 17–19, 22, *plates 6, 9–11*; Laurentian, 5–6; and mineralization, 39, 255; Oligocene, 11, 63–64;

points of interest showing, 244, 249, 252–53, 254, 256–57, 265, 271–74, *276*, 278, 280–81, 284; rift-related, 35–36; in Wheeler Geologic Area, 258–59
Voles, 141; heather (*Phenacomys intermedius*), 185
V-Rock, 293, *294*

Wagon Wheel Gap, 28, 257; fluorspar deposit, 41, 50
Wagon Wheel Gap Fluorspar Mine, 50
Wagon Wheel Gap Hot Spring, 257–58
Wanakah Formation, 289
Wannamaker Creek camp, 222
Warman, Cy, 237
Wason Park Tuff, 29, 254
Wasps, 176
Water, 95, 131, 132, 173; mining pollution, 196–97
Water boatman, 179
Water budgets, 85(table), 86, 95, *plate 21*; and stream flow, 87–88
Water quality, 113; influences on, 115–17; mining and, 196–97; monitoring of, 117–18
Watersheds, 80, 83(table), 84, *plates 16, 18*; and lake chemistry, 115–18; precipitation, 86–87, *plate 19*; stream flow, 87–89, *plate 20*; and topographic relief, 81–82
Water storage, 114
Wells, research, 91–93
Weminuche Wilderness Area, 117, 126, 167, 197, 203, 204, 247
West Benino Creek, 222
West Elk Breccia, 269
Western State College, 269

West Fork Glacier, 250
West Fork Valley, 250–51, 252
West Willow Creek, mines on, 43, 44, 47
Wetlands, types of, 129–31. *See also* Fens
WGA. *See* Wheeler Geologic Area
Wheeler, George M., 258
Wheeler Geologic Area (WGA), 258–59, *259*
Wheeler Geologic Monument, *30*, 30–31
Whitecross, 207
White Dome Lake, copepod zooplankton from, *123*
Wightman Fork, 52, 255
Wildfire. *See* Fires
Williams, Bill, 289; as Frémont's guide, 217, 218, 219, 222, 223, 225, 228
Williams Creek Reservoir, 132
Willow: mountain (*Salix monticola*), 131; plane leaf (*Salix planifolia*), 131
Willow Mountain member, 23
Windy Point, *289*
Wolf Creek Canyon, 289–90
Wolf Creek Pass, 143, 161, 169, 187, *250*, 253
Wolf Creek Ski Area, 253
Wolf Creek Valley, 289
Wolf Creek Village, 253
Woods Lake, 118, 125

Xanthoria spp., 140, 147

Yavapai orogeny, 6
Yellow Jacket Creek, 246
Yellow Jacket Summit, 246
Yellowstone caldera, 244

Zooplankton, 121–22, *122, 123*, 124, 125